Phytomedicines of Europe

ACS SYMPOSIUM SERIES 691

Phytomedicines of Europe

Chemistry and Biological Activity

Larry D. Lawson, EDITOR
Murdock Madaus Schwabe

Rudolf Bauer, EDITOR
Heinrich-Heine-Universität Düsseldorf

American Chemical Society, Washington, DC

Chemistry Library
UNIV OF CALIFORNIA

Library of Congress Cataloging-in-Publication Data

Phytomedicines of Europe : chemistry and biological activity / Larry D.
Lawson, Rudolf Bauer, editors.

   p.   cm.—(ACS symposium series, ISSN 0097–6156; 691)

"212th ACS National Meeting, Orlando, FL, August 25–29, 1996. Division of
Agricultural and Food Chemistry"—T.p. verso.

Includes bibliographical references and indexes.

ISBN 0–8412–3559–7

   1. Herbs—Therapeutic use—Europe—Congresses. 2. Materia medica,
vegetable—Europe—Congresses. 3. Medicinal plants—Europe—Congresses.

   I. Lawson, Larry D. II. Bauer, Rudolf, 1956–. III. American Chemical
Society. Division of Agricultural and Food Chemistry. IV. American Chemical
Society. Meeting (212th : 1996 : Orlando, Fla.) V. Series.

RM666.H33P49 1998
615′.321—dc21   98–10368
                                                    CIP

This book is printed on acid-free, recycled paper.

Foreword

THE ACS SYMPOSIUM SERIES was first published in 1974 to provide a mechanism for publishing symposia quickly in book form. The purpose of the series is to publish timely, comprehensive books developed from ACS sponsored symposia based on current scientific research. Occasionally, books are developed from symposia sponsored by other organizations when the topic is of keen interest to the chemistry audience.

Before agreeing to publish a book, the proposed table of contents is reviewed for appropriate and comprehensive coverage and for interest to the audience. Some papers may be excluded in order to better focus the book; others may be added to provide comprehensiveness. When appropriate, overview or introductory chapters are added. Drafts of chapters are peer-reviewed prior to final acceptance or rejection, and manuscripts are prepared in camera-ready format.

As a rule, only original research papers and original review papers are included in the volumes. Verbatim reproductions of previously published papers are not accepted.

ACS BOOKS DEPARTMENT

Contents

INDEXES

Preface

PLANTS WERE THE MAIN SOURCES OF MEDICINES throughout the world until the development in Western society of chemical, pharmacological, and toxicological methods for preparing effective and reasonably safe monomolecular synthetic drugs, a high percentage of which were developed from lead compounds discovered in plants. Although the development of modern synthetic drugs nearly ended the use of—and pharmacological research on—medicinally used plants (herbs) in the United States, this was not the case for Europe, especially for Germany, due to continued use by physicians and to favorable government regulations that encouraged research. Consequently, most of the research on the herbs that have only recently become popular in the United States has taken place in Europe. Europe's leading role in herb research is further evidenced by the long-time existence of scientific societies like the Society of Medicinal Plant Research (Gesellschaft für Arzneipflanzenforschung, founded in 1953) and the Phytochemical Society of Europe (founded in 1957)—both of whose annual meetings are held in English, although rarely attended by Americans—and several well-respected European scientific journals on medicinal plants, like *Planta Medica, Phytotheraphy Research, Phytomedicine, Fitoterapia,* and *Zeitschrift für Phytotherapie.*

Because of the lack of U.S. scientific interest in medicinal plants, much of the published research and reviews on herbs have been predominantly written in German. Therefore, the purpose of this ACS symposium was to gather together for the first time in the United States many of the leading European investigators and authorities on medicinal plants to review and discuss the current scientific status or "state of the art" of many of the more popularly used herbs, with respect to both the pharmacological evidence and the identification of the active or important components of the herbs. The reader will discover in the following chapters that there is good experimental and clinical support for the medicinal and preventative effects of many herbs, although for some popular herbs more research is still needed. Furthermore, it will be discovered that there are some plants which have good clinical support of efficacy, but for which the active compounds and/or mechanism of action have remained elusive.

The titles of the book and the symposium are not meant to imply that the plants presented in this book originated in Europe or that they are popular only in Europe. For example, echinacea and saw palmetto were originally used in the

ix

United States, and garlic has been used extensively by many cultures throughout the world. Instead, the title emphasizes the fact that these plants have been investigated by modern research, primarily in Europe, and have been popularly used there as medicines for many decades.

The book is divided into three sections: Perspective, Specific Effects, and Specific Plants. The Perspective section deals with subjects that are important to all herbs, such as status in the United States and European markets, status in the German Commission E monographs, and regulatory concerns. The Specific Effects section presents a few specific disease states and how they might be affected by a variety of plants. The largest is the Specific Plants section, which contains reviews of the biological activity and chemistry of 12 popular herbs. Unfortunately, reviews of the research on many other popular and effective herbs could not be included due to the time limitation for the symposium. However, as the editors, we feel that what is presented in the book will give the reader a good overview of the scientific status of medicinal plants as well as of some of the challenges and concerns unique to pharmacological research with plants.

Lastly, the editors would like to thank A. Douglas Kinghorn of the University of Illinois-Chicago and former chairman of the Agricultural and Natural Products Chemistry section of the American Chemical Society for inviting us to organize this symposium and book, and those colleagues who have critically reviewed the various chapters. We want to express very special appreciation to the following sponsors, who made this symposium possible:

Bioforce AG (Switzerland)
Bionorica GmbH (Germany)
Capsugel (South Carolina)
East Earth Herbs (Oregon)
Enzymatic Therapy (Wisconsin)
Finzelberg's GmbH (Germany)
Flachsmann AG (Germany)
Hevert Arzneimittel (Germany)
Klosterfrau GmbH (Germany)
Lichtwer Pharma (Germany)
Madaus AG (Germany)

Müggenberg GmbH (Germany)
Murdock Madaus Schwabe (Utah)
Pharmaton SA (Switzerland)
Phytopharm Ltd. (England)
Pure-Gar (Washington)
Salus–Haus (Germany)
Schaper & Brümmer (Germany)
Schönenberger GmbH (Germany)
Schwabe GmbH (Germany)
SmithKline Beecham (Pennsylvania)

LARRY D. LAWSON
Murdock Madaus Schwabe
Nature's Way Products, Inc.
10 Mountain Springs Parkway
Springville, UT 84663

RUDOLF BAUER
Institute of Pharmaceutical Biology
Heinrich-Heine-Universität
  Düsseldorf
Universitässtr. 1
D-40225 Düsseldorf, Germany

PERSPECTIVE

Chapter 1

Importance of European Phytomedicinals in the American Market: An Overview

Varro E. Tyler

Department of Medicinal Chemistry and Molecular Pharmacology, School of Pharmacy and Pharmacal Sciences, Purdue University, West Lafayette, IN 47907-1333

Detailed comparisons of practices on both continents reveal that European phytomedicine has had a profound influence on American herbal medicine. Of the ten best-selling herbs in the United States today, six (echinacea, garlic, ginseng, ginkgo, saw palmetto, and eleuthero) have been popularized primarily as a result of European research. This situation has come about as a result of a more favorable regulatory climate in European countries that permits botanicals to be marketed as drugs based on traditional use and/or reasonable evidence of efficacy. In the United States, demand for unrealistic amounts of evidence regarding their efficacy has discouraged research on these generally unpatentable products and relegated them to the status of dietary supplements. Following comparisons of both systems, suggestions for improving the regulation of phytomedicinals here are provided.

As political debates over various aspects of the European Union and a projected common currency continually remind us, in most respects the countries of Europe are not homogeneous, and this applies to their development and utilization of phytomedicinals. This review will not provide detailed information regarding the laws and regulations governing the approval and use of these products in the various countries of Europe. That ever-changing topic is more suitable for an attorney than for a scientist. All of the countries there allow the sale of such products as "traditional remedies" with little or no scientific or clinical evidence supporting safety and efficacy. Likewise, all of them have provisions for classifying phytomedicinals as conventional drugs. The standards applied to permit such categorization differ widely from country to country.

Instead of delving into laws and regulations, the purpose of this review is to provide background information on European phytomedicinal research that has established the safety and efficacy of a number of botanicals which have, in consequence, become very popular in the United States. For reasons that will become clear, little such research is conducted in this country, so most of the well-proven herbal products have their origin in European research. Several of them are prepared from native American plants, but their popularity is based upon information derived from studies in Europe. When it comes to phytomedicines, America owes a large debt to the Old World.

Scientific Status of Phytomedicines in European Nations

Observations, both direct and indirect, have led to the conclusion that the United Kingdom is the least advanced of the European nations with respect to the science of phytomedicine. In that regard, it is nearly—but not quite—as underdeveloped as the United States. While some of the publications emanating from the U.K. appear to be science based—the *British Herbal Compendium*, vol. 1, is a good example (1)—many others are the work of herbalists who have relatively little regard for the scientific method. The largest selection of books related to botanical medicine observed recently in London was not in Foyles or Blackwells but in Watkins Books Ltd., a firm advertising itself as "Mystical and Occult Specialists." This would seem to categorize the field as something of a nonscience in that country.

British herbalists still recommend the internal use of pokeroot for various conditions, report high concentrations of "estrogens" in certain plants, and show a disregard for proper morphologic and taxonomic nomenclature (2). With the exception of feverfew, no herbal remedy has entered the American market in recent years based on British research, and as a matter of fact, several erroneous reports on herbal toxicity have appeared in the literature as a result of lack of quality control in the U.K. Probably the most persistent of these is a study by MacGregor *et al.* (3) postulating the hepatotoxicity of scullcap and valerian in an herbal mixture. In fact, neither of these botanicals has ever been demonstrated to possess toxic properties when consumed in normal amounts. The problem arose because in Britain *Teucrium canadense* L. or American germander is almost always substituted for scullcap (4). Species of *Teucrium* contain constituents, probably furano *neo*-clerodane diterpenoids, that are transformed in the body by cytochromes P450 into hepatotoxic metabolites (5). Nevertheless, the erroneous report of scullcap and valerian toxicity continues to appear in the herbal literature.

At the other end of the spectrum among European countries interested in herbal medicine or phytomedicine is Germany. There, because of a favorable regulatory system allowing botanicals to be marketed with "reasonable" proof of safety and efficacy, research on and use of such products has flourished. The crowning achievement of the German system is the evaluation of more than 300 individual herbs and publication of therapeutic monographs devoted to them by Commission E of the Bundesinstitut für Arzneimittel und Medizinprodukte (BfArM) (6). Much of the research on which these monographs are based has either been conducted in-house by German phytomedicine producers or sponsored by them in German university laboratories. It is no exaggeration to declare that the system of phytomedicinal use developed in Germany is the envy of all other advanced nations.

Germany also accounts for the largest share of the European Union herbal drug market with sales of $3 billion of an annual $6 billion total. France occupies second place with sales of $1.6 billion representing a 26.5% market share. Next is Italy with $600 million in sales, followed by all others with combined total sales of some $800 million (7).

In terms of quantity and quality of scientific research, most European nations seem to fall somewhere between the large volume of excellent research conducted in Germany and the relatively nonscientific efforts emanating from Britain. Italy produces, and certainly publishes, a disproportionate share of original research due to the presence of a very large phytomedicine manufacturing company in Milan. However, taken as a whole, European phytomedicine is dominated by German phytomedicine, so it is not surprising that it has been the dominating influence on scientific phytomedicine in the United States as well.

Table I lists the ten best-selling herbs in selected American health food stores in 1995 (8). From this, it will be observed that six of the ten have been popularized largely on the basis of the result of European research. Considering that no modern research has been conducted anywhere on one of the herbs— namely, goldenseal—this two-thirds domination of European research influence on the best-selling herbs in the American market is indeed appreciable.

Table I. Source of Research on Best-Selling U.S. Phytomedicinals

Plant	Market Share (%)	Location
1. Echinacea (*Echinacea* spp.)	9.9	Europe (Germany)
2. Garlic (*Allium sativum*)	9.8	Europe and U.S.
3. Goldenseal (*Hydrastis canadensis*)	7.0	nowhere
4. Ginseng (*Panax* spp.)	5.9	Europe and Asia
5. Ginkgo (*Ginkgo biloba*)	4.5	Europe (Germany)
6. Saw Palmetto (*Serenoa repens*)	4.4	Europe (Germany)
7. Aloe Gel (*Aloe barbadensis*)	4.3	United States
8. Ephedra (*Ephedra* spp.)	3.5	United States
9. Eleuthero (*Eleutherococcus senticosus*)	3.1	Europe and Asia
10. Cranberry (*Vaccinium macrocarpon*)	3.0	United States

Table II. Source of Research on Other Phytomedicinals Popular in the U.S.

Plant	Location
11. Feverfew (*Tantacetum parthenium*)	England
12. Milk Thistle (*Silybum marianum*)	Germany
13. St. John's Wort (*Hypericum perforatum*)	Germany
14. Valerian (*Valeriana officinalis*)	Germany

Top Ten Herbs in the United States

Echinacea. Echinacea is a native American plant represented by several species of that genus growing primarily in the Midwestern United States, and the drug was introduced into medicinal practice by the firm of Lloyd Brothers in Cincinnati about 1880. However, the resurgence of use of the botanical in this country during recent years must be attributed to modern European research. So many different research groups have been involved in determining the immunostimulatory properties of echinacea and its various constituents that it is impossible to establish a firm priority of discovery. The literature has been summarized in considerable detail by Hobbs (9), by Foster (10), and by Bauer and Wagner (11).

For our purposes, the three studies by Beuscher *et al.* in 1977, 1978, and 1980 on the proprietary herbal combination Esberitox form a convenient starting point (12-14). Basically, these describe the influence of echinacea on cellular and humoral immunity in rats. In 1981, Wagner and Proksch began publication of a series of papers on the immunostimulant properties of echinacea that eventually concluded that high molecular weight polysaccharides were the compounds responsible for the principal activity (15). However, Bauer *et al.* later reported that the lipophilic fraction of echinacea extracts tended to be more active than the polar fraction. They found the isobutylamides and the polyenes to be the most active components, using the carbon-clearance and granulocyte tests in vivo (16).

No modern scientific or clinical studies on echinacea have been conducted in the United States, so the popularity of this herb as a useful preventive agent or therapeutic treatment for colds and influenza must be attributed directly to these and other European studies. That a phytomedicine could occupy a 9.9% share of the U.S. market without a single study having been carried out here is little short of amazing.

Garlic. Research on garlic must be viewed as both an American and a European effort. An American group headed by Cavallito first identified allicin as the antibacterial principle and determined its structure (17). However, it remained for Stoll and Seebeck in Switzerland to isolate, identify, and synthesize the stable oxygenated sulfur amino acid alliin that serves as the precursor of allicin (18). Then, in 1984, Block *et al.* in this country identified ajoene as a potent antithrombotic agent that is derivable from allicin (19). As a folk medicine for cardiovascular problems, garlic has long been popular in both Asia and Europe, but most of the clinical trials of garlic powder in coated tablets have been conducted with the German product Kwai (20). The utility of tablet coating in formulating an active, relatively odor-free dosage form of garlic is thus a European innovation.

Garlic is sold in a wide variety of dosage forms in the United States, some of which lack the ability to yield allicin following consumption and are of doubtful efficacy. The botanical has a market share of 9.8% and is used for a wide variety of effects ranging from antibacterial to hypotensive to hypolipidemic. Reduction of blood lipids, including both cholesterol and triglycerides, probably accounts for the majority of use both here and abroad. Those remain the best proven therapeutic effects of garlic (21).

Goldenseal. That goldenseal—a drug which has never been the subject of a modern scientific study, let alone a randomized clinical trial—should command a 7% share of the U.S. herb market is indeed an anomaly. The botanical apparently owes some of its popularity to an erroneous perception by members of the counterculture that goldenseal masks urine tests for heroin and other illicit drugs. Foster has identified the original source of this misconception—the novel *Stringtown on the Pike* by John Uri Lloyd (*22*). As a matter of fact, the drug is more apt to cause false positives than to obscure the presence of narcotics.

Goldenseal is not widely used in Europe, which helps explain why there has been so little scientific study of it. The herb is not monographed in *Hager's Handbuch der Pharmazeutischen Praxis*, 5th ed., nor by the German Commission E. Advocates claim it has immunostimulatory effects (*23*), but these have never been verified. For this purpose, the botanical is often combined with echinacea, so such an effect would not be unexpected.

The most recent review of the pharmacology and therapeutics of goldenseal was published in 1950 (*24*). It noted that little work had been done on the plant in the preceding 40 years. Reasons for the extreme popularity in the United States of this little-investigated botanical remain enigmatic.

Ginseng. Ginseng is the fourth best-selling herb in the United States with a 5.9% market share. Presumably, this statistic includes both the Asian and American species. Asian ginseng root has been used in China for more than 2,000 years, primarily as an aphrodisiac, probably based on the Doctrine of Signatures and its often humanoid form. The herb made its first appearance in Europe in the twelfth century, having been taken there by Marco Polo (*25*).

Over the years, numerous studies by many investigators in the Asian Rim nations, Russia, and Europe have clarified the chemistry of the saponin glycosides contained in ginseng and established with reasonable certainty their activity as performance and endurance enhancers in small animals. Almost all authorities agree that proof of efficacy as an adaptogen or tonic in humans requires additional clinical testing (*26*).

The European contribution to ginseng research that has greatly influenced the American market was the development of Ginsana brand, a product standardized to contain 4% of total ginsenosides. Said by its manufacturer to be a product of 15 years of Swiss research, the first such standardized extract dosage form was made available here in 1983 (*27*). Since that time, many additional commercial organizations have begun to market similar preparations in the United States. European—specifically, Swiss—research on ginseng may be said to be responsible for the botanical's current availability in dosage forms containing specific amounts of a standardized extract.

Ginkgo. Ginkgo biloba leaf extract is a relatively recent medical discovery used primarily to treat peripheral circulatory insufficiencies. Although the seed or nut of the tree had long been used in eastern Asia as a minor article of food and digestive aid, use of the leaf was much less common. In 1436, external application of the leaves was proposed for a variety of skin conditions; then, in 1505, they were first recommended for internal use. Modern Chinese medicine employs the leaves to "benefit the brain" (28).

Exactly how and when ginkgo came to the attention of European researchers has apparently not been recorded in detail. Schilcher has noted that its pharmacological action and therapeutic activity were discovered during a systematic investigation, but no details of the process were presented (29). An inquiry directed to the Schwabe organization in Germany in 1993 (30) elicited a response which indicated that a historical review of the discovery was in preparation, but it apparently has not yet appeared. Foster has concluded that ginkgo leaf extracts are more closely associated with European phytomedicine than with Chinese traditional medicine, but postulates that reports of use of the leaf tea in China stimulated the chemical, pharmacological, and clinical research resulting in the development of the standardized extract in Europe (31).

As a result of the European research, ginkgo biloba extract preparations standardized to contain 24% flavonoids and 6% ginkgolides entered the American market in the mid-1980s and since became the number five best-seller with a 4.5% market share. Products sold here are prepared from extracts imported from Germany, Italy, and Asia. A very large ginkgo plantation containing some 10 million trees, pruned to shrub height to facilitate harvest of the leaves, exists near Sumter, South Carolina. However, after collection and drying, the leaves are all shipped abroad for processing.

Saw Palmetto. Occupying sixth place in herbal popularity in the United States with 4.4% of the total market, saw palmetto is another native American plant. It was first introduced into medicine by a Dr. Reed in 1877 (32). By 1898, Felter and Lloyd were recommending it for a variety of conditions, including "hypertrophy of the prostate" (33). Current use in this country in the treatment of benign prostatic hyperplasia (BPH) is nevertheless the result of European research.

Harnischfeger and Stolze have summarized the pertinent literature which shows that an interest in the constituents of the fruit was developing in German laboratories in the 1960s (34). Then in 1981, Wagner and Flachsbarth (35) reported on a new anti-inflammatory polysaccharide from saw palmetto, and this was soon followed up by pharmacologic and clinical studies in France, Germany, and other European countries.

One of the most recent of these studies emanating from Belgium reported the results of an open trial with 505 patients (36). After 90 days of treatment with saw palmetto extract, 88% of the patients and 88% of the attending physicians reported significant symptomatic improvement with a low incidence (5%) of side effects. Saw palmetto has become a very popular therapy for BPH in the U.S. in recent years because of its relatively low cost and absence of some of the undesirable side effects (e.g., impotency) attendant to the use of conventional prescription medications for this condition.

Aloe Gel, Ephedra, and Cranberry. Although unrelated in most respects, aloe gel, ephedra, and cranberry—the 7th, 8th, and 10th best-selling botanicals in the U.S.—may be considered as a group because their use here has not been appreciably influenced by European research. Most of the studies dealing with the wound-healing properties of aloe gel, the bronchodilating and vasoconstricting properties of ephedra, and the anti-infective properties of the cranberry originated in America (37). Most of the worldwide use of these herbal products also takes place here.

Europeans should be grateful that the widespread chronic misuse and abuse of ephedra as a putative anorectic agent, performance enhancer, or euphoriant have not occurred to any appreciable extent on their continent. Ephedra, often sold under its Chinese name ma huang, has been reported to cause more adverse reactions and more deaths than any other herb sold in the United States (38). In April 1996, the Food and Drug Administration issued a warning to consumers not to buy products containing high doses of ephedra. Aloe gel, on the other hand, shows considerable promise in promoting wound healing when applied externally, and cranberry juice has become a very popular preventive agent for urinary tract infections (39).

Eleuthero. Usually marketed in the United States under the more glamorous designation "Siberian ginseng," eleuthero is the ninth best selling herb with 3.1% of the market here. As noted by Farnsworth et al., most of the experimental studies on this plant have been conducted in Russia, so at least a portion of the work must be considered to be of European origin (40).

An extract of the plant was first marketed in the United States in 1971 under the name Siberian Ginseng Liquid Extract. The plant is not a true ginseng, but that name was considered commercially advantageous and was justified on the basis that it did belong to the same plant family and, for that reason, had the same physiological activity (37). If this logic were extended, the consumption of deadly nightshade fruits could be justified on the basis that they belong to the same family as tomatoes.

Long-known chemical constituents were also renamed as eleutherosides to mimic the nomenclature of ginsenosides, the active principles of ginseng. This was done in spite of the fact that none was a triterpenoid saponin remotely related in structure to the ginsenosides. For example, ß-sitosterol glucoside or daucosterol was renamed eleutheroside A; syringin was designated eleutheroside B; isofraxidin glucoside was called eleutheroside B_1, and so on.

A detailed analysis of the studies conducted primarily in Russia on the ability of eleuthero to improve work output and enhance athletic performance reveals that most of the trials were not double blinded and lacked adequate controls. Haas has concluded that the only effect of eleuthero that has been adequately documented is that of an immunomodulator (41). In spite of its questionable reputation, the herb has been approved by German Commission E as a tonic. Additional randomized clinical trials are needed to confirm this evaluation.

Other Popular Herbs in the United States

There are a number of herbs that have become popularized in the United States as a result of European research but which do not fall in the top ten sales category. All of the ones mentioned subsequently are in the top twenty-five sales listing. These include feverfew, milk thistle, St. John's wort, and valerian. See Table II.

Feverfew. Long utilized in folk medicine as a treatment for headache, this plant was introduced into modern American herbal medicine as a result of clinical studies conducted in Britain in the 1980s. These showed that relatively small amounts (ca. 60-82 mg.) of feverfew leaf ingested daily reduced both the frequency and severity of migraine attacks. The sesquiterpene lactone parthenolide has been postulated to be the active principle, apparently acting as a serotonin antagonist (42). However, this has still not been verified by studies utilizing that chemical in pure form. A tentative relationship between pharmacological potency and total sesquiterpene lactone content has been established (43).

The quality of feverfew with respect to its parthenolide content has been shown to be extremely variable. One study of various plant samples and commercial products showed variations ranging from zero to 1.03% parthenolide (44). The Canadian government has established a minimum content requirement of 0.2%; French authorities recommend 0.1%. Feverfew preparations standardized at the 0.2% level are now commonly available in the United States.

Milk Thistle. Although the milk thistle plant had been recommended for the treatment of liver ailments since the time of Pliny the Elder (23-79 A.D.), it remained for Wagner *et al.* in Germany to isolate from the fruits in 1965 a mixture of active principles that was characterized in 1968 as silymarin, thereby establishing a sound scientific basis for its use (45). Silymarin, a complex mixture of flavonolignans, was subjected to extensive pharmacological study in Germany and was shown to exert a hepatoprotective effect in laboratory animals exposed to a variety of toxic compounds. Subsequent clinical trials established its utility in treating alcoholic liver disease (cirrhosis) and hepatitis (46).

Several mechanisms of action are apparently involved. Silymarin alters the permeability of liver cell membranes, thereby protecting them against the entry of toxic substances. It also stimulates protein synthesis, accelerating the regeneration process and the production of hepatocytes. In addition, silymarin acts as a free-radical scavenger and antioxidant. The flavonolignans are poorly soluble in water and are poorly absorbed from the gastrointestinal tract, so they must be administered in the form of highly concentrated extracts of the fruit.

St. John's Wort. St. John's wort is another plant valued by the ancients, primarily to relieve inflammation and promote healing. More recently, an extract of the leaves and flowering top of the plant have come to be valued in Europe for the treatment of mild depression and for nervous unrest and sleep disturbances as well (47). The antidepressive activity has been demonstrated in clinical trials, but the identity of specific active constituents has not been determined. Possibly various flavonoid derivatives are responsible.

Various mechanisms of action of St. John's wort extract have been postulated. The postulated mechanism of monoamine oxidase (MAO) inhibition has not been confirmed, and neither inhibition of catechol-*O*-methyltransferase (COMT) nor modulatory effects of a hormonal nature on the cytokines can presently be excluded (48). Obviously, additional research is required to establish with certainty the identity of the active constituents of this plant and their mode of action.

Valerian. Although the sedative properties of valerian have been recognized for 2000 years, the identity of its active constituents and their mode of action are still unknown. This is the case, in spite of fairly extensive chemical, pharmacological, and clinical studies of the herb conducted in Europe. The sesquiterpene valerenic acid was shown to inhibit gamma-aminobutyric acid (GABA) breakdown, resulting in an increased level of that compound which is associated with a decrease in central nervous system (CNS) activity. However, it is not possible to explain the activity of valerian extracts on the basis of the small amount present of this or any other postulated active constituent (*49*).

More recently, attempts have been made to explain the action of valerian extracts by determining their affinity for different sites at the GABA-A chloride channel receptor complex or for the peripheral benzodiazepine receptors. At the moment, such findings remain largely of theoretical interest.

The objective effects of valerian on sleep latency do compare favorably with those of the benzodiazepines. The botanical's efficacy, coupled with a reported lack of acute toxicity and minimal side effects, renders it a useful sleep aid (*50*). It is definitely worthy of further study to establish with certainty the nature of its active principles and their mechanism of action.

Summary and Conclusions

This review of sources of research responsible for the introduction and popularization of a variety of useful phytomedicinals in the American market has shown that, with few exceptions, the pertinent studies are of European origin. While this might be considered as a reflection of the lesser quality of American science, such is not the case. Rather, it results from the strict regulatory system in this country that demands an unreasonable investment in efficacy testing for unpatentable remedies. In those countries where more reasonable standards apply, progress has been substantial. In the United States, where sale of phytomedicines as dietary supplements is now very permissive but sale of drugs is highly restricted, most manufacturers choose to market them in the former category. This means essentially no investment in scientific or clinical studies, so the only available data derive from research abroad. In terms of phytomedicinal research, our country is indeed a very backward nation (*51*).

There is an urgent need to change this system. In the capitated health-care plans of the future, only approved drugs will likely be employed. Phytomedicines could play a very significant role in such a system because of their reduced cost and minimal side effects, if an enlightened regulatory system were implemented based on a "reasonably certain" standard of efficacy. Such a system would benefit the public health in many ways, including the additional research findings that would be bound to result from its implementation (*52*).

At the present time, scientific phytomedicine in the United States owes its existence to European research. That debt would soon be repaid if a reasonable system of regulations of herbs and phytomedicines were implemented here.

Literature Cited

1. Bradley, P.R., Ed. *British Herbal Compendium*, vol. 1; British Herbal Medicine Association: Bournemouth, England, 1992; 239 pp.
2. Ody, P. *The Complete Medicinal Herbal*; Dorling Kindersley: New York, NY, 1993, 192 pp.
3. MacGregor, F.B.; Abernethy, V.E.; Dahabra, S.; Colbden, I; Hayes, P.C. *Brit. Med. J.* **1989**, *299*, 1156-1157.
4. Foster, S. personal communication, December 31, 1991.
5. Loeper, J.; Descatoire, V.; Letteron, P.; Moulis, C.; DeGott, C.; Dansette, P.; Fau, D.; Pessayre, D. *Gastroenterology* **1994**, *106*, 464-472.
6. Blumenthal, M.; Hall, T.; Rister, R.S. *German Commission E Monographs*; American Botanical Council: Austin, TX, in press.
7. European Scientific Cooperative for Phytotherapy (ESCOP), *European Phytotelegram* **1991**, No. 3, 15-16.
8. Brevoort, P. *HerbalGram* **1996**: No. 36, 49-57.
9. Hobbs, C. *The Echinacea Handbook*; Eclectic Medical Publications: Portland, OR, 1989; 118 pp.
10. Foster, S. *Echinacea: Nature's Immune Enhancer*; Healing Arts Press: Rochester, VT, 1991, 150 pp.
11. Bauer, R.; Wagner, H. *Echinacea*; Wissenschaftliche Verlagsgesellschaft: Stuttgart, Germany, 1990, 182 pp.
12. Beuscher, N.; Beuscher, H.; Otto, B.; Schäfer, B. *Arzneim. Forsch.* **1977**; *27*, 1655-1660.
13. Beuscher, N.; Beuscher, H.; Schäfer, B. *Arzneim. Forsch.* **1978**; *28*, 2242-2246.
14. Beuscher, N. *Arzneim. Forsch.* **1980**; *30*, 821-825.
15. Wagner H.; Proksch, H. *Z. Angew. Phytother.* **1981**; *2*, 166-171.
16. Bauer, R.; Jurcic, K.; Puhlmann, J.; Wagner, H. *Arzneim. Forsch.* **1988**; *38*, 276-281.
17. Cavallito, C.J.; Buck, J.S.; Suter, C.M. *J. Am. Chem. Soc.* **1944**; *66*, 1952-1954.
18. Stoll, A.; Seebeck, E. *Experientia* **1947**; *3*, 114-115.
19. Block, E.; Ahmad, S.; Jain, M.K.; Crecely, R.W.; Apitz-Castro, R.; Cruz, M.R. *J. Am. Chem. Soc.* **1984**; *106*, 8295-8296.
20. Reuter, H.D.; Koch, H.P.; Lawson, L.D. In *Garlic: The Science and Therapeutic Application of Allium sativum L. and Related Species*; Koch, H.P.; Lawson, L.D., Eds.; Williams & Wilkins: Baltimore, MD, 1996, p. 143.
21. Tyler, V.E. *J. Am. Pharm. Assoc.* **1996**; NS*36*, 29-37.
22. Foster, S. *HerbalGram* **1989**; No. 21, 7, 35.
23. Werbach, M.R.; Murray, M.T. *Botanical Influences on Illness*; Third Line Press: Tarzana, CA, 1994, pp. 23-24.
24. Shideman, F.E. *Comm. Nat. Form. Bull.* **1950**; *18*, 3-19.
25. Huang, K.C. *The Pharmacology of Chinese Herbs*; CRC Press: Boca Raton, FL, 1993, p. 22.
26. Bahrke, M.S.; Morgan, W.P. *Sports Med.* **1994**; *18*, 229-248.
27. Horwitz, S. *Harper's Bazaar* **1984**; *117*(3), 32.
28. Del Tredici, P. *Arnoldia* **1991**; *51*, 2-15.
29. Schilcher, H. *Z. Phytother.* **1988**; *9*, 119-127.
30. Tyler, V.E. personal communication to H. Jaggy, November 8, 1993.
31. Foster, S.; Chongxi, Y. *Herbal Emissaries*; Healing Arts Press: Rochester, VT, 1992, p. 258.

12

32. Lloyd, J.U. Origin and History of All the Pharmacopeial Vegetable Drugs, Chemicals and Preparations, vol. 1; The Caxton Press: Cincinnati, 1921, pp. 277-278.
33. Felter, H.W.; Lloyd, J.U. *King's American Dispensatory,* 18th ed., 3rd rev., vol. II; Eclectic Medical Publications: Portland, OR., reprint 1983 (original 1898), pp. 1750-1752.
34. Harnischfeger, G.; Stolze, H. *Z. Phytother.* **1989**; *10*, 71-76.
35. Wagner, H. Flachsbarth, H. *Planta Med.* **1981**; *41*, 244-251.
36. Braeckman, J. *Curr. Ther. Res.* **1994**; *55*, 776-785.
37. Tyler, V.E. *Herbs of Choice: The Therapeutic Use of Phytomedicinals*; Pharmaceutical Products Press: Binghamton, NY, 1994, 209 pp.
38. Blumenthal, M. *HerbalGram* **1996**; No. 36, 21-23, 73.
39. Tyler, V.E. *The Honest Herbal,* 3rd ed.; Pharmaceutical Products Press: Binghamton, NY, 1993, pp. 25-28, 101-102.
40. Farnsworth, N.R.; Kinghorn, A.D.; Soejarto, D.D.; Waller, D.P. In *Economic and Medicinal Plant Research,* Vol. 1; Wagner, H.; Hikino, H.; Farnsworth, N.R., Eds.; Academic Press: Orlando, FL, 1985, pp. 155-215.
41. Haas, H. *Arzneipflanzenkunde*; B.I. Wissenschaftsverlag: Mannheim, Germany, 1991, pp. 135-137.
42. Awang, D.V.C. *Can. Pharm. J.* **1989**; 122, 166-169.
43. Barsby, R.W.J.; Salan, U.; Knight, D.W.; Hoult, J.R.S. *Planta Med.* **1993**; 59, 20-25.
44. Heptinstall, S.; Awang, D.V.C.; Dawson, B.A.; Kindack, D.; Knight, D.W.; May, J. *J. Pharm. Pharmacol.* **1992**; *44*, 391-395.
45. Wagner, H.; Hörhammer, L.; Münster, R. *Arzneim. Forsch.* **1968**; *18*, 688-696.
46. Morazzoni, P.; Bombardelli, E. *Fitoterapia* **1995**; *66*, 3-42.
47. Harrer, G.; Sommer, H. *Phytomedicine* **1994**; 1, 3-8.
48. Bombardelli, E.; Morazzoni, P. *Fitoterapia* **1995**; *66*, 43-68.
49. Krieglstein, J.; Grusla, D. *Dtsch. Apoth. Z.* **1988**; *128*, 2041-2046.
50. Morazzoni, P.; Bombardelli, E. *Fitotherapia* **1995**; *66*, 99-112.
51. Tyler, V.E. *Nutr. Forum* **1992**; *9*, 41-45.
52. Tyler, V.E. *Herbs Health* **1996**; *1*, 10-12.

Chapter 2

Status of Phytopharmaceuticals Within the European Market

R. Anton and B. Kuballa

Pharmacognosy Department, Faculty of Pharmacy, University Louis Pasteur
B.P. 24, 67401 Illkirch Cédex, France

After a survey of the different European regulations, the German and French positions on phytopharmaceuticals are presented and the main quality, efficacy, and safety criteria are discussed. Although the conceptual approaches are distinct, these two regulations are particularly up-to-date in the field of phytomedicines and could be of first importance for a new European harmonization. Different official institutions, such as the European pharmacopoeia, WHO, etc., are playing a very important role for the future European assessments.

It is well-known that plants, drugs, extracts, or natural derivatives have given rise to a large number of preparations used by humans as remedies since ancient times. In French, the terms "boisson, potion, poison" have the same etymology. Nowadays, plants can be classified into three categories:
- absolutely safe plants used in nutrition
- plants known for their toxic properties but having pharmacological applications, such as molecular model references for studying the mechanism of action of new drugs, such as aconitine, an anti-arrhythmic model
- plants used in therapy for which two types of raw materials have been developed: (1) industrial extraction of pure active compounds for use in severe pathologies such as morphine, digoxine, vinblastine, taxol, etc., and (2) crude herbs and extracts for use in healthcare products, OTC products, and phytomedicines for treating minor pathologies and to maintain a healthy balance.
 This last type of herbal remedy will only be discussed in a pharmaceutical context, in order to understand the problems of European harmonization. France and Germany have two separate official interpretations. It will be interesting to compare these different positions in order to develop a new common European legislation.

Present Definitions of Herbal Remedies and Plant Drug Preparations

Comprehension of the use of herbal remedies is not easy because of their very complex composition, which can also vary considerably. Therefore, it is important to provide definitions and standardization of such preparations.

The following definitions are official and are generally agreed upon by all countries:

• Herbal remedies are medicinal products containing as active ingredients exclusively plant material and/or plant drug preparations.

• Herbal drugs are sometimes whole plants, but more often they are parts of plants (roots, barks, flowering tops, leaves, flowers, fruit, seeds) including exsudates (oleoresins, gums, latex, catechu, etc.).

• Plant drug preparations can be powders, extracts, tinctures, fatty or essential oils, or expressed juices. Fractionation, purification or concentration processes may also be employed. Chemically defined isolated constituents (menthol, etc.) or mixtures are not herbal drug preparations.

• Constituents with known therapeutic activity are chemically defined groups of substances known to contribute to the therapeutic activity.

• Markers are chemically defined constituents used for control purposes.

• Homeopathic preparations are excluded.

Aspects of the European Phytomedicine Market

Presently, about 1400 herbal drugs are used in different EC countries and about 200 are the most important ones. Consequently, herbal remedies have economic relevance. About $6 billion in sales of herbal remedies in the European Union were reported in 1994, indicating that every citizen of the European Union spends $17.4 per year for herbal remedies. By 1998, this amount is expected to increase by 8% per year.

We can classify European OTC consumers into two categories: either they use phytomedicines as a self-medication or they prefer natural products, considering them as an alternative life-style. On the contrary, the term "herbal remedies" is not automatically linked to an OTC status. In Germany, and more rarely in France, a small proportion of medical prescriptions consists of herbal remedies.

European phytomedicines can be categorized as follows: cardiovascular 27.2%, respiratory 15.3%, digestive and tonics 14.4% each, hypnotic/sedative 9.3%, topicals 7.4%, others 12%. In France, about 30,000 tons of herbal plants are consumed every year. In general terms, according to the Directives 65/65 EEC and 75/318/EEC, pharmaceutical products require pre-marketing approval before gaining access to the market. Requirements for the documentation of quality, safety and efficacy, the dossier and the expert reports are laid down in Directive 91/507/EEC.

However, major problems or conflicts in the assessment of medical interest arise from harmonization of old products, generic applications (similar or identical products), divergent medical information depending on the origin and type of national authorization and diverse opinions on the criteria applied, especially in the case of safety and efficacy, as well as with label claims, such as "well-established use" and "acceptable safety". An EC note for guidance, "quality of herbal remedies", provides a well-defined framework for quality testing *(1)*.

The Legal Status of Phytomedicines in Germany

Herbal remedies constitute an important part of the pharmaceutical market, especially in the field of non-prescription medicines. In the sector of herbal OTCs, Germany holds the biggest share with $2.5 billion US in comparison with France ($1.6 billion) and Italy ($600 million).

The Federal Health Office (Bundesgesundheitsamt or BGA), renamed in 1994 as the Federal Institute for Drugs and Medical Devices (Bundesinstitut für Arzneimittel und Medizinprodukte or BfArM), is responsible for the assessment of medicines.

From 1978 to 1994, a multidisciplinary group of experts (called Commission E) established a review of more than 300 medicinal plants present on the German market. These detailed technical monographs have been elaborated on the use of each drug, the conception of which is quite different from the French approach.

In 1984, the Commission E proposed the following additional criteria for the evaluation of long term experience with herbal drugs. Without new controlled clinical trials, evidence of safety and efficacy is accepted as plausible.

- if an herbal drug is mentioned in the standard literature or in well-documented review articles, or
- if there are clinical trials which are not conclusive alone but are supported by supplementary experimental data, or
- if there is well-documented knowledge on traditional use that is supported by significant experimental studies.

A positive monograph mainly includes title of monograph, nature of the active compound (herbal drug), definition of the plant drug, indications, contra-indications, side-effects, interactions, dosage, mode of administration, length of use, precautions and pharmacological activities.

Negative monographs have also been elaborated, comprising title of the monograph, nature of active principle (plant drug), definition of the herbal drug, claimed indications, possible risks (contra-indications, side effects, interactions, effect on fertility/pregnancy) and pharmacological activities. An interesting official judgment is provided, giving a clear and useful attitude for industry: "Given the lack of evidence of efficacy, the therapeutic use of the drug is not recommended" or "Given the lack of evidence of efficacy and considering the risks induced by the drug, the therapeutical use is no longer acceptable".

In Germany, phytomedicines can be prepared from very active drug material. This is allowed in France if a complete dossier (pharmaceutical, toxicological and clinical sections) is elaborated, although for some of these plants the corresponding monographs exist in the French pharmacopoeia. These phytomedicines are not considered to be very safe. The side effects of the active ingredients are not well-known and we consider in France that the margin between therapeutic activity and toxicity is often too narrow. No clinical efficacy has been demonstrated with the present assessment criteria. Moreover, synthetic medicines with less side effects and a more precise therapeutic impact are available on the market. Besides, it is not evident that some medicines presently sold in Germany could be developed in France.

Here are some examples of monographs of plants which are not recommended in France for herbal remedies: *Adonidis herba - Chelidonii herba - Convallariae herba - Dulcamarae stipites- Myristica fragans - Petasites rhizoma - Podophylli peltati rhizoma* and *resina - Ruta graveolens - Scillae bulbus - Visci albi herba - Yohimbehe cortex* ....

In Germany, revised texts of the other monographs have been recently discussed for actualization. Some of the new texts clearly focus on specific extracts or preparations of the plants and not on the plant as a whole (*Echinaceae pallidae radix, Hippocastani semen, Crataegus sp.*). This is a general trend, and for this reason, a large number of phytomedicines cannot meet the scientific knowledge present in such a monograph and have very little chance of being granted a re-registration and of being further marketed. Therefore, fewer additional clinical studies will be carried out.

Finally, it should be noted that 4840 herbal medicinal products present on the German market, have been subject to pharmacological actions since 1978. Many of them had to be withdrawn from the market, or their use had to be restricted.

Criteria of French Legal Requirements

A group of experts was created in 1980 at the Ministry of Health in Paris. The goals of the group were threefold:
- to avoid anarchy on the market
- to define a high level of pharmaceutical quality because herbal remedies are recognized as medicines on their own
- to consider only one type of license for highly sophisticated medical drugs as phytomedicines.

In addition, the market nowadays is always flooded with masses of drug preparations from different origins. Thus, it is by all means imperative to define the high level of pharmaceutical quality.

As an example, the percentage of ginsenosides found in some Ginseng preparations on the market varies from 0.5 to 8% or more. Numerous commercial forms derived from garlic, in which both initial content and stability may vary considerably, are sold on a parallel market. How can we imagine in these conditions that the therapeutic impact will be identical for all these preparations?

The major criteria taken into consideration were the consistency of quality, safety and efficacy. The benefit/risk ratio has to be optimal. The applicant has to provide all the evidence which may be useful in supporting the application, but a high degree of flexibility for innovation is permitted in the limits of the above mentioned criteria.

A long period of use is important, including traditional indications, traditional preparations, and historical proof of the widespread traditional use. All the therapeutic indications of this drug can be claimed in a special category preceded by the term "traditionally used for". In this case, the same level of requirements as for other pharmaceutical specialties is required in the pharmaceutical documentation.

The entrepreneur must know his phytomedicine as well as possible and elaborate a complete control monograph on the basis of fingerprints to evaluate the quality, to describe the stability of the finished products, and to refer to the toxicological documentation. A consistent quality for products of plant origin can only be assured if the starting material is defined rigorously and in detail. The control of all steps of the manufacturing process is necessary to assure a consistent quality of the phytomedicine.

New commercial forms can be admitted, but the more the traditional use of the drug that is claimed, the less the toxicological and clinical documentation that will be necessary. It has been accepted that the product in question should have a well established medicinal use with an acceptable level of safety based on a well established documentation. In this case, it is not necessary to develop any clinical trial. Nevertheless, it is evident from a formal point of view that the clinical tests available in the literature are incomplete or not in accordance with today's state of the art (lack of data like possible effect on the reproduction toxicity, genotoxicity, carcinogenicity, etc.).

The recommended therapeutic indications are presented in the form of a list selected through literature search, including scientific references, and a rational point of view concerning the chemical composition of the drug as well as current pharmacological, toxicological and clinical data.

Certain uses have been excluded because they correspond to conditions for which it could be dangerous not to use therapies in which efficacy has been established in accordance with currently existing criteria.

In conclusion, a reasonable list of indications, has only been developed for minor pathologies. For each medicinal product, the number of therapeutic indications must be limited to two, chosen from the following list of main indications: non-pathological cardiac excitability; circulatory disorders - capillary fragility - hemorrhoids; wounds - itching of the scalp - emollient and antipruritic - burns - buttock erythema; toothache; gastrointestinal system - colitis - diarrhea; painful menstruations; gallbladder ailments; colds - influenza; appetite - weight gain - asthenia; weight reduction; headache; sleep disorders and neurotic states; eye irritation; cough - bronchitis - cold; minor articular pains; mouth-wash; renal system.

Currently, scientific discussions takes are under way to remove some plants on the list and to add others. A new version of the "notice to applicants" will be published at the end of 1997. In the field of clinical applications, some indications will be restricted, such as "prostatic disease" and "acne", which in the minds of physicians are pathologies which require more efficient treatments.

French Procedures for Registration of Traditional Drugs: "Notice to Applicants for Marketing Authorization" B.O. 90/22 bis (2)

Content of the Pharmaceutical Dossier. These elements summarize the French approach as described in the official 1990 text "Notice to applicants for marketing authorization". The French Ministry of Health decided to authorize some phytomedicines, and it adapted conditions for the application for marketing authorization, which have been defined for some plant drugs and preparations.

A list of 174 plants was initially established, on the basis of all scientific information available in the botanical, chemical, pharmacological, toxicological, and clinical fields. An additional list of 31 other species known for their laxative effect was also added. All these drugs are safe, and in some rare cases, a maximum dose of active ingredients is imposed (*Fucus, Valeriana*). Here are the major points concerning the official requirements assuring the best quality.

Concerning the quality of the starting material, it must be stated whether or not the active ingredients are described in the pharmacopoeia. The botanical nomenclature and origin of the plant, the manufacturing process and the qualitative and quantitative control methods for the intermediate and final products must also be described. The control procedure monograph should be based on recently described compounds using up-to-date analytical techniques (HPLC or GC fingerprints).

Validation of all the techniques must be provided, including the stability of the finished product. Furthermore, new standards must be addressed concerning pesticide residues, microbiology, solvents, and toxic metals.

If the active ingredients are described in the European or French pharmacopoeias, the manufacturer has to follow all the standards described in the monograph. If no monograph is available, a special text should be prepared by the manufacturer in the same manner as the monographs of the European pharmacopoeia.

The pharmaceutical dossier is elaborated in accordance with a provision of EC directives. Each part involves specifications, routine control tests, and scientific data relevant to the efficacy, safety and stability of the product.

Different technical steps in obtaining the plant drug, the herbal drug derived preparation, and the finished product must be clearly described. This dossier must be systematically updated in the event of a modification which may affect the quality of the finished product.

Toxicological and Clinical Documentation. A toxicological evaluation depends on the available knowledge or the traditional use of the drug, the form of administration, and the possible presence of undesirable constituents in the drug.

Three categories have been defined. The first concerns drugs for herbal teas (infusions), aqueous extracts, aqueous alcoholic extracts of low strength (< 30% alcohol v/v), traditional tinctures or aqueous alcoholic extracts of high strength (more than 30% v/v) listed in the French and/or European pharmacopoeias and laxative plants listed in the pharmacopoeias. For these types of preparations, no toxicological study is required. The second consists of an abridged toxicological study for powders of whole plants, tinctures and alcoholic extracts of non-traditional use, and unlisted combinations of drugs. This evaluation includes an acute oral toxicity trial in rats and a four-week oral toxicity trial in rats, taking into account behavior, growth, hematological, biochemical and histological (15 organs) parameters (1 group with the maximum dose that can be administered). The third category requires a complete dossier for unlisted plant medicinal products with non traditional use.

No clinical evaluation or activity is required if the plant and the indications claimed are listed.

For certain drugs, details concerning the dosage must be scrupulously followed, taking traditional use into account. For instance, in the case of herbal teas 250 mL to 1 L of tea per day must not be exceeded. For aqueous pharmaceutical forms, the daily dosage corresponds to that of the tea while for other preparations, the traditionally used dosage is taken into account. Furthermore, it is evident that the therapeutic activity may be influenced by other constituents in the preparation, distinct from the active ingredients. For example, tannins, flavonoids and saponins may enhance, prevent or prolong absorption. Consequently, every assessment of efficacy must be based on the individual herbal drug preparation, resulting in a complex situation. The above propositions have the advantages of simplifying the prescribing and the administration of herbal remedies.

Regulations in Other European Countries *(3)*

United Kingdom. No license is required when herbal medicines are used and prescribed by herbal practitioners who, in general, prepare their own medicines or obtain them from specialist suppliers. Furthermore, no license is required if no written recommendation is given to the consumer.

The review process for licensed phytomedicines was completed in 1990 and an information sheet for license holders was published in October 1985 by the Medicines Control Agency (MCA). The labels must include statements like "an herbal remedy traditionally used for the symptomatic relief of ..." and "if symptoms persist consult your doctor".

In December 1995 a new guideline "A guide to what is a medicinal product" was published by the MCA. It tries to give examples to clarify the borderline between medicinal products and, for instance, cosmetics or foodstuffs.

Ireland. The Medical Preparations Regulations of 1984 provide a licensing system for all medicines. Products containing herbal ingredients are considered to be medicinal preparations under the regulations when the labeling or associated literature makes any preventative, curative or remedial claims, or if any of the herbal ingredients present are recognized as having medicinal properties. This list includes 100 herbs.

A report of the Food Safety Advisory Committee, dated December 1993, mentions herbal extracts, herbal teas and essential oils, as being subject to authorization by the regulatory authority prior to marketing. A regulatory framework is necessary to ensure that no potentially toxic plants are marketed as herbal teas. If the herbs have medicinal benefits, they should be regarded as medicines and, therefore, regulated by the National Drugs Advisory Board.

Norway. Herbal products are not classified as medicines unless they have been registered through the marketing authorization process with the Norwegian Medicines Control Authority. If herbal remedies are not granted a marketing authorization, they cannot make any medicinal claim. In January 1994 a guideline for the registration of natural remedies was published which describes the requirements for a bibliographic application for efficacy and safety. These guidelines describe medicinal products containing substances which are suitable for self-medication and have a documentation of traditional use in Europe or North America. Thirty-two references (monographs, textbooks) are provided. If the traditional use is not documented in one of the listed references, a full toxicological and clinical dossier must be submitted.

Denmark. The Executive Order on Natural Remedies issued by the Ministry of Health on 21 September 1992 provides conditions for the registration of natural remedies. Proof of quality, safety, and efficacy must be given. A bibliographic application for therapeutic use is accepted.

Sweden. In 1994 the Medicinal Products Agency issued a guideline, "Information on application for authorization to market natural remedies". Under certain conditions, a simplified application procedure may be possible in accordance with Directive 65/65/EEC which describes a bibliographic application. Scientific literature is sufficient for preparations with well-established use. If satisfactory proof for the safety is not provided, it should be established by means of clinical trials and/or pharmacological and toxicological studies.

Finland. Administrative regulation 9/93 takes into account the provisions of directives 65/65 EEC. If the herbal medicine or its active ingredient has been used for a long time in Europe or in countries close to Europe with regard to their healthcare traditions, the safety and proposed indications for the product can be explained by information available in scientific literature. If this is not possible, an application for a marketing authorization with a full dossier is necessary.

Belgium. The regulation of the Health Ministry, dated 10 February 1995, describes the requirements for herbal medicines. A list of indications for which the different groups of plants are traditionally used is available, and a simplified procedure can be used if reference is made to the plants listed in lists I-XIX. Combination products are not accepted if they contain more than three plants from one list or if the plants belong to different lists.

Switzerland. Medicinal products are classified in several lists, depending on toxicity, indications, and active ingredients. For phytomedicines, lists C (sale limited to pharmacies) and D (pharmacies and drugstores) are of interest. Herbal products must be registered as medicines with documented quality, safety and efficacy. Their indications can be claimed on established traditional knowledge or on efficacy data.
Since 1992, requirements have been set for phytomedicines. A leaflet is intended to guarantee the correct and safe use of the medicine and to give the relevant information to the consumer. If no clinical data on the therapeutic efficacy is available, it must be stated that the properties have traditionally been attributed to the plant. If efficacy has been clinically proven, the leaflet must mention that the product is effective under certain conditions.

Italy. On 8 January 1981 a guideline was issued by the Italian Health Authority classifying herbal products as health food products or as medicines. Plants used for nutritional purposes, sold outside pharmacies, do not need an approval and are not allowed to claim therapeutic effects. Herbal products with therapeutic claims, particular pharmaceutical activity, or possibly, toxic effects are considered as medicines and are allowed to be sold only through pharmacies. Mixtures of herbs (herbal teas or similar preparations) with brand name and/or therapeutic indications have to be registered as medicinal product. With respect to the pharmacological, toxicological and clinical documentation, a bibliographic application in accordance with the Directive 65/65/EEC is possible.

Spain. A special registration for medicinal plants was established by the Ministerial Order of 3 October 1973. Products consisting exclusively of whole or powdered medicinal plants, or parts thereof, must be registered with pharmacological and analytical documents, including the proof of the indications, dosage, and the analytical methods. Preparations containing extracts, tinctures, distillates and other galenical preparations are regarded as medicines which have to fulfill all the requirements of the Spanish Drug Law.

Portugal. Herbal products entered a specific Portuguese legislation in 1993 on health products, which comprises cosmetics, medicinal plants, dietetic products with therapeutic use, and homeopathics. Phytomedicines are subject to the same registration requirements as chemically based medicines. Specific rules on herbal medicines are not yet part of the Portuguese drug law.

Greece. A new regulation was published 1 April 1994 by the Greek Ministry of Health, setting forth that herbal medicines are medicines which contain as active ingredients only plants or preparations thereof. It is not applicable for plant products used as foods. If indications are claimed, they are regarded as medicines and have to fulfill the requirements of the European regulation. The proof of efficacy depends on the indications claimed. In case of minor ailments, the traditional use has to be taken into account with lower requirements. If the traditional use cannot be shown, efficacy has to be proven and a toxicological expert report is necessary. A joint meeting of the Eastern Mediterranean regions was held in 1993 in order to examine problems about phytomedicines in this part of the world *(4)*.

Examples of Specific Problems for Phytomedicines

The quality assessment of herbal remedies is a precondition before developing pharmacological and clinical studies *(5)*. Thus a lot of conditions have to be respected. A standardized cultivation is necessary. As a matter of fact, the active ingredients vary qualitatively and quantitatively because the starting raw material is a living organism, subjected to a variety of parameters, including botanical, physiological, and ecological factors. Variables, such as the nature of soil, the type of plant (wild or cultivated), the use of fertilizers, the harvesting period, the hours of sunshine, etc., are of primary importance.

It is also important to know the exact origin of the starting material, and a contract from the supplier must be included in the documentation. This allows the supplier to be located in a possible case of toxicity. This aspect is of particular importance for those countries, which until now, have not had full control of the collection of wild plant material.

Thus, standardization of cultivation involves at least genetically selected seeds, the location of cultures, clean ecological environment, and well-known harvesting conditions. Assessment of quality does not pose fundamental problems in practice because of the long scientific tradition in pharmacognosy.

On the other hand, the active ingredients must be well-known. If the marker is well-known, classical chemical techniques can be used that are generally available in the literature. This applies, for instance, to alkaloids (Boldo, *Fumaria,* etc.), xanthine derivatives (caffeine, theophylline), anthracene derivatives (*Cassia senna, Frangula* etc.), phenolic constituents (*Crataegus, Ginkgo,* etc.), and essential oils (Labiaceae, Asteraceae).

The most well-known techniques are based on HPLC or GC fingerprints and offer many advantages: precise identification of the drug, evaluation of composition according to the year and origin of the batch, evaluation of possible modification of the chemical composition, stability of the drug and its preparations, standardization of the amount of active ingredients, and evaluation of constituents responsible for possible side effects.

Nevertheless, unknown active ingredients with therapeutic activity constitute the common case in phytotherapy. Therefore, one must choose and define adequate markers as well as appropriate identification and dosage techniques. In general, TLC, GLC and HPLC are the best ones.

As an example of standardization, Ginkgo extracts require HPLC analysis of two major types of compounds: polyphenols (free phenolic acids, flavonol glycosides, and coumaric esters of flavonols) and terpenes (ginkgolides and bilobalide). It also requires a colorimetric analysis of the proanthocyanidins. Of course, all these techniques have to be validated.

Analytical requirements are related to standardization and normalization. Standardization has to be established in each case, and the manufacturer is free to choose the markers with which to standardize the preparation. As far as possible, these markers have to be in correlation with the specifications of the extract. Normalization, however, is linked with precise amounts of known active ingredients and generally uses inert material such as lactose to define a constant dosage of active principles. Thus, normalization is meaningful only if the isolated constituent represents the active ingredient; for example, the amount of sennosides in *Cassia senna* or silymarin in *Silybum marianum* etc.

Stability tests must be carried out on three batches of the same formulation under normal test conditions (25° C over 6 or 12 months) and under accelerated test conditions (40° C over a minimum of 6 months). Analytical test procedures involve the assay (methods and validation) and the determination of degradation products;

The labeling and packaging should enable the public to be informed about the amount of the herb to use per dosage unit and how to prepare the tea. The amount of time the herb needs to be in the water is defined for each case. The type of extract and the content of constituents with therapeutic effects must be stated. Moreover, the package size should be suitable for a reasonable course of treatment, not exceeding 2 or 3 weeks.

In conclusion, a clear definition of the herbal drug preparation is necessary and comprehensive patient information is crucial, since herbal remedies are often used for self-medication. Therefore, propositions concerning rules for labeling and packaging have been elaborated. For plant drugs, the exact botanical name, the specification of the state of the drug and the percentage of active ingredients must be stated. For extracts, information must be given on the name of the drug, the drug/extract ratio, the name of the solvent used, the percentage of active ingredients and whether other constituents have been added. In the case of purified dried extracts, it is best to declare classes of compounds, such as "anthocyanoside complex", "lipid-soluble fraction", or "total alkaloids".

For fresh plants, comparative chromatograms (TLC, HPLC and GC profiles) between the fresh herb, the preparation obtained from it, and the corresponding dried herb must be furnished. When preparations are derived from plants used as foods, the pharmaceutical dossier must be complete, but the toxicological dossier is not required.

Fixed combinations are a well-known part of phytotherapy, but to avoid a completely chaotic situation and a revival of preparations such as the ancient theriak, a list of drugs authorized in combinations is given. For example: cardiac erethism-neurosedative; spasmodic colitisdiarrhoea; renal elimination-diet; aperitive-asthenia; cough-bronchitis; antalgy-articular ache. Nevertheless, further proposed combinations can be taken into consideration if the presented argumentation is coherent.

The Future of Phytomedicines

An Increase in Phytovigilance. The temptation for search new exotic species and to introduce them on the market is very high. In fact, preparation of the dossier, especially regarding safety and clinical evidence, is more difficult. This is particularly true if the place of the plants in the botanical classification is more original (species belonging to unknown botanical families); if the application of the drug is more unusual (uncommon pharmaceutical forms), and if the biosynthesis leads to original secondary metabolites the pharmacotoxicology of which is unknown.

Furthermore, preparation of more and more potent extracts containing higher levels of active ingredients is a common goal for all manufacturers. Such is the case for *Valeriana*, Ginseng preparations etc. The search for more active medicines leads to the occurrence of side effects that are observed in clinical trials. In France, such observations have recently been described.

Outstanding toxicological data on plant drugs have been published *(6, 7)*. Seven cases of acute cytotoxic hepatitis were found in 1991 after the use of phytomedicines containing *Teucrium chamaedrys*. Since 1984, 26 cases of hepatitis have occurred after using this plant. Commercial preparations containing *T. chamaedrys* have been removed from the market, and prescription preparations has been forbidden. This plant now undergoes a particular legislation.

Other cases have been observed in Belgium with a preparation of Chinese herbs, *Magnolia officinalis* and *Stephania tetrandra* contaminated by the toxic herb, *Aristolochia fung chi*. Other accidental plant mixtures are unfortunately mentioned every year.

Some plants used in Europe (*Symphytum sp., Senecio sp., Petasites sp., Eupatorium sp., Tussilago sp.* etc.) contain alkaloids characterized by a pyrrolizidine nucleus. The presence of a double bond in position 1-2 and of an ester in position 9 confers a particular hepatotoxicity to these alkaloids. Their toxicity increases with the degree of esterification, with a macrocyclic diester being more toxic than a monoester.

Furthermore, the longer the acid chain is, the more toxic the molecule will be. These alkaloids are responsible for cellular liver necrosis, veno-occlusive diseases, fetal toxicity, etc. Their metabolites are alkylating agents for biological targets, which explains their genotoxicity. In fact, even with sensitive analytical techniques such as TLC, HPLC or CPG, it is very difficult to demonstrate the absence of this type of alkaloids in preparations such as tinctures. It is advisable to eradicate the use of such drugs, especially since the benefit/risk ratio is not very high.

Numerous other examples are also available in the literature. For instance, sesquiterpene lactones present in Asteraceae, Apiaceae, Amaranthaceae, Aristolochiaceae, Frullaniaceae, Lauraceae, Magnoliaceae, Menispermaceae and Polygonaceae can give rise to allergies. Polyines (Apiaceae, Araliaceae, Asteraceae, Euphorbiaceae, Fabaceae, Rutaceae, Solanaceae), ß-carboline derivatives (Cyperaceae, Fabaceae, Polygonaceae, Rubiaceae), furochromones (Rutaceae), furocoumarines (Apiaceae, Asteraceae, Fabaceae, Moraceae, Orchidaceae, Rutaceae, Solanaceae), and lignanes (Apiaceae, Amaryllidaceae, Asteraceae, Polygonaceae, Rutaceae, Solanaceae, Zygophyllaceae) can be phototoxic. A risk of cancer is not excluded with aristolochic acid, ptaquiloside, safrole, ß-asarone, etc. Safrole, estragole and essential oils such as *Artemisia dracunculus,* etc. can be genotoxic after metabolic activation.

At the European level, lists of herbal drugs that pose more severe risks without proven benefit have been published *(7)*. These include herbal drugs with serious risks and without any accepted benefit, drugs with toxic principles where a more detailed discussion concerning the benefit/risk ratio is necessary, preparations for external use, and herbal drugs with cardiac glycosides, alkaloids etc.

New Sophisticated Manufacturing Processes. Supercritical fluid extraction is a method of major interest nowadays, with numerous advantages in the manufacturing process. Thus, we can eliminate undesired lipophilic compounds such as pesticides. If this new manufacturing process is applied to obtain new types of extracts, such extracts are very different from those of traditionally used products. Hence, they constitute a new type of future generation plant extracts. For the moment, because the specific extracts represent a part of the whole plant with no traditional use, they must be considered as new products, requiring toxicological and clinical documentation.

Some Examples of Leading Commercial Extracts. Among the large variety of drugs sold on the European market, some of the leader examples will be discussed.

Ginkgo biloba is the main extract presently on the market. It is very specific in terms of preparation and composition and requires about ten different industrial steps to prepare the form used to treat vascular diseases. This type of extract has given rise to many pharmaco-toxicological and clinical trials and to numerous scientific papers in the world. It seems that the indications claimed are well-established, and clinical evidence has begun to be demonstrated on the basis of serious criteria.

Indeed, a great diversity of phytopharmaceutical preparations containing Ginkgo leaf extracts have recently flooded the market. Almost all of them claim the same therapeutic properties without having the same chemical profile, because the nature of the extraction is quite different, resulting in extracts with other specifications. This is the reason why in France we cannot accept the transposition of the therapeutic evidence to any given extract, especially since the initial extract is very well standardized. Therefore, a French generic extract is impossible.

Ginseng is a plant consumed worldwide with a long tradition of use. The various methods of preparation give rise to a large quantity of different forms. It is very difficult to fix a very strict chemical profile in relation to the therapeutic properties, considering the fact that some ginsenosides have stimulating activities, while others have a depressive effect. This drug has been included in the positive list, including a limit for the daily dosage.

Hypericum is a well-known plant used in Germany as antidepressant. In France, the problem was recently discussed. It was concluded that depression has to be considered as a major pathology and without new clinical proofs versus placebo and versus well-known reference chemical drugs, this type of indication cannot be allowed in France.

Echinacea also constitutes an important drug on the market and is used as an immunostimulant. The main problem is that the use of this plant is quite recent (about 50 years), and the traditional use by the Indian population of the US is not very credible. Moreover, the situation is complicated by the fact that the roots or aerial parts of about three different species are distinctly used. The chemical composition is well-known and very complicated (polysaccharides, alkylamides, polyphenols, etc.) on a qualitative and quantitative level. Many pharmacological tests have been set up, but clinical tests have not been established with strong protocols and the present results are questionable. This plant has not been included in the French list.

In conclusion, the list of the plants proposed by the French Ministry of Health is not definitive. New proposals may be formulated by the applicant, provided that the application is supported by a scientific dossier with up-to-date references, proof of safe use is furnished, and evidence of long term use is shown. In addition, closely related varieties, subspecies, and species may be proposed but the more they differ from those on the list or from the chemotypes usually employed, the more evidence the applicant must provide to support the new product. Moreover, the more the preparation differs from the usual form of administration (most often an herbal tea), the higher will be the level of evidence for safety. However, extracts obtained with solvents of different polarities may be recombined in order to reconstitute the initial balance of constituents in the drug.

The Role of Some Official Institutions

WHO. The notion of herbal remedies has a very broad international basis and a definition was established during the sixth international conference of drug regulatory authorities in Ottawa (Oct. 1991) *(8)*. Its objectives were *to promote collaboration between national drug regulatory authorities, to forge a consensus on matters of mutual interest, to facilitate timely and adequate exchange of technical information and to discuss contemporary issues of international relevance.*

WHO defined herbal remedies in terms close to the European definition: *"Finished, labeled medicinal products that contain as active ingredients aerial or underground parts of plants, or other plant material, or combinations thereof, whether in the crude state or as plant preparations. Plant material includes juices, gums, fatty oils, essential oils, and any other substances of this nature. Herbal medicines may contain excipients in addition to the active ingredients. Medicines containing plant material combined with chemically defined active substances, including chemically defined, isolated constituents of plants, are not considered to be herbal medicines".*

A quite different but non-negligible role is played by this international organization concerning the assessment of herbal medicines. The aim of a guideline published in 1991 is the following: *to assist national regulatory authorities, scientific organizations, and manufacturers to undertake an assessment of the documentation/submission/dossiers in respect of such products.*

The content of this guideline gives information on pharmaceutical assessment (GMP), crude plant material, plant preparations, finished product, stability, safety assessment, toxicological studies, documentation of safety based on experience, assessment of efficacy and intended us and product information for the consumer and promotion.

A recent traditional medicine program has been discussed in order to publish new data sheets about the most commonly used medicinal plants in the world. But the objectives of this work are quite different, giving complete information on worldwide distributed plants which can be used by local populations under well-known conditions (pharmaceutical form, dosage ...) instead of expensive synthetic drugs.

ESCOP (European Scientific Cooperative on Phytotherapy). In order to provide scientifically based assistance for a harmonious assessment of phytomedicines, ESCOP was founded in 1989. Having defined phytomedicines as *"medicinal products containing as active ingredients only plants, parts of plants or plant material or combinations thereof whether in the crude or processed state"*, ESCOP may play a key role for a European concurrence. The general purpose of ESCOP is *to develop a coordinated scientific framework, to promote the acceptance of phytomedicines, to support and initiate clinical and experimental research, and to improve international accumulation of scientific and practical knowledge.* Its main objective is to establish criteria that will contribute to the acceptance of phytotherapy on a European level.

Since November 1992, the ESCOP scientific committee has prepared original European monographs in the form of SPCs (Summary of Product Characteristics). The first six monographs, with a very precise technical content, have been submitted to the Committee for Proprietary Medicinal Products (CPMP). At the same time, an important work is being carried out concerning a priority list of some 50 plants in order to obtain the possibility of mutual recognition. The purpose of an SPC is to gather the best scientific information and to represent the fruit of cooperation between scientists of different disciplines and different countries. An SPC includes information on therapeutic indications, dosage, and pharmacological properties. Other paragraphs give many details on pharmacodynamics, pharmacokinetic properties, and preclinical safety data, each statement being supported by scientific references. For example, SPCs have been elaborated on *Sennae folium, Sennae acutifoliae fructus, Sennae angustifoliae fructus, Frangulae cortex, Matricariae flos, Valerianae radix*, etc.

Further cooperation among phytotherapy associations is being developed, and about 2500 persons are estimated to be involved in this organization. In the field of efficacy and safety, a trend will appear in the future, that will require more clinical data for phytomedicines. Therefore, the EU-Biomed Research Program (European Commission Biomedical and Health) has been set up to support ESCOP's work. It will provide research expertise on the determination of European standards for safe pharmacovigilance studies and the effective use of phytomedicines.

European Pharmacopoeia. This pharmacopoeia was created in 1963 and now represents 25 European countries. It consists of 16 groups of experts and has published more than 700 monographs.

One of the main purposes of the pharmacopoeia is to define assurance of the quality of the raw material. This is accomplished by describing control security, batch consistency, and batch safety. It employs analytical methods that are provided in the pharmacopoeial monograph, which consists of the following sections: the "definition" gives the content of active ingredients of the drug; the section "characters" comprises information that should not be regarded as constituting analytical requirements; the chapter "identification" includes all the mandatory tests performed to establish the identity of the drug; and the "assay" includes spectroscopic methods that may be used if a set of similar molecules of the same chemical class are thought to be responsible for the biological activity. GC or HPLC methods are used to quantitate precise active constituents. Information must also be given about the storage of raw material.

Before preparing new monographs, the most important questions are related to acceptance of new species, the choice of reference substances for TLC (the natural substances present in the drug generally are expensive) and the HPLC and GC fingerprints. The greatest difficulty lies in defining acceptable ranges and, sometimes, upper content limits of active constituents or negative markers when their toxicity is known.

Some Comments on the Assessment of Phytotherapy at a European Level

Different points of view have been discussed with respect to official aspects *(9-10)* and industrial concerns *(11-13)*. A consensus was established in 1988, as described in "Quality of herbal remedies" *(1)*. In 1994, the CPMP adopted final core-SPCs for four laxatives, but since then, no concrete progress has been apparent. However, the situation has made some progress: new procedures for EU marketing authorizations have been elaborated, and there is a new agency to coordinate activities. Nevertheless, some European countries have discovered that herbal medicinal products are on their markets and that the problem linked with the anarchy has been underestimated. But presently, it seems that a European harmonization of herbal remedies is not a first priority. In addition, divergent national points of view and traditions exist in different European member states, such as concerning the labeling, "traditionally used" products. On the other hand, for some persons, the label "traditional" could be seen as discrediting, especially as this term is not used in chemically defined substances with comparable conditions. It seems that this label is not acceptable on the basis of EC directives and might be used only for products limited to national markets such as the recent proposal of a committee on phytomedicines in London in 1995. A proposal for the assessment of phytomedicines clarifies the position of phytomedicines under the scope of 65/65/EEC *(14)*.

On a national level, phytomedicines are purely based upon national experience or tradition. The marketing authorization is subject to following requirements: quality, proofs of safety and efficacy set by national authorities, positive plant list, and positive indication list. A statement concerning the "traditional use" must be provided.

On a national and/or European level, phytochemicals are recognized by scientific standards. The quality must be proven and the requirements on safety and efficacy are covered by specific bibliographical references or monographs/SPCs or clinical trials according to an appropriate model. Concerning the area of use, the application should be supported by a monograph/SPC and/or indication and by clinical trials. Other main problems lie in the interpretation of terms like "well-established use" and "acceptable safety" and the apparent lack of guidance in this way.

A European solution has to be found because the exclusion of herbal remedies would lead to the paradoxical situation that homeopathic remedies would be accepted while herbal remedies would be excluded! Such an approach would be neither reasonable nor acceptable.

Some meetings from European experts are planned before the end of 1997 to discuss the different approaches in terms of national regulations and to find new coherent points of view in order to elaborate a European common legislation.

Conclusion

The potential of nature is very vast, because of its incomparable diversity and the numerous ethnopharmacological data currently available. A fantastic hope for the future may be concretely developed in two directions: (1) in terms of research, the constant discovery of new leader compounds gives rise to the industrial production of highly sophisticated molecules against severe pathologies; and (2) in terms of public health, the development of our environmental resources for producing new products such as herbal remedies. In each case, scientists and specialists of different disciplines need to develop a very close collaboration, because natural products are characterized by very specific and complex problems. A platform is needed to discuss these problems, to exchange experiences with the assessment of herbal drugs, and to create a special group facilitating mutual recognition.

We cannot accept the popular saying, "All that is natural is safe". The development of new drugs precisely requires quality, safety, and efficacy limits. Their harmonization should be a priority for the future, in order to set up a European free common market.

Literature Cited

1. Commission of the European Communities. The Rules governing Medicinal Products in the European Community, November 1988, Volume III. Note for Guidance: Quality of Herbal Remedies.
2. French Ministry of Health. French Procedures for Registration of Traditional Drugs. Notice to Applicants for Marketing Authorization. B.O. 90/22 bis, 1990.
3. Future Regulation of Herbal Products in Europe and Worldwide. AESGP Annual Meeting Istanbul, 29 May-1 June 1996.

4. Drug regulatory authorities of the Eastern Mediterranean regions. Report on the Joint WHO/DSE Regional Meeting Tunis, Tunisia, 2-8 November 1993.

5. Bauer, R.;Tittel, G. *Phytomedicine* **1996**, *2* (3), 193-198.

6. De Smet, P.A.G.M. In *Adverse Effects of Herbal Drugs*;De Smet, P.A.G.M.;Keller, K.;Hänsel, R.;Chandler, R.F. Eds. Springer Verlag: Berlin, **1992**, vol. 1, pp 1-72.

7. De Smet, P.A.G.M. In *Adverse Effects of Herbal Drugs*;De Smet, P.A.G.M.;Keller, K.;Hänsel, R.;Chandler, R.F. Eds. Springer Verlag: Berlin, **1993**, vol. 2, pp 1-90.

8. WHO. *Report of the Sixth International Conference of Drug Regulatory Authorities in Ottawa*, Canada, October 1991, pp 73-80.

9. Keller, K. *European Phytotelegram* **1994** (6), 40-47.

10. Keller, K. *Proceedings of the 3rd International Conference on Pharmacopoeias and Quality Control of Drugs*, Rome, November 1992, pp 243-250.

11. Steinhoff, B. *Br. J. Phytother.* **1993-94**, *3* (4), 190-193.

12. Steinhoff, B. *Br. J. Phytother.* **1993-94**, *3* (2), 76-80.

13. Grünwald, J.;Büttel, K. *Pharm. Industrie* **1996**, *58*, 209-214.

14. Proposal for the assessment of phytomedicines, AESG members meeting, London 31 January 1995, pp 36-37.

Chapter 3

The German Commission E Monograph System for Phytomedicines: A Model for Regulatory Reform in the United States

Mark Blumenthal

American Botanical Council, P.O. Box 201660, Austin, TX 78720

No longer relegated solely to the shelves of health food stores, herbs have become mainstream products and constitute the fastest growing category in drugstores. This requires an appropriate regulatory mechanism to review and evaluate the quality, safety, and efficacy of these natural products. The German Commission E stands out as the most rational in the world. In Germany herbs are routinely prescribed by physicians and sold in pharmacies. The Commission E system consists of a panel of experts convened under the auspices of the former Federal Health Agency which reviewed bibliographic information on approximately 300 herbs sold in Germany, including historical and traditional use, chemical studies, pharmacological and toxicological research, clinical studies, epidemiological data, and patient case files. Safety was assessed on a doctrine of absolute certainty, but efficacy was based on a more relaxed standard—a doctrine of reasonable certainty. The monographs, published in the German Federal Gazette, offer therapeutic guidelines including approved uses, contraindications, side effects, interactions with other drugs, dosage information, and related data. The American Botanical Council has translated and published all Commission E monographs into English to provide health professionals and the general public with accurate, authoritative assessments of herbal products.

One of the biggest questions facing the herb industry, consumers, Congress, and regulators is what kind of rules are appropriate for the regulation of herbal products. Specifically, the biggest question lies with the issue of how to transmit the health and therapeutic properties of herbs to health consumers.

The recent debate and eventual passage of the Dietary Supplement Health and Education Act of 1994 (DSHEA) was marked by intense debate surrounding the issue of health claims and how such health benefits of dietary supplements would be approved by the Food and Drug Administration (FDA). The Agency was concerned about the prospect that manufacturers wanted too lenient a standard for making *health claims* (long-term, dietary associated preventive claims as contrasted to *therapeutic or drug claims* which are more short-term and deal more with transient symptoms).

On the other hand, industry and many consumer groups were concerned about the possible overzealous regulation of dietary supplements by FDA and the agency's attempt to classify as drugs any high potency supplements that it deemed intended for therapeutic uses, as evidenced by proposed rules FDA issued in June 1994. As a result of DSHEA, American health consumers now have assurance that many of the supplements they ingest in the belief that these products convey a health benefit will be protected from over-regulation by the FDA. The new law allows manufacturers to provide safety information and warnings that formerly were restricted to drug labeling. DSHEA also allows manufacturers to make limited "structure and function" claims as to how an herbal product affects the structure or function of the body so long as the claim is not therapeutic in nature and does not imply the treatment or cure of a pathology.

Despite the gains claimed by the dietary supplement consumers and industry in the passage of the DSHEA, the new law does not deal with the widespread use of herbs and phytomedicines for their therapeutic benefits. A new regulatory framework is still needed to create an appropriate set of criteria and procedures for the approval of herbs as therapeutic agents or drugs.

Unfortunately, the current system for the assessment of ingredients for use in nonprescription or OTC (over-the-counter) drugs is appropriate for only a few phytomedicines. In 1992, a group of European and American phytomedicine manufacturers and marketers (the European-American Phytomedicines Coalition, EAPC) filed a petition with the FDA that requested that Agency to classify as *old drugs* herbal medicines or phytomedicines marketed in Europe for a "material time" and for a "material extent." These would then fall under the FDA OTC Drug Review, rather than being classed as *new drugs* requiring the expensive and time-consuming New Drug Application (NDA) process established for most conventional drugs *(1)*. On October 3, 1996, FDA published an Advanced Notice of Proposed Rulemaking which indicated that the Agency might be more willing to accept foreign market data in consideration of proposed OTC drug status for an ingredient; however, a number of conditions were suggested that may limit the likelihood that leading European phytomedicines might be reviewed *(2)*. The proposed regulations are being reviewed by FDA in light of public comments made to the Agency *(2)*.

The American Botanical Council, a nonprofit research and education organization, has instituted a review of six major industrialized nations and how they approach the challenge of regulating herbs and phytomedicines as nonprescription medicines. In this review (called the Traditional Medicine Research Project), one particular model stands out, the German Commission E. This Commission is an independent division of the former Federal Health Agency, now the German Federal Institute for Drugs and Medical Devices (analogous to the FDA), that *actively* collected information on herbal medicines and evaluated them in relation to their safety and efficacy. These evaluations were published in the form of brief monographs in the German Federal Gazette (analogous to the *Federal Register*) and were intended as package inserts for herbal drugs sold in pharmacies.

The method of approval is quite interesting. Commission E reviewed information about each herb's history of use and chemistry, pharmacological, and experimental studies on animals, human clinical studies if available, epidemiological studies, and even subjective evaluations by patients and physicians from clinical experience. Emphasis was placed on safety when efficacy judgments were made. According to Varro E. Tyler, Distinguished Professor Emeritus of Pharmacognosy at Purdue University and long the senior author of the textbook *Pharmacognosy,* has noted, "Commission E evaluates efficacy based on a doctrine of reasonable certainty," as contrasted with our FDA's insistence on a "doctrine of absolute proof" based on information that *is passively* submitted to FDA from drug manufacturers *(3).*

The Commission E was established in 1978 and has since published over 400 monographs (including revisions) covering over 300 herbs and herb combinations sold in Germany. The Commission includes people with expertise in various aspects of medicinal plant research and use: physicians, pharmacists, pharmacologists, and toxicologists, as well as representatives of the pharmaceutical industry and lay persons. Of these monographs, about 200 are positive; that is, they approve the use of an herb for particular use or uses. About 100 are negative assessments, usually based on either lack of sufficient data to approve actions and/or significant toxicity concerns.

Herbs and medicinal plant products have always been popular and widely respected in Germany. In his introduction to the American Botanical Council's forthcoming English translations of these monographs, Tyler writes *(3):*

> The therapeutic use of herbs and phytomedicines has always been very popular in Germany. About 600-700 different plant drugs are currently sold there, singly and in combination, in *Apotheken* (pharmacies), *Drogerien* (drugstores), *Reformhäuser* (health food stores), and *Märkte* (markets). In addition to the self-selection of herbal products by consumers, about 70% of the physicians in general practice prescribe the thousands of registered herbal remedies, and a significant portion of the $1.7 billion annual sales (a conservative estimate) is paid for by government health insurance. In 1988, 5.4 million prescriptions were written for a single phytomedicine, *Ginkgo biloba* extract, a figure that does not include the substantial over-the-counter sales of the product.
>
> In view of this significant role which phytomedicines play in Germany, it is only natural that the government there would develop a mechanism to assure users of their safety and efficacy. The process is unique. For various reasons, even other advanced nations have not yet chosen to emulate it. But it is worthy of imitation, and it is probably only a matter of time before consumers in other countries are able to benefit from the German experience.

A review of the first 285 monographs revealed that most have a positive assessment, while 66% mention various risks, and 58 have no plausible evidence of efficacy resulting in a negative assessment. The monographs list known contraindications for the various herbs. Some of the more frequently mentioned contraindications include allergy to active constituent (63 monographs), restricted use during pregnancy and/or lactation (24 monographs), gallstones (15 monographs) and inflammatory kidney disease (7 monographs). Thirty-five monographs limit the period of use to reduce side effects. Types of side effects include gastrointestinal disorders (35 monographs), allergic reactions - mostly skin reactions with some more severe generalized reactions (30 monographs), photosensitivity (5 monographs). In addition, 7 monographs mention that the herb may influence the absorption of other drugs taken simultaneously *(4, 5)*.

The American Botanical Council has undertaken the translation and publication in English of the German Commission E Monographs. All monographs published in Germany through 1994 have been translated and will be published in the fall of 1997. Multiple cross references and indices are being created to increase the usefulness and ease of use of the monographs *(6)*.

Prof. Tyler has some glowing words for the Commission E Monographs: "They represent the most accurate information available in the entire world on the safety and efficacy of herbs and phytomedicines. As such, they are worthy of careful study by anyone interested in any type of drug therapy. Ignorance of the Commission E monographs is ignorance of a substantial segment of modern medicine" *(3)*.

Approved Uses and Dosage

The following are examples of the approved uses for some of the leading phytomedicines sold in the U.S. according to the Commission E Monographs, with the suggested doses. Regarding doses, all begin with the phrase "Unless otherwise prescribed" and then the specific general dose recommendation is given. Also, the phrase "Equivalent preparations" is usually added in reference to the practice of prescribing and using preparations which are generally equivalent in active constituents to the drug form and dosage prescribed, e.g., the use of a water infusion or tincture which is equivalent to a stated dose of dried herb (drug). Other sections of the monographs (e.g., Contraindications, Side Effects, Mode of Administration, Duration of Application, Actions) are not included here. For a full text of each monograph, see Blumenthal et al. 1997 *(6)*.

Asian Ginseng root *(Panax ginseng)*
• Tonic for invigoration and fortification in times of fatigue and debilitation or declining capacity for work and concentration.
• Convalescence.
Daily dosage: 1–2 g of root; equivalent preparations.

Chaste Tree fruits *(Vitex agnus-castus)*
• Irregularities of the menstrual cycle.
• Premenstrual complaints.
• Mastodyma.
  Note: In case of feeling of tension and swelling of the breasts and at disturbances of menstruation, a physician should be consulted for diagnosis.
Daily dosage: aqueous-alcoholic extracts corresponding to 30–40 mg of the drug.

German Chamomile flower *(Matricaria recutita)*
• Externally: Skin and mucous membrane inflammations, as well as bacterial skin diseases, including those of the oral cavity and gums. Inflammations and irritations of the respiratory tract (inhalations). Ano-genital inflammation (baths and irrigation).

• Internally: Gastrointestinal spasms and inflammatory diseases of the gastrointestinal tract.

Daily dosage: Boiling water (ca. 150 ml) is poured over a heaping tablespoon of chamomile (ca. 3 g), covered, and after 5–10 minutes passed through a tea strainer. Unless otherwise prescribed, for gastrointestinal complaints a cup of the freshly prepared tea is drunk three to four times a day between meals. For inflammation of the mucous membranes of the mouth and throat, the freshly prepared tea is used as a wash or gargle.

External: for poultices and rinses, 3–10% infusions; as a bath additive, 50 g to 10 liters (approx. 2.5 gal.) water; semi-solid formulations with preparations corresponding to 3–10% herb.

Echinacea purpurea herb *(Echinacea purpurea)*
• Internally: Supportive therapy for colds and chronic infections of the respiratory tract and lower urinary tract.
• Externally: For use on poorly healing wounds and chronic ulcerations.

Internal daily dosage: 6–9 ml expressed juice. (This refers to fresh juice of above-ground parts.) [Dosage is also offered for injectible preparations.]

External dosage: Semi-solid preparations containing at least 15% pressed juice.

Echinacea pallida root *(Echinacea pallida)*
• Supportive therapy for influenza-like infections.

Daily dosage: Tincture (1:5) with 50% (v/v) ethanol from native dry extract (50% ethanol, 7–11:1), corresponding to 900 mg herb.

Eleuthero (Siberian) Ginseng rhizome *(Eleutherococcus senticosus)*
• Tonic for invigoration and fortification in times of fatigue and debilitation or declining capacity for work and concentration.
• Convalescence.

Daily dosage: 2–3 g of root.

Garlic *(Allium sativum)*
• Supports dietary measures to reduce serum cholesterol levels.
• Preventative measure for age-related vascular changes.

Daily dosage: 4 g fresh garlic; equivalent preparations.

Ginger root *(Zingiber officinale)*
• Indigestion.
• Prevention of motion sickness.

Daily dosage: 2–4 g rhizome; equivalent preparations.

Ginkgo biloba leaf extract *(Ginkgo biloba)*
[This preparation is a concentrated dry extract (35–67:1, average 50:1) extracted with acetone and water. The preparation is standardized at 22 to 27% flavone glycosides and 5–7% terpene lactones, of which approximately 2.8–3.4% consists of ginkgolides

A, B, and C, as well as approximately 2.6–3.2% bilobalide. Ginkgolic acid should be less than 5 ppm. The given ranges include manufacturing and analytical variances.]
Uses:
• (a) For symptomatic treatment of disturbed performance in organic brain syndrome within the regimen of a therapeutic concept in cases of demential syndromes with the following principal symptoms: memory deficits, disturbances in concentration, depressive emotional condition, dizziness, tinnitus, and headache. The primary target groups are dementia syndromes, including primary degenerative dementia, vascular dementia, and mixed forms of both.

Note: Prior to starting treatment with ginkgo extract, clarification should be obtained as to whether the pathological symptoms encountered are not based on an underlying disease requiring a specific treatment.
• (b) Improvement of pain-free walking distance in peripheral arterial occlusive disease in Stage II of Fontaine (claudicatio intermittens) in a regimen of physical therapeutic measures, in particular walking exercise.
• (c) Vertigo and tinnitus of vascular and involutional origin.
Daily dosages:
Indication (a): 120–240 mg extract in 2 or 3 doses.
Indications (b) and (c): 160 mg extract in 2 or 3 doses.

Hawthorn leaf with flower *(Crataegus monogyna; C. laevigata)*
• Decreasing cardiac output as described in functional level II by the NYHA (New York Heart Association).
Daily dosage: 160–900 native, water, ethanol extract (ethanol 45% v/v or methanol 70% v/v, drug-extract ratio = 4–7:1, with defined flavonoid or procyanidin content), corresponding to 30–168.7 mg procyanidins, calculated as epicatechin, or 3.5–19.8 mg flavonoids, calculated as hyperoside in accordance with DAB 10 (German Pharmacopoeia 10th ed.).

Horse Chestnut seed *(Aesculus hippocastanum)*
(This drug is defined as a dry extract adjusted to 16–20% triterpene glycosides, calculated as anhydrous aescin.)
• Treatment of complaints found in pathological conditions of the veins of the legs (chronic venous insufficiency), for example pain and a sensation of heaviness in the legs, nocturnal systremma (cramps in the calves), pruritis, and swelling of the legs.

Note: Other non-invasive treatment measures prescribed by a physician, such as leg compresses, wearing of supportive elastic stockings, or cold water applications, must be observed under all circumstances.
Daily dosage: 100 mg aescin (escin) corresponding to 250–312.5 mg extract two times per day in delayed release form.

Kava Kava root *(Piper methysticum)*
• Conditions of nervous anxiety, stress, and restlessness.
Daily dosage: Herb and preparations equivalent to 60–120 mg kava pyrones.

Milk Thistle fruit *(Silybum marianum)*
• Crude drug (dried seeds): indigestion.

• Standardized extract: toxic liver damage and supportive treatment of chronic inflammatory diseases of the liver and liver cirrhosis.
Daily dosage of crude drug: 12–15 g ; of standardized preparation: 200–400 mg of silymarin, calculated as silibinin.

Saw Palmetto fruit *(Serenoa repens)*
• Urination problems in benign prostate hyperplasia stages I and 2.
Daily dosage: 1–2 g sabal berry or 320 mg lipophilic ingredients extracted with lipophilic solvents (hexane or ethanol 90% v/v).

St. John's wort herb *(Hypericum perforatum)*
• Internally: For psychovegetative disorders, depression, anxiety and/or nervous restlessness. Oily hypericum preparations for indigestion.
• Externally: Oily hypericum preparations for treatment in post-therapy of acute and contused injuries, myalgia and first degree burns.
Daily dosage for internal use: 2–4 g of drug or 0.2–1.0 mg of total hypericin in other forms of drug application.

Valerian root *(Valeriana officinalis)*
• Restlessness, sleeping disorders based on nervous conditions.
Dosage:
Infusions (teas): 2–3 g drug per cup, once to several times per day.
Tincture: ½–1 teaspoon (2–5 ml), once to several times per day.
Extracts: Amount equivalent to 2–3 g of drug, once to several times per day.
External use: 100 g for one full bath; equivalent preparations.

Literature Cited

1. Pinco, R.G. and L.D. Israelsen. European-American Phytomedicines Coalition Citizen Petition to Amend FDA's OTC Drug Review Policy Regarding Foreign Ingredients. July 24, 1992.
2. Food and Drug Administration. Eligibility Criteria for Considering Additional Conditions in the Over-the-Counter Drug Monograph System Request for Information and Comments. *Federal Register* Vol. 61 (193): 51625-31. Oct. 3, 1996.
3. Tyler, V.E. In *German Commission E Monographs: Therapeutic Monographs on Medicinal Plants for Human Use;* Blumenthal, M.,; Gruenwald, J.; Hall, T.; Riggins, C.W.; Rister, R.S. (eds.); Klein, S.; Rister, R.S. (transl.).; American Botanical Council: Austin, TX, 1997.
4. Bergner, P. *HerbalGram* **1994,** *30:* 17 & 64.
5. Keller, K. *Zeitschrift Phytotherapie* **1992,** *13; 116-120* (English).
6. Blumenthal, M.; Gruenwald, J.; Hall, T.; Riggins, C.W.; Rister, R.S. (eds.); Klein, S.; RS. Rister (transl.). *German Commission E Monographs: Therapeutic Monographs on Medicinal Plants for Human Use.* American Botanical Council: Austin, TX, 1997.

Chapter 4

Botanicals: A Current Regulatory Perspective for the United States

Loren D. Israelsen

Utah Natural Products Alliance, 2046 E. Murray-Holladay Road, Suite 204,
Salt Lake City, UT 84117

Herbs enjoy a unique regulatory status because of the Dietary
Supplement Health and Education Act of 1994 (DSHEA). Scientific
literature may now be used in connection with the sale of herbs to
consumers. A new class of claims called Statements of Nutritional
Support are being used to promote the physiological effects and
benefits of herbs. This has resulted in an extraordinary growth of herb
sales in the U. S. Important issues are now being debated concerning
the establishment of pharmacopeial standards for herbs, the OTC
status of herbs such as valerian and ginger, the major changes in new
drug review standards for botanicals, and the protection of proprietary
research on standardized plant extracts. This chapter will present an
overview of these issues, with detailed analysis of current law and
emerging policies.

Botanicals have always been a challenge to regulators. The very scope of these products
and their wide and varied uses make legal definitions and regulatory policy difficult. For
many reasons, the FDA has either ignored or attacked botanicals for over 50 years.
Following World War II, the rise of synthetic drugs, patent protection and great success
in conquering feared diseases were key factors that molded FDA's structure, priorities
and expertise. The Center for Drug Evaluation and Research (CDER) employs 1475
persons (1), many of whom hold Ph.D.'s or M.D.'s, while only a handful have academic
or research experience in pharmacognosy or related disciplines. Botanicals, once the core
of businesses such as Parke Davis, Merck, Upjohn and Lilly, became reception room
artifacts and the trade of ephemera collectors. Between 1900 and 1960, the vast majority
of botanicals lost their USP status. Without industrial support and patent protection,
research funds disappeared. Unable to meet stricter new drug approval requirements,
botanicals were relegated to haphazard review under the 1972 OTC drug review system.
Today, only a handful of botanicals of commercial importance are recognized as OTC

active ingredients such as capsaicin (Category I), senna, cascara sagrada (Category III) and slippery elm bark (a crude drug).

On the food side, the 1958 Generally Recognized as Safe (GRAS) list recognized a number of botanicals as Generally Recognized As Safe. However, this recognition extended largely to the flavoring uses of botanicals for alcoholic beverages. Where would gin be without juniper berries? Later in 1974, the agency published a compliance policy guideline which purported to classify selected herbs as safe, unsafe, or of undetermined safety. This was done in response to a field inquiry following a complaint about the use of herbs by the White Woman's Magic Society of Houston, Texas. This document listed 27 herbs as unsafe, and between 1974 and 1986 this was the principal basis for enforcement action against herbs by FDA. The industry in turn widely criticized this document as lacking scientific legitimacy or factual accuracy.

Passage of the Nutrition Labeling Education Act (NLEA) in 1990 brought renewed hope that botanicals would enjoy an improved status, particularly with respect to the new health claim provisions which, for the first time, allowed disease/nutrient relationship claims. FDA's subsequent rule-making proposed to treat botanicals as non-nutrients barring any chance of health-claim approvals. FDA rejected extensive industry comments objecting to this proposal. This was the last straw. With no other options available, the Utah botanical industry which represents nearly $1 billion in annual sales, joined forces and laid out a set of legislative principles they believed were necessary to resolve these regulatory attacks. In June of 1992, Senator Orrin Hatch introduced the Health Freedom Act, modeled after the Utah template. This legislation passed in 1994 as the Dietary Supplement Health and Education Act (DSHEA). It became the largest constituent issue before the U.S. Congress since the Vietnam War. Now widely regarded as the model for consumer-driven campaigns, the dietary supplement industry accomplished the nearly impossible. So stunning was this victory, that renowned food and drug lawyer Peter Barton Hutt remarked,

> "They therefore went to Congress and succeeded in obtaining what was without question the single worst legislative defeat that FDA has ever encountered in its entire history...This was an extraordinary and humiliating defeat for FDA...Thus, the impact of the 1994 statute is enormous."*(2)*

Passage of the DSHEA signaled two things:
(1) A fundamental reform of FDA's haphazard and ineffective regulation of botanicals, and
(2) Dietary supplement consumers are a political and economic force to be reckoned with.

Central to the bill (and a constant point of negotiating friction) was the status and regulation of botanicals. I have "reconstructed" the DSHEA to focus exclusively on how it defines and now treats botanicals. While many of these provisions also apply to other dietary ingredients, review of the law with an eye toward botanicals provides a useful and concise picture of the new regulatory framework for this specific class of products.

Botanical Summary of DSHEA

1)   (Section 3) Definition.   The term "dietary supplement" means an herb or other botanical or a concentrate, constituent, extract or combination of any botanical that is intended for ingestion as a tablet, capsule or liquid form and is not represented for use as a conventional food or as a sole item of a meal or the diet and is labeled as a dietary supplement. This includes new drugs which were marketed as botanicals prior to such approval but does not include a botanical approved as a new drug or authorized for investigation as a new drug for which substantial clinical investigations have begun and made public and which was not before such time marketed as a dietary supplement.

2)   Botanicals are Not Food Additives.

3)   (Section 4). A botanical is considered unsafe if it presents a significant or unreasonable risk of illness or injury under conditions of use recommended or suggested in labeling, or it is a new botanical for which there is inadequate information to provide reasonable assurance that such botanical does not present a significant or unreasonable risk of illness or injury. In any preceding under this paragraph, the United States shall have the burden of proof to show that a botanical is adulterated.

4)   (Section 5) Botanical Supplement Claims. A publication including an article, a chapter in a book or an official abstract of a peer-reviewed scientific publication that appears in an article and was prepared by the author or editors of the publication which is reprinted in its entirety shall not be defined as labeling when used in connection with the sale of botanicals to consumers when, (1) it is not false or misleading, (2) does not promote a particular manufacturer or brand of botanical, (3) is displayed or presented with other items on the same subject matter so as to present a balanced view of the available scientific information on a botanical, and (4) if displayed in an establishment, is physically separate from the botanical and does not have appended to it any information by sticker or other method. This provision shall not apply to or restrict a retailer of botanicals from selling books or publications as a part of their business.

5)   Section 6. A statement for a botanical may be made if:

(a)   The statement describes the role of a botanical intended to affect the structure or function in humans, characterizes the documented mechanism by which a botanical acts to maintain such structure or function or describes general well-being from consumption of a botanical;

(b)   The manufacturer of the botanical has substantiation that such a statement is truthful and non-misleading, and

(c)   The statement contains prominently displayed and in bold-faced type the following: "This statement has not been evaluated by the Food and Drug Administration. This product is not intended to diagnose, treat, cure or prevent any disease." Such statements cannot claim to diagnose, mitigate, treat, cure or prevent any disease. The manufacturer of a botanical making such a statement shall notify the secretary (FDA) within 30 days after first marketing such a statement.

6)   Botanical dietary supplements are misbranded if:

(a)   It is a botanical dietary supplement and the label or labeling of the botanical fails to list the name of each ingredient, the quantity of such ingredients, or, if a proprietary blend, the total quantity of all ingredients.

(b)   The label or labeling of the botanical fails to identify the product by the term "dietary supplement."

(c) The label or labeling of the botanical fails to identify any part of the plant from which the ingredient is derived.

(d) The botanical is:

❏ Covered by the specifications of an official compendium;

❏ Is represented as conforming to such official compendium and fails to do so; or

❏ The botanical is not covered by an official compendium and fails to have the identity and strength that it represents to have, or

❏ It fails to meet the quality, purity or compositional specifications based on validated assays or other appropriate methods that it is represented to meet.

7) (Section 8). A botanical shall be deemed adulterated unless:

(a) The product contains only botanicals which have been present in the food supply as an article used for food in a form in which the food has not been chemically altered, or

(b) There is a history of use or other evidence of safety establishing that a new* botanical, when used under conditions recommended in the labeling, will reasonably be expected to be safe and at least 75 days before introduction into commerce to provide the secretary with information on which the manufacturer has concluded that the botanical will reasonably be expected to be safe.

*A new botanical means one that was not marketed in the United States before October 15, 1994.

8) (Section 9) Good Manufacturing Practices. A botanical is unsafe if it is prepared, packed or held under conditions that do not meet current good manufacturing practices. The secretary may by regulation prescribe good manufacturing practices for botanicals which shall be modeled after current GMP's for food and may not impose standards for which there is no current and generally available analytical methodology.

9) (Section 10). A botanical is not a drug solely because its label or labeling contains a statement of nutritional support. Also, a botanical shall not be deemed misbranded if its label or labeling contains directions or conditions of use or warnings.

10) (Section 12) Commission on Dietary Supplement Labels. A commission shall conduct a study on and provide recommendations for the regulation of label claims and statements for botanicals including the use of literature in connection with the sale of botanicals and procedures for evaluation of such claims. The commission shall evaluate how best to provide truthful, scientifically valid and not misleading information to consumers.

11) (Section 13) Office of Dietary Supplements. The secretary shall establish an Office of Dietary Supplements (ODS) within the National Institutes of Health (NIH). The purposes of the office are as follows:

(a) To explore the potential role of botanicals to improve health care.

(b) To promote scientific study of the benefits of botanicals in maintaining health and preventing chronic disease.

(c) The Director of the ODS shall:

❏ Conduct and coordinate scientific research relating to botanicals

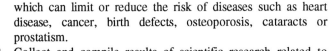

which can limit or reduce the risk of diseases such as heart disease, cancer, birth defects, osteoporosis, cataracts or prostatism.

❑ Collect and compile results of scientific research related to botanicals.

❑ Serve as a principle advisor to the NIH, the Centers for Disease Control and Prevention (CDCP) and the Commissioner of FDA on issues relating to botanicals including botanical regulations, the safety of botanicals, scientific issues arising in connection with the labeling and composition of botanicals and shall compile a database of scientific research on botanicals.

As can be seen, the DSHEA lays down a new and unique approach to defining and regulating botanicals. I am unaware of any other regulatory system or approach like this in the world. It is nearly two years since the DSHEA became law. A great deal of work has been done, both by industry and by FDA, to implement and to further develop various aspects of this law. The following are important areas of general interest.

1) <u>Statements of Nutritional Support</u>. The FDA has received over 1,000 notices for statements of nutritional support, the majority of which are of botanical origin. Unfortunately, no accompanying sales data are presently available to determine how successful this new class of claims is in the marketplace.

2) <u>Nutrition and Other Labeling Regulations</u>. The FDA has issued proposed regulations setting out formats for botanical dietary supplement labels, nutrition labeling, etc. After careful review, the botanical industry identified several serious concerns and filed extensive comments objecting to various aspects of these regulations.

3) <u>Good Manufacturing Practices</u>. The agency has recently published the industry's draft GMP's as an Advanced Notice of Proposed Rulemaking (Docket No. 96N-0417). Following a public comment period, it is expected these regulations will proceed through formal rule-making and be officially adopted. This is seen as a very favorable step for both FDA and the botanical industry. Important issues remain with regard to how nutrients and botanicals will be treated with respect to identification of raw materials, microbiological contamination, appropriate analytical techniques, etc.

4) <u>The Commission on Dietary Supplement Labels</u>. This Commission was a political compromise when the Senate Labor and Human Resources Committee could not agree on how dietary supplements should fit into the larger NLEA health claims scheme. The Commission is composed of seven members. Members with expertise in botanicals include Dr. Norman Farnsworth, University of Illinois at Chicago, and Robert McCaleb, President, Herb Research Foundation. The Commission has held field hearings and received extensive comments from interested parties and is now preparing a report of their findings and recommendations for the President and the Congress. The Commission hopes to have their report completed by June of 1997. The Commission appears to have been persuaded that botanicals should be distinguished from nutrients and accorded separate status as either herbal remedies, traditional medicines or similar. Many believe this is appropriate as it pertains to claims structures but not at the expense of losing dietary supplement status with respect to other provisions of the DSHEA.

5) The Office of Dietary Supplements (ODS). The ODS is now headed by Dr. Bernadette Marriott. Initially, its focus has been on nutrients. However, more recently, the ODS has begun to focus attention on the role of botanicals in health care and is establishing a priority list of botanicals based on sales and on therapeutic importance in order to identify appropriate research and public education priorities. Much of the important work regarding the long-term place of botanicals in the U.S. marketplace is yet to be decided. Many important decisions will be made over the next 18 months that will greatly affect both the requirements for the manufacture and sale of these products but also allow the claims, creation of research incentives or recognition of botanicals as new prescription or OTC drugs.

Other Developments Affecting the Regulatory Status of Botanicals

1) U.S. Pharmacopeia. In July 1996 the *U.S. Pharmacopeia (USP)* held an open conference to discuss approaches for establishment of quality standards and use information for botanicals. Four draft monographs on valerian, ginger, garlic and ginkgo have been published in the pharmacopoeial forum for comment. Further monographs are expected to be published as well. The establishment of *USP* standards for botanicals would serve three important purposes:

❑ To legitimize and distinguish quality botanicals in the mind of the public and professionals.
❑ Open the way for FDA drug review and approval of botanicals as drugs (traditional medicines) for which *USP* standards may exist.
❑ Become an authoritative source of usage information for health care professionals.
❑ Set the stage for reimbursement or health insurance coverage for botanicals.
❑ Create incentives for compliance with GMP's in order to meet *USP* standards.

2) FDA. There are signs of activity within FDA. The agency has acted as a co-sponsor of several conferences on botanicals over the last 18 months and is now rumored to be organizing a botanical review committee within the Center for Drug Evaluation and Research (CDER). The role of such a committee remains unclear but signals an intent to devote greater resources to the botanical area.

The agency has also hinted that it will soon publish a notice in the *Federal Register* responding to the 1992 citizen petition of the European American Phytomedicine Coalition (EAPC) which seeks recognition of well-established phytomedicines as old drugs in the U.S. for OTC drug review purposes.

Summary and Conclusions

For many years, botanicals held a very unsure and tenuous place in both American society and FDA regulation. The recent resurgence of botanicals for dietary and medical use has, in part, led to passage of the DSHEA which affected important changes with respect to the regulation of botanicals. This new status as dietary supplements creates both opportunity and also many new questions.

Ultimately, botanicals in the U.S. may hold no less than five levels of regulatory recognition. They are:

- ❏ New Prescription Drugs
- ❏ OTC Drugs
- ❏ Traditional Medicines
- ❏ Dietary Supplements
- ❏ Conventional Foods

Depending on the scientific and commercial resources available, various companies may elect to market botanicals in one or more of these categories.

Barring an unforeseen reversal of law, botanicals as dietary supplements will continue to enjoy impressive growth, and as the American consumer becomes more familiar with the various dietary and medicinal benefits of these products, they will continue to include botanicals as a part of their life both in the kitchen and in the medicine chest.

Literature Cited

1. FDA Almanac, Fiscal Year 1995.
2. Guide to U.S. Food Labeling Law, February 1996.

SPECIFIC EFFECTS

Chapter 5

Search for New Plant Constituents with Antiasthmatic and Antihypertonic Activity

H. Wagner

Institute of Pharmaceutical Biology, University of Munich, Karlstrasse 29,
D–80333 Munich, Germany

The development of target-directed *in vitro* screening methods (specific enzymes, primary cell cultures and receptors) has considerably stimulated the search for new potential drugs from plants. Inspite of the fact that the results obtained *in vitro* often have no counterpart *in vivo* and have to be confirmed in experimental animal studies, this new approach opens up the possibility of designing leading structures for the development of new drugs. This review summarizes results obtained in the author's research laboratory in recent years in an extensive program, which comprises the following areas: <u>allergic asthma and cardiovascular diseases, in particular, high blood pressure.</u> For the screening of isolated compounds and the monitoring of plant extract fractionations key enzymes and primary cell cultures of humans and animals as well as some *in vivo* models were used. The drugs for screening were selected primarily following ethnopharmacological guidelines. The results of structure-activity relationship studies are discussed with regard to their therapeutic relevance.

Antiinflammatory drugs and those for the treatment of cardiovascular diseases rank second and third on the list of new drugs wanted for rational therapy, respectively. Although we possess a great number of powerful synthetic drugs we must confess that we do not have enough causally and rationally applicable drugs. In addition a great number of them have the handicap of causing severe adverse effects and being too toxic. In this context, it might also be worth stressing that from about 2000 registered human diseases and disorders, only about 40% can be cured. Bearing this in mind, it is essential to utilize the almost limitless resources provided by the plethora of potential medicinal plants of traditional medicine for an overall and thorough screening. However, only in a few exceptional cases will a screening program such as this result in a highly effective drug. Most of the new plant constituents found by a bioguided fractionation can be regarded as leading structures only, which may serve as starting templates for synthetic modification with the further possibility of optimizing bioavailability and pharmacokinetic properties and, hence, the efficiency of this plant constituent. It is an interesting coincidence that a battery of pharmacological *in vitro* methods has recently become available, which readily lend themselves to automation. They comprise the use of enzymes, cells or receptors. In most cases these target-directed methods are specific, and have practical and other important implications with regard to operational time, costs and especially medical ethics. The only disadvantage of these programs is that the results obtained *in vitro* often have no counterpart *in vivo*. In other words, in each case the *in vivo* efficacy must be proven and confirmed in experimental animal studies. Furthermore, we must keep in mind that due to the complex pathophysiology of many diseases it is not always possible to find and select the adequate target model.

An obvious advantage of this target-directed screening program is the chance of finding the major active principles of a medicinal plant, which would make it possible to standardize plant extracts, a requirement for reproducible pharmacological and clinical studies.

In the following we present the most important results of our screening program for new potent antiasthmatic and antihypertonic drugs over the past 10 years.

Screening for Antiasthmatic Plant Constituents

Methodology. The complex pathophysiology of allergic asthma and the many possible sites of attack for antiasthmatic drugs, as shown in Figure1, suggested that we should start screening the plant extracts with an *in vivo* model, and use the *in vitro* methods later for investigating the mechanism of action of the isolated compounds. We used a noninvasive whole-body computer-aided plethysmographic lung function model according to Dorsch (*1*). This very sensitve method measures the degree of bronchial obstruction and records not only the parameter 'compressed air', but also the airway resistance thorax gas volume, maximum flow rates, tidal volume, breathing frequency, and I/E (inspiration/expiration) ratio. The substances, or extract fractions under investigation, were applied either orally or as an aerosol to the guinea pigs sensitized to ovalbumin 1 h or 30 min prior to inhalation challenges with PAF (platelet activating factor) or ovalbumin (*2*).

Since it is generally accepted that allergic asthma is generated as a consequence of inflammatory processes, all or some of the following *in vitro* effects on mediator systems and cell functions were measured: cyclooxygenase, 5-lipoxygenase, chemotaxis, biosynthesis of LTB_4 and LTC_4, thromboxane, biosynthesis, chemoluminescence and histamine release. In some cases the *in vivo*-rat paw edema model in mice was also carried out.

Results of the Screening Program. From the many plant drugs used in traditional medicine of different countries for the treatment of allergic asthma, three drugs were selected for thorough screening: *Allium cepa*, *Picrorhiza kurroa*, and *Galphimia glauca*.

Allium cepa. Onion juice has been recommended for centuries for the treatment of pain and swelling after bee or wasp stings. In homeopathy *Allium* D_4 is used as an expectorant, antitussive and as a remedy against allergic rhinitis and bronchitis. A

48

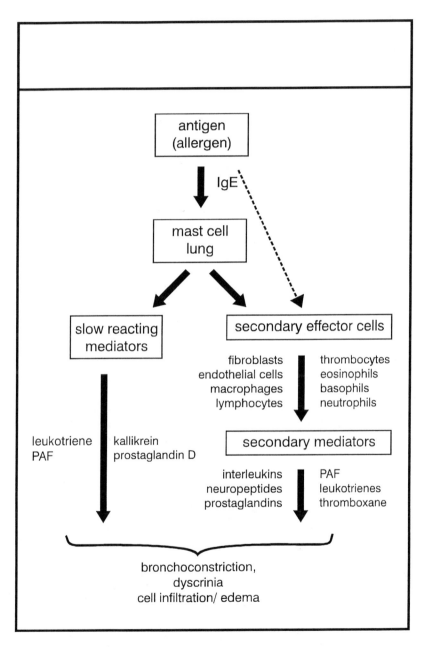

Figure 1. Pathophysiology of allergic asthma.

preliminary study carried out with onion juice administered subcutaneously to one asthma patient revealed a marked reduction of the immediate and late bronchial reaction. In another study with adult voluntccrs, topical onion treatment (intradermal injection of an ethanolic onion extract) prior of the application of anti IgE showed a strong reduction of the cutaneous inflammation in a manner similar to the inflammation after bee or wasp stings. Edema formation was especially reduced (3).

Using the bioguided fractionation method we were able to localize the antiasthmatic principle of the onion juice in the lipophilic $CHCl_3$-fraction. 13 compounds were isolated from two active subfractions using MPLC (medium performance liquid chromatography). All compounds were structurally elucidated by modern spectroscopic methods. They could be identified as trans and cis methylsulphinothioic acid-S-1-propenylester (**1**), cis- and trans-n-propylsulphinothioic acid-S-1-propenyl ester, (**2**), n-propylsulphinothioic acid-S-n-propyl ester (**3**) and trans-5-ethyl-4,6,7-trithia-2-decene-4-S-oxide (**4**) trans, trans and trans, cis 5-ethyl-4,6,7-trithia-2,8-decadiene-4-S-oxide (**6**) and as the diastereoisomers of the latter three compounds (**4**). Five compounds (**1-3**) belong to the class of α,ß-unsaturated thiosulphinates. The other novel unsaturated α-sulphinyldisulphides (**4-6**) were called "cepaenes".

These compounds are not present in the intact bulb. They are generated from sulphur-containing amino acids upon squeezing or slicing of the bulb, by enzymatic catalysis and sequential reactions. Most of them are relatively unstable. As shown in some animal experiments, the thiosulphinates of onion are the major antiasthmatic principles of freshly prepared onion extracts *in vivo* (5). At concentrations of 10 and 100 mg/kg, they were able to reduce the allergen or PAF induced bronchial obstruction by 30-60%. *In vitro* they were found to be potent dual inhibitors of cyclooxygenase and 5-lipoxygenase (6). The cepaenes also very effectively inhibit cyclooxygenase and 5-lipoxygenase *in vitro* (7). Whether the cepaenes also exhibit antiasthmatic activity *in vivo*, could not be evaluated due to a shortage of enough pure substance.

In Table I the *in vitro* effects of some isolated and synthetic thiosulphinates as well as cepaenes on various mediator systems and cell functions are shown (8).

Picrorhiza kurroa. This plant is a small herb with tuberous roots that grows in the Himalayan area at altitudes of 3000 - 5000 m. In Ayurvedic medicine it is used mainly for the treatment of liver and lung diseases, but also for chronic dysentery and other complaints. Controlled trials performed under different arthritic conditions showed marked improvement of clinical symptoms (9,10).

In a preceding *in vitro* investigation, *Picrorhiza* extracts were found to inhibit histamine release *in vitro* (11) and to enhance the bronchodilating effects of sympathomimatic amines (12). The bioguided fractionation of an ethanolic extract led to the isolation of some new triterpene derivatives (cucurbitacins) and, in the more polar fraction, of a phenylglycoside identified as androsine (**7**)(13,14). Of all the isolated compounds, only androsine was found to inhibit experimentally induced bronchial obstruction by 60-70% when applied in doses of 10 mg/kg. The inhalation of the same compound at a concentration of 0.5 mg/kg was much more effective.

This result prompted us to perform detailed structure-activity relationship studies in the class of acetophenones. From more than 25 synthesized or isolated acetophenones, aglycons and glycosides with a varying substitution pattern, 3,5-dimethoxy-4-

$$R_1-S(O)-S-R_2$$

1
R$_1$:Me R$_2$:CH=CH—Me *(trans)*
R$_1$:Me R$_2$:CH=CH—Me *(cis)*

2
R$_1$:nPr R$_2$:CH=CH—Me *(cis)*
R$_1$:nPr R$_2$:CH=CH—Me *(trans)*

3 R$_1$:nPr R$_2$:CH$_2$—CH$_2$—Me

4

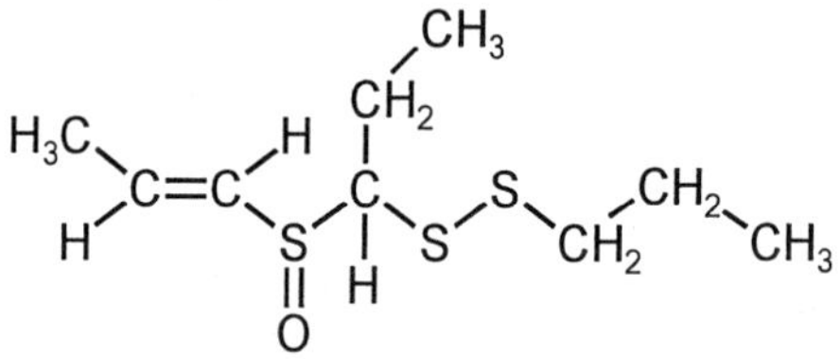

5

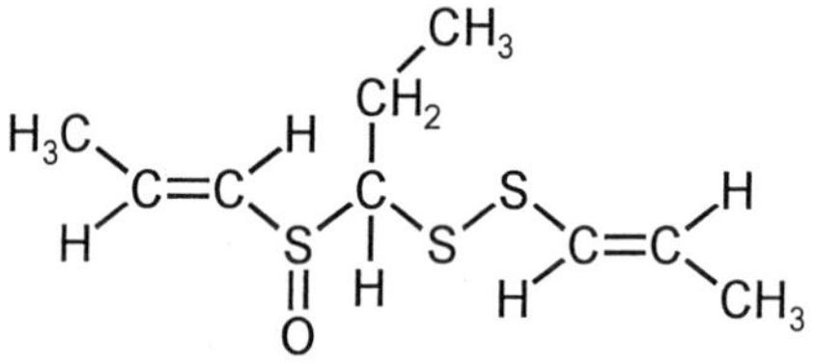

6

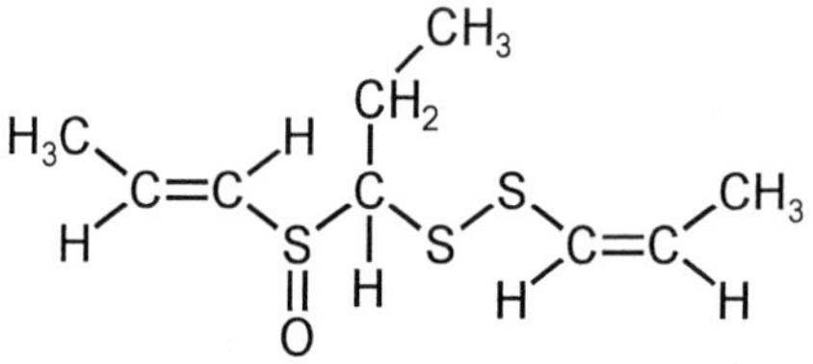

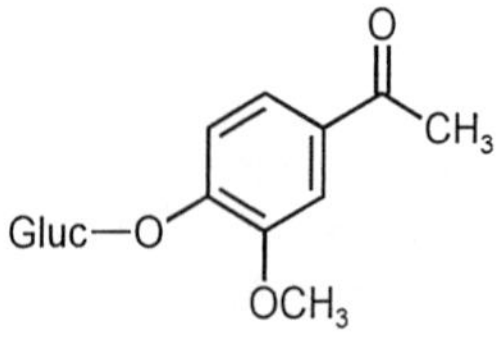

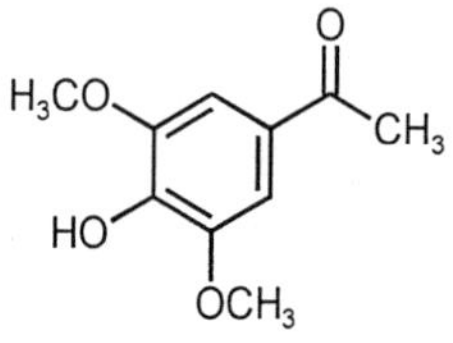

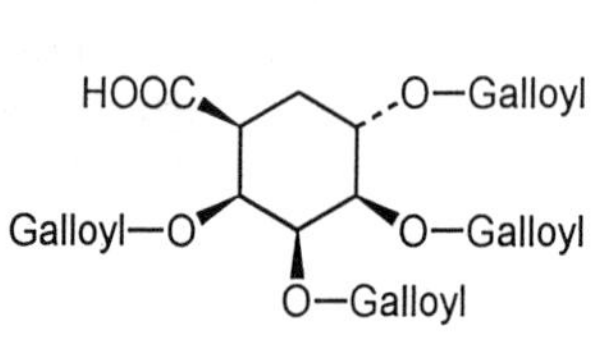

7

8

9

hydroxyacetophenone (acetosyringenin **8**) was found to be the most active compound (*15-16*). It prevented allergen- and PAF-induced bronchial obstruction by more than 80% after a single dose of 0.5 mg applied by inhalation aerosol, or 10 mg/kg applied orally (Figure 2).

To summarize, it may be concluded that aglycones are more effective than glycosides when applied by inhalation. Acetophenones with hydroxyl groups at the para position to the acetyl group and an additional methoxy group at the 3 or 5 position showed by far the best effects (*16*). From the bioguided fractionation of *Picrorhiza kurroa* extract, it can be concluded that the extracts might not be useful for a rational therapy of allergic asthma - firstly because of the low concentration of the acetophenone in the extract and, secondly, because of the presence of the relatively toxic cucurbitacins.

Galphimia glauca. The plant, abundant in Central and South America, is prescribed by curanderos of some tribes in the rain forest against pollinosis. The first studies on patients with pollinosis were promising (*17*), but no information was available on the mode of action of the active compounds. The oral pretreatment of guinea pigs with 320 mg/kg of the alcoholic *G. glauca* extract prevented bronchial obstruction to the inhalation of ovalbumin and PAF. This effect was observed not only 1 and 12 h after a single dose but even after 4 days (*18*).

By bioguided fractionation we were able to isolate and identify gallic acid and ellagic acid, in addition to methylgallate, the rare tetragalloyl-quinic acid (**9**) and the flavonoids hyperoside, isoquercitrin, kaempferol-glucoside and quercetin/kaempferol-glucoside-gallate.

After a single oral dose of 5 mg/kg tetragalloyl quinic acid showed significant effects (by 60-80%), whereas gallic acid, methyl-gallate and quercetin were much less active (45 mg/kg)(*19*). Surprisingly, the flavonol-acyl-glycoside was found to exert a strong effect on the classical complement model (*20*). We also observed an inhibitory effect on histamine release from human polymorphonuclear leukocytes. Since many flavonoids are known as inhibitors of cyclooxygenase as well as lipoxygenase, and the gallates as potent oxygen radical scavangers and antioxidants, it might be justified to suggest a different mechanism, most likely acting in combination, as being involved in the antiasthmatic effect of the Galphimia extract.

Results of Other Laboratories. In traditional medicine of China, Japan and India (Ayurvedic-medicine) many other plants are claimed to be useful in the treatment of allergic asthma (Table II). Whereas, for most of them the clinical proof according to GCP (Good Clinical Practice) is lacking, many pharmacological *in vitro* and *in vivo* investigations have been carried out which seem to confirm their antiasthmatic potential. Many isolated substances have also been screened for antiallergic and antiasthmatic activity.

Sankawa and Chun (*21*) screened 20 drugs from traditional Chinese medicine using the passive cutaneous anaphylaxis test (type 1 immediate response). Ten plant extracts showed good inhibition effects (100-300 mg/kg *i.v.*). In an *in vitro* mast cell model, highly methoxylated flavonoids isolated from *Citrus aurantium*, e.g. nobiletin, 3,4,6,7,8,,3',4'- heptamethoxy flavone, tangeretin and others (*22*), were found to be good inhibitors of histamine release. Flavonoids, a coumarin, a neolignan and sesquiterpenes

Table 1. Effects of thiosulphinates (TS), isolated from onion or synthesized, on mediator systems and cell functions in vitro.

	A	B	C	D	E	F	G	H
Dimethyl-TS[a]	0	0	0	0	0	↓		↓
Methyl-propyl-TS[a,b]				0	0			↓↓
Dipropyl-TS[a]	↓↓	↓	↓	0	0		↓↓	
Diallyl-TS	↓	↓↓	↓↓	↓↓↓	↓↓		↓	↓↓↓
Methyl-phenyl-TS				↓↓↓	↓↓		↓	
Phenyl-methyl-TS				↓↓↓	↓↓	↓↓	↓↓↓	↓
Diphenyl-TS	0	↓↓	↓↓	↓↓↓	↓↓	↓↓		↓↓
Methyl-l-propenyl-TS[a]				↓↓↓	↓↓↓			↓↓↓
Propyl-l-propenyl-TS[a]				↓↓↓	↓↓↓			

↓↓↓ = Reduction by more than 30% at concentrations of 1μM.
↓↓ = Reduction by less than 30% at 1 μM.
↓ = Reduction at 100 μM.
[a] Identified in *Allium cepa* L.
[b] Isomers.

A = Histamine release
B = Biosynthesis of LTB4
C = Biosynthesis of LTC4
D = 5-Lipoxygenase
E = Cyclooxygenase
F = Thromboxane biosynthesis
G = Chemotaxis
H = Chemiluminescence

Table II Antiasthmatic and antiallergic drugs of the traditional medicines of Asia.

Adhatoda vasica	India
Asarum sagittaioides	Korea
Bupleurum falcatum	China
Centipeda minima	Korea
Cinnamomum cassia	China/Japan
Ganoderma lucida	China/Japan
Glycyrrhiza glabra	China
Magnolia species	Japan
Picrorhiza kurroa	India
Scutellaria baicalensis	China
Tylophora indica	India

with antiallergic properties were likewise isolated from *Magnolia salicifolia* and *Centipeda minima* (*21*). Baicalein from *Scutellaria baicalensis* has been reported to exhibit antiallergic activity on the basis of its inhibition of hypersensitive immediate response (*23*) and the thrombocyte lipoxygenase (*24*). From Ayurvedic medicine two other plants with antiasthmatic activity are known: *Tylophora antiasthmatica* and *Adhatoda vasica*. For the first plant, a non-anticholinergic and antihistamine mechanism of action, which might be due to the *Tylophora*-alkaloids has been reported (*25*). Comprehensive pharmacological investigations have been described also for *Adhatoda vasica*, whose two major alkaloids vasicine and vasicinone showed bronchodilatory activity comparable to that of theophylline (*26*).

Conclusions. To summarize all the results of screening, *i.e.* those available in the literature as well as obtained by ourselves, it may be concluded that none of the active compounds isolated has the pharmacological profile of a betasympathicomimetic, nor that of an anticholinergic or a xanthin derivative. There is strong evidence that most of them interfere with defined mediators or cell functions which play a key role in the pathological processes of inflammation.

Screening for Antihypertensive Drugs

Although a great number of synthetic antihypertensive drugs are available, such as ACE-I + II inhibitors, Ca-antagonists, ß-blockers and diuretics, publications reporting several adverse effects of some of them are increasing. Furthermore, as far as the heart and blood circulation diseases are concerned, we know only a limited number of relevant targets, e.g. enzymes, receptors, and their proteins. In the case of Ca-channel blockers, the channels are not clearly defined structurally. Therefore, the screening of natural products relying on bioassays is still valid.

In the search for new antihypertensive plant compounds, we again followed the ethnopharmacological approach, and selected plants which in traditional medicine have gained a high reputation in the treatment of hypertension, coronary heart diseases, heart arhythmia and myocardial infarct. Our screening program aimed at new compounds with ACE-inhibiting- and Ca-antagonistic activity.

Plant ACE-inhibitors. Within the enzyme cascade of the renin-angiotensin system, the angiotensin converting enzyme (ACE) plays an important role in the regulation of blood pressure. Acting as a dipeptidyl carboxypeptidase, the enzyme liberates angiotensin II, an octapeptide with strong vasopressor activity from angiotensin I (a decapeptide) which, in turn, is a product of the action of renin. In addition to directly causing vasoconstriction, angiotensin II stimulates the synthesis and release of aldosterone, which also increases the blood pressure by promoting sodium and water retention (*27*). If the formation of angiotensin II and thus the inactivation of vasodilatory kinins are suppressed by selective ACE-inhibitiors, the results are a decrease in blood pressure, natriuresis and diuresis (*28*).

Methodology. We developed a new *in vitro* assay by using the chromophore- and fluorophore-labelled tripeptide dansyltriglycine as substrate and ACE from rabbit lung.

The enzyme cleaves the substrate into dansylglycine and diglycine. The extent of cleavage is determined by separation of the split products on reversed phase high performance liquid chromatography and quantification (29)(Figure 3).

This target-directed enzyme assay is very specific, highly sensitive, reproducible, and lends itself very easily to automation. The screening, bioguided fractionation and isolation of the active compounds of about 30 plant extracts led to compounds belonging to two major classes - the procyanidins and flavonoids. Additional potent compounds could be detected in the class of peptides.

Procyanidins. Within this class the oligomeric procyanidins and the C-4/C-6 linked dimeric compounds, with inhibition rates of 36-60% at concentrations of 0.17 and 0.33 mg/ml respectively, were more active than the C-4/C-8-linked dimeric forms (Table III) (30). This finding was in agreement with other results obtained from procyanidins of *Vitis vinifera* and *Cupressus sempervirens* (31). Pure procyanidins isolated from *Lespedeza capitata* (32), *Cistus clusii* and *Amelanchier ovalis* (30) showed moderate (50-90% inhibition at 0.20-0.33 mg/ml) or good activity (30-90% inhibition at 0.15 - 0.2 mg/mL). Monomeric flavan-3-ol and (+)-catechin were inactive. The procyanidin fraction of *Crataegus oxyacantha*, a plant used for the treatment of moderate heart insufficiency, showed also an ACE-inhibiting activity (33). This finding is of interest in light of a clinical trial performed with an *Crataegus* extract, against the synthetic captopril, a well known ACE inhibitor. In this trial the *Crataegus* extract revealed clinical results very similar to that of captopril (34).

From the Alchemy Molecular Modelling program (Tripos Inc.) used for calculating three-dimensional models of the energy minimized structures, we can conclude that the *in vitro* activity of procyanidins might be due to the formation of chelate complexes between the electron-rich heterocyclic oxygen and hydroxy functionalities of different monomeric units with the zinc atom of the enzyme (30).

Flavonoids. Within this class of compounds, some derivatives of 8-hydroxychromane and -chromone, e.g. mesquitol (**10**), teracacidin or gossypetin, and also polyhydroxyflavones with a phenolic hydroxyl group in positions 2' or 4' of the B-ring (e.g. in morin (**11**) and amentoflavone) (Table IV) inhibited ACE to 60-90% at a concentration of 0.33 mg/mL (30). It is remarkable that all active compounds are characterized by a phenolic hydroxyl group in the vicinity of a heterocyclic oxygen atom. Therefore, one may surmise that the *in vitro* activity of these compounds is due to the generation of chelate complexes with the zinc atom within the active centre of the ACE. Such a mechanism has also been proposed for the microbial ACE inhibitor phenacein (**12**) (IC_{50} 0.0001 mg/mL ~ 390 nmol/l (35).

Peptides. Among the plant peptides tested, the tripeptide glutathione showed an IC_{50} of 0.0032 mg/mL. By using the computer-aided active analogue approach we could show that the glutathione molecule possesses a similar C-terminal structure which fits into the active centre of the enzyme receptor. The same mechanism of interaction can also be suggested for the peptides, phytochelatin and the metallothionins, which have a primary structure of repetitive γ-glutamylcysteinyl units (36,37) and thus the capacity to bind to heavy metal ions (Table IV).

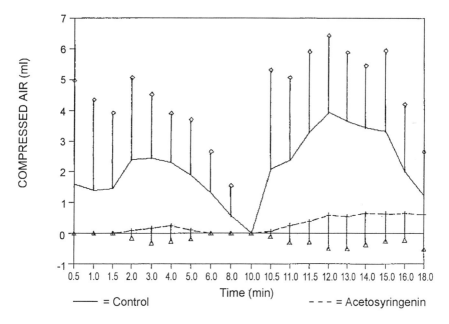

Figure 2. Prevention of allergen (ovalbumin) induced bronchial obstruction (inhalation of ovalbumin at 0 and 10 min) in 10 guinea pigs by the inhalation of 0.5 mg acetosyringenin 30 min prior to the inhalation challenge. Percent inhibition of bronchial obstruction at the first allergen challenge (time 0): $83 \pm 27\%$; and at the second challenge: $86 \pm 24\%$ ($p<0.001$).

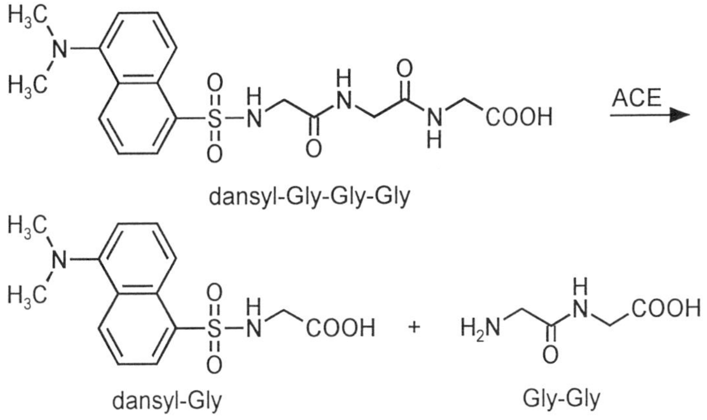

dansyl-Gly-Gly-Gly

dansyl-Gly   +   Gly-Gly

Figure 3. ACE-catalyzed cleavage of dansyltriglycine.

Table III. ACE-inhibitory activity (% ± s.d.) of compounds isolated from *Lespedeza capitata, Amelanchier ovalis, and Cistus clusii* (means of two different experiments).

Lespedeza capitata		
Dimeric procyanidin B_3	0.33 mg/mL	62.1 ± 1.4
Dimeric procyanidin B_1	0.33 mg/mL	58.2 ± 2.1
Dimeric procyanidin B_6	0.17 mg/mL	36.7 ± 0.3
Trimeric procyanidin C_1	0.17 mg/mL	43.8 ± 3.8
Amelanchier ovalis		
Dimeric procyanidin B_2	0.33 mg/mL	65.9 ± 1.6
Dimeric procyanidin B_4	0.33 mg/mL	61.8 ± 2.2
Cistus clusii		
Pentameric prodelphinidin	0.14 mg/mL	IC_{50}
Proanthocyanidins		
[4,8]-3,4-trans-(-)-fiseti-nidol(+)-catechin	0.33 mg/mL	59.2 ± 1.4
[4,8]-3,4-cis-(-)-fiseti-nidol-(+)-catechin	0.33 mg/mL	66.4 ± 3.1
[4,6]-3,4 cis-(-)-fiseti-nidol-(+)-catechin	0.17 mg/mL	61.8 ± 0.7

Table IV. Screening results for ACE-inhibitory activity (% ± s.d.) with flavonoids, peptides, and *Allium ursinum* (means of two different experiments).

Flavonoids		
Morin	0.33 mg/mL	63.7 ± 2.3
(+)-Mesquitol	0.33 mg/mL	77.6 ± 2.7
(-)-Teracacidin	0.33 mg/mL	69.4 ± 1.3
Amentoflavone	0.33 mg/mL	62.5 ± 1.1
Gossypeptin	0.33 mg/mL	68.4 ± 2.3
Peptides		
Glutathione [γ-Glu-Cys-Gly]	0.0032 mg/mL	IC_{50}
Phytochelatin [(γ-Glu-Cys)-Gly; n=2-11]	0.034 mg/mL	IC_{50}
Metallothionein-fragment [Lys-Cys-Thr-Cys-Cys-Ala]	0.0013 mg/mL	IC_{50}
Allium ursinum		
γ -L-glutamyl-(+)-S-allyl -L-cysteine sulphoxide	0.03 mg/mL	48.7 ± 1.4

While screening *Allium ursinum* (wild garlic) extracts, we detected a marked ACE-inhibiting activity in a very polar fraction. The bioguided fractionation and isolation led to γ-glutamyl-(+)-S-allyl-L-cysteine sulphoxide, which exhibited an ACE-inhibiting effect of ~ 50% at a conc. of 0.03 mg/mL (Table IV).

Plant Ca-Channel Blockers. Calcium antagonistic drugs act on calcium channels by inhibiting the Ca^{2+} influx into the cell. The synthetic Ca-blockers in use can be divided into the three main groups: dihydropyridines, benzodiazepines and phenylalkylamines. The first type, such as nifedipine, verapamil, and diltiazem, for example, blocks the voltage-operated channels (VOCCs) in the membranes of the muscles like prenylamine and fendiline, but may also interact with receptor-operated Ca^{2+} channels (ROCCs), Na^+ channels or calmodulin, and are therefore classified as "nonselective" calcium-antagonists by WHO (World Health Organization).

Of the various Ca-channels which are known today (L,-N, T and P type), the "long lasting" or "slow" type channel plays the most important role for the heart and smooth muscle. Test models in which a response (increase in force, uptake of calcium or increase in $[Ca^{2+}]$) is elicited by depolarization (increase in potassium concentration) can only be the first step in determining a possible calcium-channel blocking. The only method for determing a calcium-blocking effect with certainty is the measurement of the calcium inward current through calcium channels by the patch clamp method. In this context it must be stressed that the various models available today, with e.g. isolated muscle preparations (e.g. guinea-pig, trachea, heart muscle, vascular smooth muscles or various cell lines from animal organs) can show widely different results, which render it difficult to compare the potency of the various substances tested.

In our screening we used papillary muscles from the right ventricles of guinea pig-hearts (*38*). The muscles were stimulated by square-wave pulses through punctate electrodes at their mural end. The isometric contraction curves were monitored on an oscilloscope and a high speed pen recorder. The contractions were elicited at a low contraction frequency under the influence of ryanodine (1 μM) and noradrenaline (0.3 μM). In this experiment a decrease in contraction force correlates well with the calcium-channel blocking activity of the compounds (*39*).

In our first approach we tested some essential oils, which are known for their diuretic and spasmolytic efficacy and are used in traditional medicine. We found a concentration-dependent effect on the force of contraction of papillary muscles with parsley, celery fruit and clove oil (*40*). Subsequent studies with the putative major active principles of the oils, apiol (**13**) (*40*) and eugenol (**14**) (*41*) (see Figure 4), revealed an IC_{50} of 29 μM (apiol) and 224 μM (eugenol) (Table V).

Structure activity relationship studies in the class of phenylpropanoids (Table V) showed that a certain degree of methoxylation and the substitution pattern of the benzol ring has an important influence on the magnitude of activity (*40*) (Figure 4). Nifedipine exhibited in a similar test model an IC_{50} of 0.3 μM.

Among the substances subjected to this screening program, we found some naphtho-quinones (e.g. juglone, plumbagin), terpenoids (e.g. bisabololoxide A) garlic constituents (allicin, ajoene), pyranocoumarins (visnadin) (**15**) and some alkaloids (e.g. sanguinarine-NO_3) with pronounced activity (*42*).

Other research groups have subsequently conducted similar screening programs using

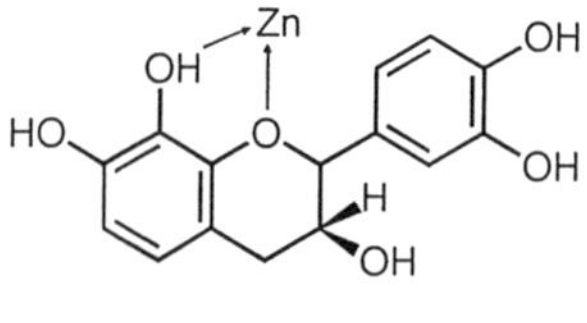

58

10

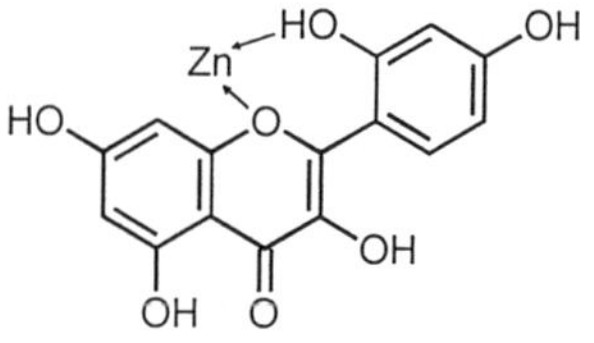

11

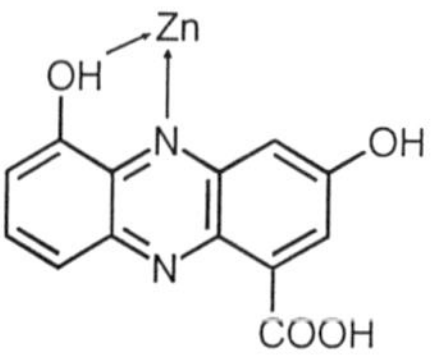

12

Table V. Calcium-channel blocking effects of various phenylpropanoids

Substance	IC_{50} (μM)	SEM[*] (μM)
apiol	29.2	2.31
allyltetramethoxybenzol	63.1	3.74
trans-anethol	75.2	9.4
α-asaron	76.0	7.77
ß-asaron	63.1	19.58
methylchavicol	257.6	19.59
myristicin	87.6	9.27
safrol	57.7	6.5
eugenol	224.0	---

[*]Standard error of the mean for ≤ six muscles.

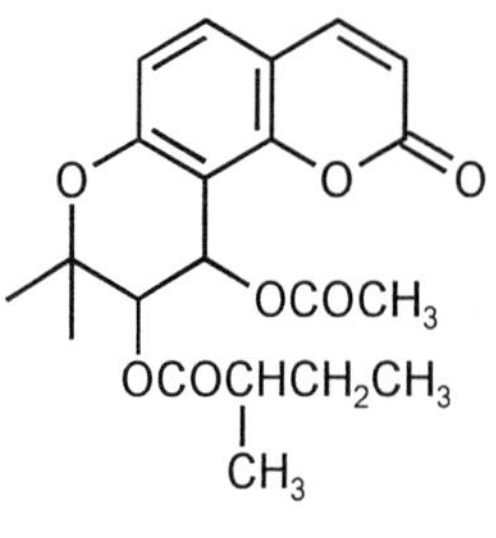

15

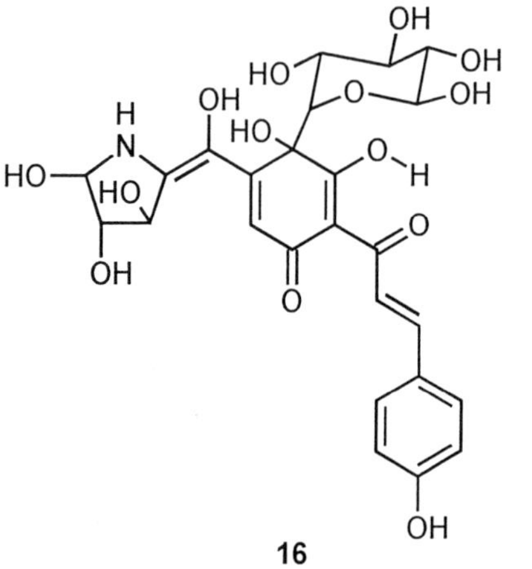

16

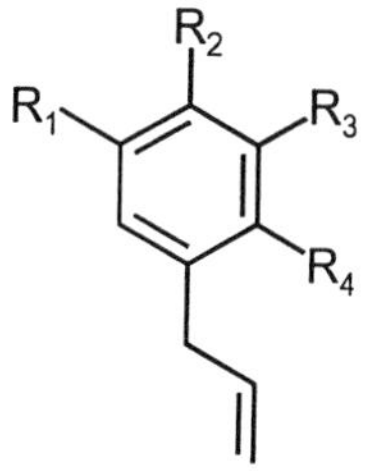

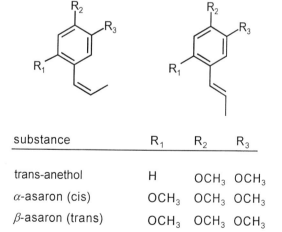

	substance	R_1	R_2	R_3	R_4
13	apiol	OCH_3	$-O-CH_2-O-$		OCH_3
	allytetra-methoxybenzol	OCH_3	OCH_3	OCH_3	OCH_3
	methylchavicol	H	OCH_3	H	H
14	eugenol	H	OH	OCH_3	H
	myristicin	OCH_3	$-O-CH_2-O-$		H
	safrol	H	$-O-CH_2-O-$		H

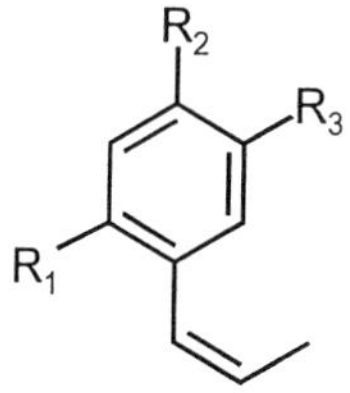

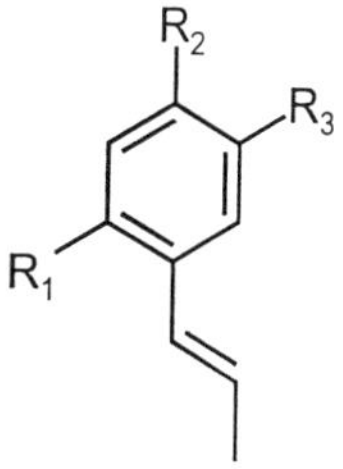

substance	R_1	R_2	R_3
trans-anethol	H	OCH_3	OCH_3
α-asaron (cis)	OCH_3	OCH_3	OCH_3
β-asaron (trans)	OCH_3	OCH_3	OCH_3

Figure 4. Structures of the phenylpropanoids tested for calcium-channel blocking.

60

rat thoracic aortas, guinea pig ileum, mouse myocardial cells, rat pituitary cells or cultured cells (*43-47*). Among the compounds listed in a review article written by Vuorela *et al.* (*48*), furanocoumarines, coumarins and some mono- and sesquiterpenoids occupy a preferred place.

Regarding the mechanism of action, it is likely that most of these compounds are real Ca-channel blockers. However, definite proof of this by means of the patch-clamp method has hitherto only been possible for a few compounds, e.g. eugenol, apiol, and tinctormine (**16**)(*48*).

Literature cited

1. Dorsch, W.; Waldherr, U.; Rosmanith, J. Pflügers Arch. **1981**, *391*, 236-241.
2. Dorsch, W.; Hess, V. *Int. Arch. Allergy Immunol.* **1992**, *99*, 496-498.
3. Dorsch, W.; Ring. I. *Allergy* **1984**, *39*, 43.
4. Bayer, Th.; Breu, W.; Seligmann, O.; Wray, V.; Wagner, H. *Phytochemistry* **1989**, 2373-2377.
5. Dorsch, W.; Wagner, H.; Bayer, Th.; Fessler, B.; Hein, G.; Ring, I.; Scheftner, R.; Sieber, W.; Strasser, Th.; Weiß, E. *Biochem. Pharmacol.* **1988**, *37*, 449.
6. Wagner, H.; Dorsch, W.; Bayer, Th.; Breu, W.; Willer, F. *Prostaglandins, Leucotrienes and Essential Fatty Acids* **1990**, *39*, 59-62.
7. Bayer, Th.; Wagner, H.; Wray, V. *Lancet* **1988**, 8616/I, 906.
8. Breu, W.; Dorsch, W. In *Economic and Medicinal Plant Research*; Farnsworth, N.; Wagner, H., Eds.; Academic Press, London, **1994**, *vol. 6*, 115.
9. Langer, I.G.; Gupta, O.P.; Atal, C.K. *Indian J. Pharmacol.* **1981**, *13*, 98-103.
10. Mogre, K.; Vora, K.; Sheth, U.K. *Ind. J. Pharmacol.* **1981**, *13*, 253-259.
11. Dorsch, W.; Stuppner, H.; Wagner, H.; Gropp, M.; Demoulin, S.; Ring, I. *Int. Arch. Allergy Appl. Immunol.* **1991**, *95*, 128-133.
12. Mahajani, S.S.; Kilkarni, R.D. *Int. Arch. Allergy Immunol.* **1977**, *53*, 137-144.
13. Stuppner, E.; Müller, P.; Wagner, H. *Phytochemistry* **1991**, *30*, 305-310.
14. Stuppner, H; Wagner, H. *Planta Med.* **1989**, *55*, 467-469.
15. Dorsch, W.; Wagner, H.; Neszmelyi, A.; Stuppner, H. *Int. Arch. Allergy and Immunology* **1992**, *99*, 493-495.
16. Dorsch, W.; Müller, A.; Christoffel, V.; Stuppner, H.; Antus, S.; Gottsegen, A.; Wagner, H. *Phytomedicine* **1994**, *1*, 47-54.
17. Wiesenauer, M. *Allgemein-Med.* **1982**, *58*, 1850-1852.
18. Dorsch, W.; Bittinger, M.; Haas, A.; Müller, A.; Kreher, B.; Wagner, H. *Int. Arch. Allergy Immunol.* **1992**, *97*, 1-7.
19. Neszmélyi, A.; Kreher, B.; Müller, A.; Dorsch, W.; Wagner, H. *Planta Med.* **1993**, *59*, 164-167.
20. Wagner, H.; Knaus, W.; Jordan, E. *Zeitschr. f. Phytother.* **1987**, *8*, 148-149.
21. Sankawa, K.; Chùllù, Y.T., In *Antiallergic substances from Chinese Medicinal Plants in Advances in Chinese Medicinal Materials Research*; Chang, H.M.; Young, H.W.; Tso, W.W.; Koo, A., Eds.; World Scientific Publ. Co., Singapore, **1985**.
22. Amellal, M.; Bronner, C.; Briancon, F.; Haag, M.; Anton, R.; Landry, Y. *Planta Med.* **1985**, *16*.
23. Koda, A.; Nagai, H.; Wada, H. *Folia Pharmacol. Japan* **1970**, *66*, 237.
24. Sokiya, O.; Okuda, H. *Biochem. Biophys. Res. Commun.* **1987**, *105*, 1090.
25. Rao, K.V.; Wilson, R.A.; Cumminga, B.J. *Pharm. Sci.,* **1971**, *60*, 1725.
26. Chopra, R.N.; Ghosh, S. *Indian J. Med. Res.* **1925**, 205-210.

27. Wyrratt, M.J. *Clin. Physiol. Biochem.* **1988**, *6*, 217-229.
28. Cushman, D.W.; Ondetti, M.A.; Chening, H.S.; Antonaccio, M.J.; Murathy, V.S.; Rubin, B. *Adv. Exp. Med. Biol.* **1980**, *130*, 199-222.
29. Elbl, G.; Wagner, H. *Planta Med.* **1991**, *57*, 137-141.
30. Wagner, H.; Elbl, G.; Lotter, H.; Guinea, M. *Pharm. Pharmacol.* **1991**, *1*, 15-18.
31. Meunier, M.T.; Villié, F.; Jouadet, M.; Bastide, J.; Bastide, P. *Planta Med.* **1987**, *53*, 12-15.
32. Wagner, H.; Elbl, G. *Planta Med.* **1992**, *58*, 297.
33. Elbl, G. *P.D.-Thesis*; **1991**, University of Munich, Germany.
34. Tauchert, M.; Ploch, M.; Hübner, W.D. *Münch. Med. Wschr.* **1994**, *136* (Suppl. 1) 27.
35. Bush, K.; Henry, P.R.; Souser-Woehlcke, M.; Trejo, W.H.; Slusarchyk, D.S. *J. Antibiot.* **1984**, *3*, 1308-1312.
36. Grill, E.; Zenk, M.H.; Winnacker, E.L. *Science* **1985**, *230*, 574-576.
37. Yoshida, A.; Kaplan, B.E.; Kimura, M. *Proc. Natl. Acad. Sci. USA* **1979**, *76*, 486-490.
38. Reiter, M.W.; Vierling, K.; Seibel, N.S. *Arch. Pharmacol.* **1984**, *325*, 159-169.
39. Sensch, O.W.; Vierling, W.; Brandt, M.; Reiter, N.S. *Arch. Pharmacol.* **1993**, *347*, R 78.
40. Neuhaus-Carlisle, K.; Vierling, W.; Wagner, H. *Pharm. Pharmacol.* **1993**, *3*, 77-79.
41. Vierling, W.; Sensch, O.; Neuhaus-Carlisle, K.; Wagner, H. *Naunyn-Schmiedeberg Arch. Pharmacol.* **1995**, *351*, R 96.
42. Neuhaus-Carlisle, K.; Vierling, W.; Wagner, H. *Phytomedicine* **1997** in press.
43. Härmälä, P.; Vuorela, H.; Hiltunen, R.; Nyiredy, Sz.; Sticher, O.; Törnquist, K.; Kaltia, S. Phytochem. Anal. **1992**, *3*, 42-48.
44. Rauwald, H.W.; Brehm, O.; Odenthal, K.P. *Phytother. Res.* **1994**, *8*, 135-140.
45. Vuorela, H.; Törnquist, K.; Sticher, O.; Hiltunen, R. *Acta Pharm. Fenn.* **1988**, *97*, 167-174.
46. Thastrup, O.; Fjalland, B.; Lemmich, I. *Acta Pharmacol. et Toxicol.* **1983**, *52*, 246-253.
47. Namba, T.; Morita, O.; Huang, S.L.; Goshima, K.; Hattori, M.; Kakinchi, N. *Planta Med.* **1988**, *54*, 277-282.
48. Vuorela, H.; Vuorela, P.; Törnquist, K.; Alaranta, S. *Phytomedicine 4* in press.
49. Meselhy, M.R.; Momose, Y.; Hatakeyama, N.; Hattori, M.; Kadota, S.; Namba, T. *Phytomedicine* **1995**, *3*, 277-281.

Chapter 6

Herbal Drugs in the Treatment of Benign Prostatic Hyperplasia

H. Schilcher

Alfred Neumann Anger 17, D-81737 München-Perlach, Germany

The natural history of benign prostate hyperplasia (BPH) is a slow enlargement of fibromuscular and epithelial structures within the gland, eventually leading to obstructive urinary symptoms experienced to some extent by most men over the age of 50. Its etiology and pathogenesis have not yet been clearly defined. Therefore, the therapy strategy is controversial. Besides the transurethral resection of the prostate, medical treatments, including alpha-receptor blocking agents, 5-alpha-reductase inhibitors, and phytomedicines, are becoming more and more important world-wide. Herbal remedies, such as *Cucurbitae peponis semen*, *Serenoae repentis fructus, Urticae radix,* and other herbal drugs, have a long tradition for successful treatment of BPH in Europe, although no exact mechanism of action nor precise classification of the active compounds are known. However, in the past decade many clinical and pharmacological studies have confirmed the empirical observations, thereby advancing these traditionally used phytomedicines to rational drugs.

Benign prostatic hyperplasia (BPH) is one of the most common causes of urination problems in men, manifesting itself as hesitancy, incomplete voiding of the bladder, terminal dribbling, urgency, frequency, and nocturia.

Pronounced BPH is found in 25% of men 40 to 50 years of age and in nearly 80% of men over 70 years of age. Despite the considerable importance of the clinical picture, particularly for general practitioners and urologists, the etiology and pathogenesis of BPH remain unclear (*1*). The ongoing controversy concerning therapeutic strategy is therefore entirely understandable. Many urologists believe that treatment should not target BPH as such but only the associated symptoms, especially the distressing obstructive and irritative urination symptoms (*2*).

Provided that the patient receives regular medical attention, the symptoms are not life-threatening, but they can substantially diminish the patient's quality of life. The strategy of "watchful waiting" until surgery becomes necessary is understandably rejected by most patients. The approximately DM 250 million (wholesale price) spent by Germans on prostate remedies in 1995 is certainly not to be seen as a "financed luxury". Surgical treatment remains the gold standard in urology because, in a scientific sense, it is still the only proven causal therapy (3). Nevertheless, there is a growing interest worldwide in conservative pharmacological measures for the patient's sake, even though uncertainty about the underlying mechanisms of action is still a source of scientific contention.

Etiology and Pathogenesis of BPH / Problems Finding Medications

According to the current state of medical knowledge, prostatic hyperplasia is due mainly to increased proliferation of the stromal components (supportive tissue) of the prostate. To a far lesser extent, glandular hypertrophy also occurs (1). Histologically, marked growth of the muscular stroma is seen. Despite numerous experimental studies, it is still unclear how this leads to palpable and easily diagnosed hyperplasia, and this lack of knowledge makes any causal therapy extremely difficult.

The recent finding that there is often no direct correlation between prostate size and symptom intensity further complicates the rational search for suitable drugs. And last but not least, the aforementioned symptoms, which are of primary concern to the patient, may also be due to concurrent abacterial prostatitis with relatively little hypertrophy of the prostate.

Of the **five** hypotheses advanced to explain the development of hyperplasia (4) (Table I), only **two** are mentioned here, since they are most amenable to experimental investigation, and since the prostate drugs used—whether synthetic or of plant origin—have a regulating effect on the postulated causes. The first is synthesis of dihydro-testosterone (DHT) from testosterone, the second a shift of the androgen-estrogen ratio due to increased synthesis of 17ß-estradiol, likewise from testosterone.

Table I. The pathogenic background of BPH. The development of benign prostatic hyperplasia is a multifactorial process and the pathogenic mechanism is triggered by various hormones and mediators.

· Increased concentration of
1. SHBG (Sexual hormone binding globulin)

· Increased enzymatic activity of
2. 5α-Reductase: Testosterone → Dihydrotestosterone
3. Aromatase : Testosterone → Estrogen

· Increased concentration of
4. Cholesterol (a precursor of testosterone and its metabolites)

· Increased concentration of
5. Prostaglandins and leukotrienes (edema and inflammation mediators)

Table II. Pharmacological Treatment of BPH with Synthetic Drugs

α_1 **Blockers (alfuzosin and terazosin)**	Selective blockade of noradrenergic α_1 receptors in the prostate, trigonum vesicae, and urethra. Blockade of noradrenergic α_1 receptors not only in the vascular system but also in other extraprostatic tissues.	Relaxation of smooth muscles in the bladder neck and prostate. Improvement of obstructive and especially irritative symptoms.
Finasteride	Competitive inhibition of 5α-reductase and thus reduction of plasma DHT levels.	Remission of hyperplasia. The reduction of the prostate volume by approx. 20% does not necessarily correlate with the intended improvement of symptoms.
Aromatase inhibitors	Prevention of the age-related shift of the androgen-estrogen ratio, i.e. inhibition of 17-β-estradiol (E_2) synthesis.	Delay of hyperplasia. Improvement of urinary flow.
Antiestrogens	Delay of the age-related shift of the androgen-estrogen ratio in favor of the androgen fraction.	Delay of hyperplasia. Improvement of urinary flow.

Pharmacological Approaches

Table II gives a general overview of the synthetic prostate drugs, which account for about 19% of prostate medications currently sold in Germany. By relaxing the smooth muscles of the prostate and bladder neck, **alpha-adrenoceptor blockers** (alfuzosin and terazosin) bring about an improvement in the "dynamic component", i.e. the obstructive and especially the irritative symptoms. Of course the side effects and the cost of treatment (DM 1.50 per day) must also be considered. Although **finasteride** has been tested in clinical studies with the longest observation times and leads to the greatest remission of hyperplasia of all the prostate drugs (6), in many patients improvement of symptoms does not correlate with a reduction in prostate volume. In addition, the responder rate is only 50%, and therapeutic success can be expected only after about six months of treatment, which costs DM 3.50 per day and is associated with an adverse-effect rate of over 10%. Because of their adverse effects, **antiestrogens** have not proven very useful. In addition, they are too expensive.

With a market share of about 80% in Europe, **herbal prostate drugs** are clearly at the top of the list of the most frequently sold prostate remedies—whether by prescription or over the counter. This superiority stems primarily from their multifactorial sites of action (Table III), their almost complete lack of adverse effects, and their much lower therapeutic costs of about DM 1.00 per day. Phytomedicines for relieving BPH symptoms can be divided into *two groups* (Tables 4 and 5).

Table III. Pharmacological Treatment BPH with Phytomedicines

Phytomedicines	**Multifactorial** sites of action:	Primary therapeutic aim is to alleviate or eliminate *symptoms.*
	a) Prostatotropic activity via effects on endocrine metabolism: weak to moderately potent, not comparable to finasteride	See: International Prostate Symptom Score (I-PSS and Boyarski Score) (WHO Consensus Report 2nd Int. Consult. on BPH, Paris 1993)
	b) Direct or indirect antiinflammatory activity: dependent on the phytomedicine and the dosage	
	c) Antiedemic or anticongestive activity.	
	d) Musculotropic activity, i.e. positive effects on the detrusor	Causal therapy - if one exists at all - is pursued only as a secondary aim.
	e) Anticonvulsive and weak sedative activity	
	f) Weak bacteriostatic activity	
	Synergistic and additive overall effects	

The first group (Table IV) comprises four crude drugs for which **positive monographs** have been drawn up by the Expert Commission E of the German Health Office in Berlin (now the Federal Institute of Pharmaceuticals and Medical Products [BfArM]). Only these four drugs are discussed below in detail.

The second group (Table V) consists of crude drugs and crude-drug preparations which are currently marketed as prostate remedies but for which **no** Commission E **monographs** are available. It is highly likely that, except for the phytosterol mixture from *Hypoxis rooperi*, all these drugs will be withdrawn from the market by the year 2004 unless their efficacy has been demonstrated by that time.

Monographed Phytomedicines for the Treatment of Urination Disturbances Associated with BPH

The four drugs listed in Table IV have two things in common: first, according to the indications specified by the Commission E monographs, they are suitable only for the treatment of the **urination disturbances** associated with BPH, i.e. they are not approved for causal therapy; and second, they may only be used in **stage I to II BPH** according to Alken (*4*) (same as stage II to III BPH according to Vahlensieck).

Table IV. Monographed Phytomedicines used to Treat BPH

1.	*Cucurbitae peponis semen* = pumpkin seeds Especially from the cultivated variety *Cucurbita pepo* Linné, convar. citrullinina Grebenscikov, var. styriaca Greb. (whole crude drug, comminuted seeds, fatty oil and extracts)
2.	*Pollinis siccum extractum* = rye pollen extract Hydrophilic and lipophilic whole extract of the pollen of *Secale cereale* Linné
3.	*Serenoae repentis fructus* = saw palmetto fruits Lipophilic extract of the fruit of *Serenoa repens* (Bartram) Small, formerly *Sabal serrulata* (Michaux) Nutall ex Schultes
4.	*Urticae radix* = nettle root Alcoholic-aqueous extract of the roots of *Urtica dioica* Linné and more rarely from *Urtica urens* Linné

Table V. Non-Monographed Phytomedicines used to Treat BPH

1.	Purified extract of the roots of ***Hypoxis rooperi*** Linné, consisting of a phytosterol mixture conforming to the requirements of the US National Formulary XIII and a free β-sitosterol and β-sitosterol-β-D-glucoside as chief constituents and pentenyl glycosides.
2.	Extracts from the bark of ***Prunus africana*** (syn. *Pygeum africanum*) containing phytosterols.
3.	Ethanolic-aqueous extracts from the bark and flowers of ***Populus*** species (e.g. *P. tremuloides, P. tremula,* etc.) containing salicyl glycosides and phenyl-substituted organic acids.
4.	Aqueous extracts from ***Epilobium*** species (*E. parviflorum* Linné and *E. augustifolium* Scop) containing flavonoids and phenyl-substituted organic acids.
5.	A further **30 crude drugs** are contained in combination preparations.

Cucurbita pepo seeds

Of the four monographed herbal prostate drugs, pumpkin seeds have the longest tradition (7). They have been used to relieve BPH symptoms, irritable bladder, and bladder complaints in general since the 16th century, especially in southern European countries, Germany, and Austria (7).

Understandably, the reports by G. Öllinger, H. Bock, A. Lonicerus, F. B. Vietz (7), and other authors do not describe the therapeutic successes in terms of a clear-cut nosology (disease classification) or via a scientifically oriented evaluation of the results. Nor are there any documented epidemiologic studies by physicians. Nevertheless, one can appreciate that on the basis of empirical medicine, modern clinical studies were not at first deemed necessary.

The conflicting positive reports from empirical medicine are evidently due to the fact that the seeds of different *Cucurbita* varieties were used medicinally. More recent phytochemical investigations have shown marked differences in the content of phyto-sterols, tocopherols, and amino acids in pumpkin seeds of different varieties (8, 9).

Reproducible scientific findings have been available only since researchers began using a phytochemically precisely defined cultivar, *Cucurbita pepo* Linné, convar. citrullina Grebenscikov, var. styriaca Greb., forma Fink (4). This *Cucurbita pepo* cultivar, referred to hereafter simply as medicinal pumpkin seeds, is distinguished by a higher content of delta-7-phytosterols, amino acids, and selenium than other pumpkin seeds. Table VI gives an overview of the constituents that make efficacy in the treatment of BPH plausible.

Table VI. Active principles in and effects of the seeds of *Cucurbita pepo* Linné, convar. citrullina Greb., var. styriaca Greb.

Constituent	Activity / Mechanism of Action
Delta-7-sterols in free form and as glycosides	Antiinflammatory Diuretic Inhibition of DHT binding to cellular receptors Reduction of DHT in prostate tissue Normalization of prostatic metabolic parameters
Selenium	Antiinflammatory
Other hitherto unidentified constituents	Antimicrobial
Linoleic acid	Precursor of prostaglandins E_2 and $F_{2\alpha}$, which are involved in the regulation of detrusor-sphincter interactions
Tocopherols (β- and γ-tocopherols)	Strengthen the connective tissue and muscle
Carotenoids	Oxygen free radical scavengers
Magnesium salts	Improvement of neuromuscular functions

Proceeding from the "DHT hypothesis" as one of the possible causes of BPH, tests were conducted on human fibroblasts to determine whether pumpkin sterols are able to displace dihydrotestosterone from androgen receptors. The experimental findings show a significant dose-dependent reduction in DHT binding to the cells in the presence of incubated pumpkin sterols (10).

These first experimental results were later supplemented and confirmed by a clinical pharmacologic study (11). Serum levels of acid phosphatase, prostate-specific antigen, bound and unbound testosterone, and SHBG were determined, as well as the DHT concentration in resected tissue (open prostatectomy). In addition, the administered pumpkin sterols were qualitatively identified in resected tissue by gas chromatography. The results of the clinical pharmacological study showed that the isolated pumpkin sterols have prostatotropic activity in comparison to an untreated control group.

Between 1962, when the first clinical observational study was published (12), and 1982, there appeared six further publications dealing with a combination pumpkin preparation (13–18). The pharmaceutical preparation tested consisted chiefly of medicinal pumpkin seeds. A placebo-controlled double-blind study was conducted in 1980 with a fixed combination of this kind (*18*). Unfortunately, only one open observational study with a single-substance pumpkin preparation has so far been performed (*19*). A placebo-controlled double-blind study complying with GCP guidelines and using the International Prostate Symptom Score (I-PSS) is currently under way following a successful pilot study (*20*). Despite the lack of GCP studies, it must be assumed on the basis of the available scientific information and the correct use and interpretation of the drug guidelines that medicinal pumpkin seeds are effective.

Pollonis siccum extractum

Pollens are also traditional prostate remedies (*21*) but do not have the same importance as pumpkin seeds. Documented empirical medical reports are likewise not available with bee-collected pollens, the composition of which can be very heterogenous (*22*). A scientific "breakthrough" was achieved with the use of a standardized whole extract (hydrophilic plus lipophilic dry extracts) from a mixture of about 90% rye pollen (*Secale cereale*) and 10% timothy (*Phleum pratense*) and corn (*Zea mays*) pollen. Hereafter, this three-pollen extract will be referred to simply as "grass-pollen extract." Normally, a pollen mixture consists of 20 to 40 different pollens (*21*).

Earlier, several experimental studies showed that the *hydrophilic* and *lipophilic* extracts differed in their activity profiles but complemented each other outstandingly in terms of the therapeutic goal. There is, therefore, a rational basis for use of the whole extract (*1*).

The *lipophilic* fraction exhibits dose-dependent noncompetitive inhibition of 5α-reductase and 3α- and 3β-hydroxysteroid dehydrogenases with unequal affinity in the epithelium and stroma. It, therefore, influences the synthesis of dihydrotestosterone and has prostatotropic activity. The activities of the DHT-metabolizing enzymes 5α-reductase, 3α-HSOR, and 3β-HSOR have been determined in, for example, mechanically separated epithelial and stromal fractions from 10 normal and 20 hyperplastic prostates.

In addition, with the help of radioactively labeled arachidonic acid, it was shown that the *lipophilic* grass-pollen extract dose-dependently inhibits cyclooxygenase and 5-lipoxygenase in sheep bladder microsomes and in leukemic leukocytes of the rat (RBL-1 cells). The IC_{50} values were similar to those of the comparative substance, diclofenac.

The *hydrophilic* fraction had no activity in this test system but did exhibit antiinflammatory and anticongestive activity in animal experiments. The experimental results make the benefit of grass-pollen therapy plausible by documenting an antiinflammatory effect, as prostatic congestion may also be due to the accompanying inflammatory syndrome (*1*).

Clinical efficacy was tested in an open multicenter study with 1894 patients of 208 general practitioners as well as in two placebo-controlled double-blind studies with 103 and 92 patients. The open observational study ran for twelve months, the double-blind studies for six months. Overall, the study results show a significant reduction in pronounced *congestion* and an improvement in urodynamic parameters.

The results of the two experimental studies and the clinical studies are detailed in the textbook and manual *Benigne Prostatopathien* (benign prostatic diseases), edited by W. Vahlensieck and G. Rutishauser (*1*).

Serenoae repentis fructus

Extracts of the dried berries of the saw palmetto (*Serenoa repens* Bartram Small; originally called Sabal) are popular folk remedies in North America, though their use is almost unknown in European traditional medicine. Saw Palmetto extracts in fixed combinations have been used increasingly since the 1950s in Germany as well, and in the past 15 years *lipophilic* extracts of saw palmetto fruit have gained increasing medical and scientific importance. On the basis of experimental and clinical studies conducted in the past ten years, saw palmetto preparations have assumed first place among the herbal prostate drugs.

Experimentally, the following effects of sabal extracts have been demonstrated: antiinflammatory, antiexudative, and antiedemic properties, as well as prostatotropic effects, e.g. inhibition of 5α-reductase and aromatase, and effects on the receptor binding of androgens (see reference *23* for overview). Both the *aqueous* extract, which consists primarily of an acidic polysaccharide (*24*), and the flavonoid-rich *alcoholic* extract (*25*) were shown to have antiinflammatory and antiedemic activity in rat paw edema and in perfused rabbit ears.

The *lipophilic* extract was also observed to have marked antiedemic activity in various animal models (*26, 27*). Dose-dependent inhibition of leukotriene B_4 and thromboxane B_2 synthesis has been demonstrated in peritoneal leukocytes of the rat, as well as a potent antiinflammatory effect on croton-oil-induced edema of the mouse ear (*28*). A CO_2 extract of saw palmetto was observed to inhibit cyclooxygenase and 5-lipoxygenase in microsomes obtained from sheep seminal vesicles (*29*).

Inhibition of 5α-reductase was demonstrated by five different groups (*28–32*) in various experimental models and also in comparison to finasteride. Both the ethanol extract and the hexane extract inhibit aromatase (*28*). Dose-dependent displacement of dihydrotestosterone from nuclear and cytosolic androgen receptors was reported by four different groups (*26, 28, 29, 32, 33*).

In an alternative *in vivo* system (human prostatic tissue transplanted to the nude mouse) the synergistic effects of DHT and estrogen were observed, as well as sabal-extract-induced inhibition or slowing of prostatic hyperplasia (*34*). Other animal experiments confirmed the antiandrogenic effects of lipophilic saw palmetto extracts (*26*).

The following clinical studies have so far been conducted:

1. A six-month multicenter, placebo-controlled study in which a significant improvement of nocturia, dysuria, urinary flow, and residual urine volume was observed (*35*).
2. A three-year prospective multicenter study with 435 patients in which significant efficacy was found using the I-PSS, even in comparison to synthetic prostate drugs (*36*).

3. Three placebo-controlled double-blind studies (*37–39*) with observation periods of three to six months, in which significant improvement was found in both subjective and objective parameters.

In conclusion, lipophilic saw palmetto extracts meet the requirements of the drug guidelines in every respect. They also have an extremely low rate of adverse reactions and a low cost compared to finasteride and other synthetic prostate drugs.

Urticae radix

Nettle extracts do not have any traditional use, and the first observations by Rückle were not made until 1950 (*40*). These were followed by experimental and clinical studies on the use of nettle extracts in BPH. Meanwhile, numerous experimental and clinical studies have been carried out to confirm the efficacy and determine the dosage (*41*).

Experimental studies have demonstrated the following effects (reviewed in *4, 23*): inhibition of aromatase using placental microsomes as an aromatase source and observing the conversion of labeled androstenedione to estradiol (*42*); inhibition of human leukocyte elastase (HLE) to demonstrate the antiinflammatory effects of a nettle extract (*43*); inhibition of complement activation as a parameter of antiinflammatory activity (*44*); investigation of interaction with SHBG, the most important androgen transport protein in the bloodstream (*45*); investigation of immunomodulating properties (*44*), especially with respect to prostatic malignancies; observation of effects of a nettle-root extract on the morphology and activity of prostate cells (*46*); and investigation of the effects on ATPase activity in BPH tissue (*47*).

A combination of *Urtica dioica*-root extract (with 30% methanol) and *Pygeum africanum-bark* extract (with methylene chloride) showed a synergistic action and significantly (P=0.05) increased 5α-reductase-inhibitory activity in concentrations as low as 0.1 mg/ml (ED_{50} 0.24mg/ml) (*54*).

In a pilot study, the efficacy of an alcoholic-aqueous nettle extract was tested in prostatic hyperplasia of the dog (*48*). All the study results point to both antiinflammatory and prostatotropic activity of hydroalcoholic nettle extracts.

Efficacy in patients was investigated between 1982 and 1991 in eight different multicenter studies (41). A significant reduction in nocturia frequency, an increase in urinary flow rate, and a reduction in residual urine volume were observed. Two placebo-controlled double-blind studies (*49, 50*) only partly confirmed the results of the observational studies. Whereas, no significant differences were found between placebo and active-treatment groups in terms of subjective symptoms, the objective parameters showed an approximately 44% increase in urinary volume compared to placebo and a 10% increase in maximum flow. Another double-blind study is currently in progress.

A special property of hydroalcoholic nettle extracts is the evidently specific interaction of *nettle lectins* with membrane receptors and the associated effects on cell differentiation and proliferation processes.

Table VII. Phytomedicines versus Synthetic Drugs in the Treatment of BPH

Phytomedicines	5α-Reductase Inhibitors and α_1-Receptor Antagonists
Multiple active substances	A single active substance
Multifactorial experimental and clinical activity profile	A single, fairly clearly defined activity principle, e.g. inhibition of DHT-ynthesis, selective α_1 blockade
Aim Improvement of symptoms	Aim Causal therapy (not yet achieved)!
Clinical effects Marked improvement of symptoms	Clinical effects With finasteride, the volume reduction does not correlate with a significant improvement of symptoms. With α_1-receptor blockers a reduction of nocturia.
Adverse effects Negligible	Adverse effects: With finasteride, up to 12% With α_1-receptor blockers: hypotension
Costs Reasonable	Costs 3 to 5 times higher!
Recommendation For routine BPH therapy, especially by general practioners.	Recommendation For *special* therapy by urologists.

Urtica dioica agglutinin (UDA) may be the major antiprostatic compound of Urtica dioica root-drug, acting by blocking the EGF-R (epidermal growth factor receptor) in prostate tissues (*53*).

Non-Monographed Crude Drugs and Crude-Drug Preparations

Within the group of non-monographed crude drugs and crude-drug preparations, the phytosterol mixture from the African *Hypoxis rooperi* LINNÉ must be mentioned at least because of its commercial importance. This mixture of natural substances consists of at least six phytosterols and goes by the name of β-sitosterol. It was medically prescribed as early as 1974 and, next to pumpkin seeds, can look back on the longest medical history.

At first sight, especially if one is unfamiliar with the experimental and clinical findings, it is difficult to appreciate the therapeutic relevance of purified and concentrated *Hypoxis rooperi* extract, as many foods contain substantial quantities of phytosterols, particularly β-sitosterol. However, experimental data (*4*), clinical observational studies on over 800 patients (*51*), and especially the most recent placebo-controlled double-blind study (*52*) indicate that the phytosterol extract of *Hypoxis rooperi* is an effective drug for the treatment of stage I to II BPH symptoms.

72

Other crude drugs and crude-drug preparations for the symptomatic treatment of BPH (Table V), the use of which is known only in empirical and folk medicine, are not considered in this paper.

Summary

In summary, even when the scientific findings are subjected to a stringently critical evaluation, it must be concluded that the denigration of herbal prostate drugs to mere placebos, as occasionally occurs, is not scientifically tenable. All in all, far more experimental and clinical studies have been conducted with herbal prostate drugs than with "fashionable" drugs, such as α_1-receptor antagonists and finasteride, since with the latter only studies of recent design are available.

A dogmatic evaluation of one drug group or another is certainly not in the patient's best interest. As the overview (Table VII) shows, both groups of substances are useful for individualized treatment of benign prostatic hyperplasia, provided that conservative pharmacological treatment of BPH is basically accepted and that the patient's quality of life is also considered.

Literature cited

1. *Benigne Prostatopathien;* Vahlensieck, W.; Rutishauser, G., Eds.; Georg Thieme Verlag: Stuttgart, New York, 1992.
2. *Prostataerkrankungen-Diagnostik, konventionelle und alternative Behandlungsverfahren;* Frohmüller, H.; Theiß, M.; Bracher, F., Eds.; Wissenschaftl. Verlagsgesellschaft: Stuttgart, 1995.
3. Whitfield, H.; *Brit. J. Urology* **1995**, *76, Supplement 1*, 1-73.
4. *Phytotherapie in der Urologie;* Schilcher, H., Ed.; Hippokrates Verlag, Stuttgart 1992.
5. Brom, S. *Dtsch. Apoth. Ztg.* **1996**, *136*, 29-36.
6. Stoner, E. et al. *Urology* **1994**, *43*, 284-290.
7. Schilcher, H. *Med. Monatsschr. Pharm.* **1996**, *19*, 178-179.
8. Sauter, M. *Phytochemische und andere Untersuchungen von Cucurbita pepo L. convar. citrullinina I.GREB. var. styriaca I.GREB. mit Hinblick auf mögliche prostatotrop wirksame Inhaltsstoffe;* Dissertation; Freie Universität Berlin, 1984.
9. Naeimi, M. *Zur Identifizierung und Phytochemie von Cucurbitae semen, insbesondere der Samen von Cucurbita pepo L. convar. citrullinina I.GREB. var. styriaca I.GREB*; Dissertation; Freie Universität Berlin 1994.
10. Schilcher, H.; Schneider, H. J. *Urologe [B]* **1990**, *30*, 62-66.
11. Schilcher, H.; Dunzendorfer, U.; Ascali, F. *Urologe [B]* **1987**, *27*, 316-319.
12. Auel, W. *Der Landarzt* **1962**, *38*, 372-373.
13. Weinkammer, E. *Physikal.-Diätetische Therapie* **1964**, *9*, 2-5.
14. Lütgen, Th. *Dtsch. Med. J.* **1970**, *21*, 3-7.
15. von Weiser, K. *Dtsch. Med. J.* **1972**, *23*, 1-7.
16. Reuter, M. A.; Loenicker, E. M.; Reuter, H. J. *Erfahrungsheilkunde* **1980**, *13*, 512-515.
17. Schoenenberger, A.; Hauri, D. *Ergebnisse der klinischen Prüfung von Prosta Caps Fink;* Urologische Klinik des Universitätsspitals Zürich, Jahresbericht, 1981.
18. Kurth, W. *extracta medica practica* **1980**, *1*, 315-324.
19. Nitsch-Fitz, R.; Egger, H.; Wutzl, H.; Maruna, H. *Erfahrungsheilkunde* **1979**, *12*, 1009-1013.

20. Smith-Kline Beecham - Fink, Germany, *personal communication*
21. Schilcher, H.; Gärtner, Ch. *Ztschr. f. Phytotherapie* **1990**, *11*, 77-80.
22. Gärtner, Ch. *Phytochemische und andere Untersuchungen von bienengesammelten Mischpollen und Rapspollen;* Dissertation; Freie Universität Berlin, 1989.
23. Koch, E. in *Phytopharmaka in Forschung und klinischer Anwendung;* Loew, D.; Rietbrock, N.,Eds.; Steinkopff-Verlag: Darmstadt, 1995, 68ff.
24. Wagner, H.; Flachsbarth, H.; Vogel, G. *Planta med.* **1981**, *41,* 252-258.
25. Hiermann, A. *Arch. Pharmazie* **1989**, *322,* 111-114.
26. Stenger, A.; Taraye, J.-P.; Carilla, E.; Delhorn, A.; Charveron, M.; Morre, M.; *Gaz. Med. France* **1982**, *89,* 2041-2048.
27. Taraye, J.-P.; Delhorn, A.; Lauressergues, E.; Stenger, A.; Barbara, M.; Bru, M. Villanova, G.; Caillol, V.; Aliaga, M. *Ann. Pharm. Franc.* **1983**, *41,* 559-570.
28. Koch, E.; Biber, A. *Urologe [B]* **1994**, *34,* 90-95.
29. Breu, W.; Hagenlocker, M.; Redl, K.; Tittel, G.; Stadler, F.; Wagner, H. *Arzneim-Forsch./ Drug Res.* **1992**, *42,* 547-551.
30. Drücker, E.-M.; Kopanski, L.; Schweikert, H. U. *Planta med.* **1989**, *55,* 587-590.
31. Liang, T.; Liao, S. *Biochem. J.* **1992**, *285,* 557-562.
32. Rhodes, L.; Primka, R. L.; Berman, C.; Vergult, G.; Gabriel, M.; Pierre-Mallce, M.; Gibelin, B. *Prostate* **1993**, *22,* 43-51.
33. Carilla, E.; Briley, M.; Fauran, F.; Sultan, C.; Duvilliers, C. *J. Steroid Biochem.* **1984**, *20,* 521-52.
34. Otto, U.; Wagner, B.; Becker, H.; Schröder, S.; Klosterhalfen, H. *Urol. Int.* **1992**, *48,* 167-170.
35. Cukier, D. *Gaz. Med.* **1984**, *34 Supplement 1,* 34-38.
36. Bach, D. *Urologe [B]* **1995**, *35,* 178-183.
37. Champault, G.; Patel, J. C.; Bonnard, A. M. *Brit. J. Clin. Pharmacol.* **1984**, *18,* 461-462.
38. Mattei, F. M.; Capone, M.; Acconcia, A. *Urologie/Nephrologie* **1990**, *2,* 346-350.
39. Helpap, B.; Oehler, U.; Weisser, H.; Bach, D.; Ebeling, L. *J. Urol. Path.* **1995**, *3,* 175-182.
40. *Brennesselwurzeltee bei beginnender BPH;* Rückle, E., Ed.; Hippokrates-Verlag: Stuttgart, 1950, 55-56.
41. *Benigne Prostatahyperplasie - 3. Klinisch-experimentelles Expertengespräch zu Fragen der benignen Prostatahyperplasie in Sevilla 1990;* Rutishauser, G., Ed.; Zuckschwerdt-Verlag: München, 1990.
42. Kraus, R.; Spiteller, G.; Bartsch, W. *Liebigs Ann. Chem.* **1991**, 335-339.
43. Koch, E.; Biber, A. in press
44. Wagner, H.; Willer, F. *Nat.-Ganzheitsmed.* **1990**, *3,* 309-312.
45. Hryb, D. J.; Khan, M. S.; Romas, N. A.; Rosner, W. *Planta med.* **1995**, *61,* 31-32.
46. Ziegler, H. *Fortschr. Med.* **1982**, *100,* 1832-1834.
47. Farnsworth, W. E. *Medical Hypothesis* **1993**, *41,* 358-362.
48. Daube, G. In *Benigne Prostatahyperplasie II;* Bauer, H. W., Ed.; Zuckschwerdt-Verlag: München, 1988, 63-66.
49. Vontobel, H. P.; Herzog, R.; Rutishauser, G.; Kres, H. *Urologe [A]* **1985**, *24,* 49-53.
50. Dathe, G.; Schmid, H. *Urologe [B]* **1987**, *27,* 223-225.
51. Albrecht, J. *Urologe [B]* **1994**, *34,* 20-25.
52. Berges, R. R.; Windeler, J.; Trampisch, H. J.; Senge, Th. *Lancet* **1995**, *345,* 1529-1532.
53. Wagner, H.; Geiger, W.N.; Boos, G.; Samtleben, R. *Phytomedicine* **1995**, *1,* 287-290
54. Hartmann, R.W.; Mark, M.; Soldati, F. *Phytomedicine* **1996**, *3,* 121-128.

Chapter 7

Plant Polysaccharides and Cancer

G. Franz

University of Regensburg, Department of Pharmaceutical Sciences,
D–93040 Regensburg, Germany

A variety of polysaccharides from different biological sources has been
shown to enhance the immune system. The immunostimulating
potential of these biopolymers is described as mainly promoting the
activity of phagocytosis of the respective macrophages and granulocytes, inducing the production of TNFα and interleukins, and, finally,
acting on the complement system. The most active polymers appear to
be branched (1→3) ß-D-glucans, which possess randomly dispersed
single ß-D-glucopyranosyl units attached to C6. A great variety of
other neutral and acidic polymers of different molecular dimensions
have been proposed as so called "anti-tumor polysaccharides." The
corresponding activities were mostly shown on allogeneic tumors such
as Sarcoma 180. However, clinical experience, has only been
documented for the fungal ß1→3 glucans, such as lentinan and
schizophyllan. The actual relevance of polysaccharide application in
cancer adjuvant therapy will be discussed.

Carbohydrates are found throughout biological systems and can be involved in many
physiological processes. However, they are rarely used as therapeutic agents. It was
surprising during the last years to learn about possible effects of a broad variety of
carbohydrate polymers for treating a wide range of diseases such as immune disorders,
tumors, diabetes, hypercholesterolemia, thrombosis, inflammation, etc. (*1*). Some of
these new fields of polysaccharide applications have been clinically approved, while
others are still controversial and have not been accepted for clinical purposes.

An essential prerequisite to attributing distinct physiological effects to a
polymer is a knowledge of its structure. Since defining the physiological activity of a
polysaccharide depends on its level of purity, it is essential to utilize appropriate
techniques for their isolation, purification and structure determination.

The previous lack of interest in the therapeutic possibilities of carbohydrates is
mostly a consequence of their complex structure, which has made both analysis and

synthesis extremely difficult. The monomer units of a polysaccharide can be joined in many different combinations, each of which may have a different biological activity. Since there may be many different types of monosaccharides in a chain and since these can further be attached in many different ways, the resulting polymers are very complex. However, in nature there are a relatively limited number of polysaccharide structures. Furthermore, the number of possible conformations is also rather restricted.

There are considerable differences in the physical properties of polysaccharides, such as solubility, rheology, gel-forming abilities, etc. Knowledge of these physical parameters are important prerequisites for possible medicinal application of these compounds.

Biological systems, including the immune system, show a distinct need for regulation. Once an immune response is initiated, mechanisms must exist to control the extent of the effect and to regulate it over time. The response should be self-limited by the removal of the stimulatory compound.

The so-called biological response modifiers (BRM) constitute a group of therapeutic agents capable of altering immunological responses. Although the mechanisms of action of immunomodulators are presumed to be very complex, part of their activity may proceed by a direct effect on the immune cells. This is a possibility for evaluating the potency of such compounds through measurements of cellular responses in vitro. The different assay methods should be adequate to measure the suppression or stimulation of the tested cellular systems. Some assays are more reliable, while others might be the result of an artifact. It is always essential to combine several test models in order to obtain a reliable picture.

In vitro tests have been shown to be very useful for evaluation of immunomodulation and for the comparison of potencies (2). Some in vitro tests correlate fairly well with in vivo animal models, but their validity for clinical use is still questionable.

According to Wagner (3), immunomodulators are compounds predominantly leading to a nonspecific stimulation of the immunological defense system. They should not affect immunological memory cells and, therefore, have only a relatively short activity potential. Since their effect fades relatively fast, they have to be administered continuously or in regular intervals. This was shown to be the case, when polysaccharides, such as lentinan or schizophyllan, are being used as oncostatic (stopping tumor growth) polysaccharides (4). When preparations containing complex higher plant polysaccharides are utilized as nonspecific immunostimulators against bacterial or viral infections, a prolonged treatment period is needed (5).

Screening Methods for Immunostimulants

For an appropriate screening to evaluate potential immunomodulating compounds, many methods have been established which allow a determination of the functional state and efficiency of the cellular and humoral nonspecific immune system. The target cells include the granulocytes, macrophages, T-lymphocytes, NK-cells, and, finally, the classical and alternative pathways to complement activation. Very often the tests comprise various in vivo infection tests with microorganisms such as *Candida*, *Listeria*, *Streptococcus* and *Staphylococcus* (6). The scope of all test systems is to define cellular targets, to establish dose/response curves, to predict in vivo activity, and to evaluate toxicity. In some cases these test models can be utilized even in preclinical trials. The more recently developed flow cytometry has improved the methodology by selecting distinct cellular systems.

It must always be seriously considered whether the data obtained with highly sophisticated in vitro models truly reflect the situation in vivo and are indeed relevant for an appropriate judgment of the efficacy of the system (7). As has been shown in the case of distinct polysaccharide structures, it was quite obvious that certain in vitro effects related to a stimulation of immune cell targets, did not correspond with the in vivo results (8). The exploration of the immunological functions by a specific drug must be followed in each case by the evaluation of the identical compound in the total complex biological system (9).

Polysaccharides as Immunomodulators

A series of new pharmacological activities of biopolymers, such as the polysaccharides, have been described (10). These polysaccharides were shown to differ in both structure and immunostimulating potencies. Since in some cases it was demonstrated that structure dependent effects against tumor cells via the immune system could be achieved, many publications appeared with a great variety of results. One of the first reviews about this topic was published as early as 1976 (11). Another overview in the field of the so-called antitumor polysaccharides appeared in 1987 (12). Later, many results came mainly from two research centers active in isolating new polysaccharides from different biological sources, evaluating the respective structures, and testing for specific immunostimulating activities (13).

The immunostimulating potential of these newly detected biopolymers was described as mainly promoting the activity of macrophage phagocytosis and granulocytes, inducing the production of TNFα and interleukins, and acting on the complement system. It became obvious that all the different polysaccharide structures described elicit an immune response, when certain structural parameters are present, such as the triple helical polymer conformation.

Isolation and Purification of Polysaccharides

Since the biological activity of such polymers largely depends on its state of purity and the chemical structure, it is essential to utilize various modern techniques for their isolation, purification and structure determination. Polysaccharides are usually isolated by precipitation from non-dialysable fractions of the appropriate extracts from biological material.

Uronic acid-containing polysaccharides mostly need dilute sodium hydroxide for their extraction. Specific salt concentrations are sometimes essential for a non-destructive extraction of certain polysaccharide types. The crude polysaccharide fractions are usually further purified by ion exchange, gel permeation, and affinity chromatography (14).

The molecular weights of the polysaccharides are determined with a broad range of molecular sieves such as the Sepharoses or Sephadexes, or by ultracentifugal analysis (15). The degree of polymerization can be further determined using periodate oxidation or GLC after reduction of the reducing end groups (16).

Structure Determination of Polysaccharides

In complex carbohydrates, the glycosidic linkages are usually determined by methylation analysis with the classical Hakomori method or a modification thereof (*17*). Identification of partially methylated alditol acetates are possible in standard procedures by GLC-MS (18). The determination of anomeric configurations of the different sugars present in a repeating unit, the detection of amino- or deoxysugars, and the identification of 0-acetyl groups and other non-sugar substituents can be routinely performed by ^{1}H and ^{13}C NMR spectroscopy. However, the amount of information obtained decreases with the increasing size and complexity of the polymers (*19*).

Fungal Antitumor Polysaccharides - Lentinan and Schizophyllan

The most important problems in cancer research are to increase the survival time of a patient and to prevent tumor recurrence after surgical resection. Cytocidal anticancer chemotherapeutics may have detrimental side effects and destroy the host's defense mechanisms. In addition, a possible induction of new tumors by cytotoxic agents might occur (*20*). On the other hand, there is good evidence for the existence of intrinsic resistance to cancer. The examples are an equilibrium state with proliferation, regression in a small amount of cancer cells, and spontaneous regression of cancer. An increase in this resistance is one of the goals for finding new anticancer drugs. One of the most interesting problems in cancer treatment is the recurrence after surgical resection of the primary tumor, which is caused by the development of micrometastasis. Unfortunately, postoperative chemotherapy has the disadvantages of insufficient effectiveness and a high level of more or less toxic side effects.

In recent years, an increasing number of polysaccharides has been identified which possess physiological activities correlated to the immune system. Most of these polysaccharides have been isolated from *Basidiomycetes*. Furthermore, *Oomycetes* and *Ascomycetes* have been examined for their potential in providing immunological active polymers. Variations in both the structure and biological activity of *Oomycetes* antitumor glucans have been described (*21*).

Probably one of the best studied polymers with pronounced immunostimulating activities is lentinan from the mushroom *Lentinus edodes,* which mushroom has long been used in oriental folk medicine as a therapy for cancer. lentinan was shown to be a hot water soluble natural glucan with a ß-1,3 structure and ß-1,6 branches and a molecular weight range from 400,000-800,000 (*22*). In vivo tests have shown broad antitumor effects in mice against a range of transplanted or induced sarcoma, fibrosarcoma, and carcinoma. lentinan has also been demonstrated to exhibit activity against metastasis of some tumors (*23*). Other model tumors, such as B16 melanoma were less responsive to a lentinan treatment (*24*).

The optimal doses in these animal experiments were usually much less than the maximum tolerated dose, with larger doses showing a decreased anticancer activity.
The immune system seemed to be involved by activating mainly the macrophages, leading to the assumption that in vivo effects are due to a stimulation of cell-mediated immunity. The exact mechanisms of action, however, remain obscure.

The first clinical trials with lentinan co-administered with mitomycin C and 5-fluorouracil demonstrated a significant improvement of survival of advanced colorectal cancer patients (25). In another phase III clinical trial, stomach cancer patients were administered lentinan with tegafur while the patients in the control group received tegafur alone. This study showed a highly significant increase in survival of patients who received lentinan. The one year survival rate was increased from 2.9% (control) to 21.2% for the lentinan-treated patients. Of the lentinan treated patients, 2.5% were still alive after five years; whereas, all of the control patients had died after three years (26). Most interestingly, side effects were rare and neither serious nor pronounced.

The antitumor effect, mediated by stimulation of the immune system, seems to be associated with the ß-1,3-basic structure of these glucans. However laminaran, pachyman, and other similar non-cellulosic ß-glucans, which also possess ß-1,3 linkages, are devoid of a pronounced antitumor effect. These ineffective polymers have in part a monohelical tertiary structure; whereas, lentinan and related polysaccharides possess a triple helical architecture. Therefore, the possible antitumor effect is closely related to the type of glycosidic linkage, the tertiary structure, and tthe nature of the side chains (27).

A second example of a fungal polysaccharide being potent as antitumor drug is sizofilan (schizophyllan or SPG), obtained from a highly viscous culture filtrate of *Schizophyllum commune* (28). SSPG has a ß-1,3 linked glucan backbone and side chains each comprising one glucose molecule bound to the main chain through ß-1,6 glucosidic linkages at an interval of three glucose molecules. It has been confirmed by X-ray diffraction that SPG is a trimer having a triple helical structure (29). As in the case of lentinan, the antitumor effect of SPG in animals has been investigated using various transplanted syngeneic and allogeneic tumors. As a result, administration of the glucan alone and co-administration with chemotherapy or radiotherapy were found to suppress tumor growth, increase anti-metastasis, and prolong the survival rate (30).

Since no adverse reactions were observed in phase I clinical studies with SPG, phase II studies were conducted in patients with lung cancer, cancer of the digestive organs, head, neck and uterus. These studies reported that indeed SPG had antitumor effects and also improved the host's immunological parameters (28). No direct cytotoxic effects against tumor cells have been shown for SPG and lentinan. Their antitumor activity is mediated by a stimulation of the host's immune system. Therefore, the immunomodulating potencies of these glucans have been intensively investigated in order to clarify their specific mode of action. It was concluded that there are three main aspects for the mode of action of ß-1,3/1,6-glucans:

1. Helper T-cell oriented immunopotentiation.
2. Macrophage mediated nonspecific immunopotentiation.
3. Non-immunological potentiation of the host's defense mechanisms through acute phase proteins, CSF, and the complement system.

For SPG, it was shown that application of the glucan after 5-fluorouracil treatment resulted in considerably enhanced NK-cell proliferation. Furthermore, precursors of various mature cell types, including NK-cells, were found to differentiate and rapidly regenerate cells damaged by chemotherapy (31). These results suggest that administration of SPG after chemotherapy in general may be useful for cancer patients.

Phytophthora - Glucans: Structural Variations and Antitumor Activity

Phytophthora parasitica is a phytopathogenic fungus belonging to the *Oomycetes* which is known to contain noncellulosic cell wall glucans (*32*). These glucans have some similarities to those described for the *Basidiomycetes*. One main difference is the presence of prolonged side chians compared to schizophyllan and lentinan and a relatively broad molecular mass distribution of 10-25 kDa.

In earlier investigations, it was speculated that a high molecular mass of > 50 kDa is an essential prerequisite for any glucan-mediated immunomodulation. With the isolation of water soluble ß-1,3/1,6-glucans from the cell walls of *Oomycetes,* it could be clearly demonstrated that even a molecular mass as low as 10 kDa still exhibits antitumor activity comparable to that of SPG, which has an average molecular mass of ~ 450 kDa. An essential requirement seems to be a certain degree of branching of less than 33% for any effect against Sarcoma 180. The lower the degree of branching, the higher is the biological activity, as demonstrated by a comparison of the inhibition rates at a dose of 0.2 mg/kg. From these investigations, it could be concluded that a pronounced antitumor activity in different animal tumor models can be expected from low branched ß-1,3/1,6-glucans with relatively high molecular mass (*33*).

Another representative of the *Oomycetes, Pythium aphanidermatum* was also examined for possible immunomodulating glucans. It is known that this fungus contains cellulose, instead of chitin, and other noncellulosic cell wall and storage glucans (*27*). The ß-1,3/1,6-cell wall glucan was shown to have a molecular mass of 10 kDa and to be highly branched; whereas, the storage glucans with a similar basic structure and a molecular mass of 20 kDa was only slightly branched. It is interesting to note that both of these glucans were active against Sarcoma 180 in vivo, even though they have different degrees of 1,6-branching,

Antitumor Activity of Chemically Modified ß-1,3-Glucans

Besides the typical conformation of such glucans, it is interesting to evaluate the influence of different side chain sugars. For this purpose, the ß-1,3-glucans, curdlan and lichenan, were chemically modified by attaching arabinose, rhamnose, D-glucose, and gentiobiose side chains. The antitumor activity of the newly branched polysaccharides was tested against the allogeneic Sarcoma 180 (*21*). The rhamnosyl and arabinosyl curdlan derivatives were the most active, but glucosyl and gentiobiosyl curdlan derivatives actually decreased the antitumor activity. The derivatives of lichenan showed no significant activity, except for the glucosyl derivatives with a low degree of branching. These data again were the proof for the fact that a 1,3-linked backbone is advantageous for any immunostimulating activity of the ß-glucans. It was further demonstrated in these studies that the derivatives of curdlan with pronounced antitumor activity had no ordered structures, such as triple helices (*34*).

Biological Activities of Higher Plant Polysaccharides

Most of the more detailed studies in the past have been carried out with defined polysaccharides from fungi, algae, and lichens, which have at least some common structural features in that the majority are ß-1,3-glucans with side chains. Many other carbohydrate polymers are known, but these were rarely shown to be potent immunostimulators.

The situation for higher plants is much more complex, both from the chemical point of view and from the multiplicity of the different actions upon the immune system (*35*). The most important source for higher plant polysaccharides is the cell wall, which has a great variety of matrix polysaccharides. These so-called hemicelluloses may be subdivided into xylans, mannans, and galactans, according to their sugar composition.

Each group contains polysaccharides with considerable structural diversity. For the xylans, xylopyranose residues are ß-1,4-linked to which arabinose, galactose, glucose or uronic acids can be attached. Mannans are mostly associated with glucose or galactose either in the backbone or as side chains. Galactans are very abundant, mostly as arabinogalactans, where the galactopyranose residues are linked ß-1,3. In close connection, the complex group of pectins is always found. The most abundant components of the pectin-polysaccharides are the polyuronic acids, which are made of galacturonic acid residues. They are often accompanied by arabinans and galactans. The polyuronic acids are mostly composed of unbranched chains of D-galacturonic acid, in which are inserted L-rhamnopyranosyl residues.

Storage polysaccharides, such as the starch components (amylose and amylopectin), fructans, mannans, and galactomannans seem to be of lesser importance as immunostimulating biopolymers (*36*).

Due to the great chemical variety, it is obvious that the different polysaccharides cannot have the same site of interaction in the complex human immune system. Therefore, at present, it is almost impossible to draw any structure/activity relationships, as is the case for fungal glucans. It is still extremely difficult to indicate which structural features are essential for a specific immunostimulating activity. There is a great diversity in the literature in view of the different test methods, in vivo and in vitro conditions, purity of the tested compounds (LPS-contamination), and dose-dependence (*37*).

In many cases, the relationship between immunological activity in vitro for a specific polymer does not correlate with in vivo effects, such as for the antitumor effect (*2*). This might be due to the fact that the absorption of macromolecules and particles from the gut after orally application is still a matter of discussions. Whereas fungal glucans are only active after parenteral application, most of the higher plant polysaccharides are orally applied. However, an interaction with intestinal Peyer's patches could be an explanation.

Some data have demonstrated an absorption of intact macromolecules across the gut (*38*). The question is still open as to whether and how the responsible cellular systems in the gut can recognize the great variety of the different polysaccharide structures. It must be remembered that in the daily diet of plant food, the body is constantly in contact with almost identical polymers. Since it is obvious that the various polysaccharide structures can not have the identical sites of attack in the immune system, several studies were carried out in order to examine the possible polysaccharide structure-related effects (*3*). As far as macrophage stimulation and further phagocytosis is concerned, these processes seem to be correlated mainly with uronic acid-containing arabinogalactans, 4-0-methyl glucuronoxylans, and neutral xyloglucans. On the other hand, the complement-activating polymers were shown to be pectin-like macromolecules. Complement activation appears to be intrinsically associated with several immune reactions, such as the activation of macrophages and lymphocytes (37).

It is still difficult to conclude a distinct polysaccharide structure-related interaction with defined cellular systems of the immune system. Some conclusions, however, might be drawn. The highly anionic pectin-like structures can activate the complement system via the alternative pathway. This activation is mainly expressed by the rhamnified regions. It was concluded that the combination of the rhamnogalacturonan and the neutral sugar chains are responsible for this kind of interaction. Mitogenic activity is associated with increasing size of the polysaccharides tested and, again, with the anionic charge of these macromolecules (3). Antitumor activity, mostly shown in vivo with the transplanted Sarcoma 180, can only be related to ß-1,3-glucan basic structures. The rhamnification of the basic glucan chain seems to play a modifying role (21). Other neutral or acidic polysaccharides seem to be inactive in most tumor test models.

Induction of cytotoxic macrophages are correlated in vivo as well as in vitro with rhamnogalacturonans, mostly those rhamnified with arabinogalactan side chains (3). A clear cut correlation of defined structures with an increase in phagocytosis seems to be rather difficult and mainly dependent upon the test system utilized (6). In some cases, when pectic polysaccharides were tested utilizing bone macrophages, an increase of phagocytotic activity was demonstrated. However, utilizing pectin-like polymers from *Ginkgo biloba* (39), only a minor and nonspecific antitumor activity seems to be present. It is interesting to note that in these cases, where a phagocytotic induction with branched rhamnogalacturonans can be observed, that the activity is lost when the arabinogalactan side chains are cleaved and only the remaining highly acidic backbone is tested (7).

Literature Cited

1. Franz, G.; Alban, S., Kraus, J. *Macromol. Symp.* **1995**, *99*, 187-200.
2. Kraus, J.; Roßkopf, F. *Pharm. Pharmacol. Lett.* **1991**, *1*, 11-15.
3. Wagner, H. *Pure and Appl. Chem.* **1990**, 7, 1217-1222.
4. Chihara, G. *Int. J. Immunother.* **1989**, *5*, 1677-1683.
5. Jurcic, K.; Melchart, D., Hohmann, M., Bauer, R., Wagner H. Z.*Phytother.* **1989**, *10*, 67-70.
6. Wagner, H.; Jurcic, K. *Methods in Plant Biochem.* **1991**, *6*, 195-217.
7. Kraus, J.; Franz, G. In *Microbiol. Infections;* Friedmann H., Ed.; Plenum Press: New York, 1992, pp. 299-308.
8. Roßkopf, F. Dissertation; 1992; University of Regensburg, Germany.
9. Seymour, L. J. *Bioactive and Compatible Polymers* **1991**, *6*, 199-216.
10. Bohn, J.A.; BeMiller, J.N. *Carbohydr. Polymers* **1995**, *28*, 3-14.
11. Whistler, R.L.; Bushway, A.A., Singh, P.P., Nakahara, J. Tokuzen, R. *Adv. Carbohydr. Chem. Biochem.* **1976**, *32*, 235-275.
12. Witczak, Z.J.; Whistler, R.L. In *Industrial Polysaccharides*; Stivala, S.S.; Crescenzi, V., Dea, J.C.M., Eds.; Gordon and Breach Science Publ.: New York, 1987, pp. 157-173.
13. Kraus, J. *Pharmazie in unserer Zeit* **1990**, *19*, 157-165.

82

14. Sherma, J.; Shirley, C.C., Eds.; In *CRC Handbook of Chromatography: Carbohydrates*, CRC Press: Boca Raton, 1991, Vol. 4.
15. Blaschek, W. In *Polysaccharide*; Franz, G.; Ed.; Springer Verlag: Heidelberg, Berlin, New York, Tokyo, 1991, pp. 18-47.
16. Aspinall, G.O. In *The Polysaccharides*; Aspinall, G.O.; Ed.; Academic Press: New York, 1982, pp. 35-131.
17. Harris, P.J.; Henry, R.J., Blakeney, A.D., Stone, B.A. *Carbohydr. Res.* **1984**, *127*, 59-73.
18. Biermann, J.; McGinnis, G.D., Eds.; In *Analysis of Carbohydrates by GLC and MS*, CRC Press: Boca Raton, 1989.
19. Sweeley, C.C.; Nunez, H.A. *Ann. Rev. Biochem.* **1985**, *54*, 765-801.
20. Chihara, G. *Int. J. Immunother.* **1989**, *5*, 167-176.
21. Demleitner, S.; Kraus, J., Franz, G. *Carbohydr. Res.* **1992**, *226*, 239-246.
22. Chihara, G. *Rev. Immunol. Immunofarmacol.* **1984**, *4*, 85-100.
23. Aoki, T. In *Immune Modulating Agents and their Mechanisms*; Fenichel, R.L.; Chirigos, M.A., Eds.; Marcel Dekker: New York, 1984, pp. 63-77.
24. Chihara, G.; Hamuro, J., Maeda, Y.Y., Shiio, T. *Cancer Detect. Prev. Suppl.* **1987**, *1*, 423-443.
25. Taguchi, T.; Furue, M., Kondo, T., Hattori, T., Itoh, J., Ogawa, N. *Japan J. Cancer Chemother.* **1985**, *12*, 366-375.
26. Taguchi, T. *Cancer Detect Prev. Suppl.* **1987**, *1*, 333-345.
27. Blaschek, W.; Käsbauer, J., Kraus, J., Franz, G. *Carbohydr. Res.* **1992**, *231*, 293-307.
28. Furue, H. *Drugs of Today* **1987**, *23*, 335-346.
29. Tabata, K.; Ito, W., Kojima, T., Kawabat, S., Misaki, A. *Carbohydr. Res.* **1981**, *89*, 121-135.
30. Sagawara, J.; Lee, K.C., Wong, M. *Cancer Immunol. Immunother.* **1984**, *16*, 137-142.
31. Tsuchiya, Y.; Matsutani, M., Inoue, M., Sato, S., Asano, T., Yajima, M. *Cancer Immunol. Immunother.* **1991**, *34*, 17-23.
32. Fabre, J.; Bruneteau, M., Ricci, P., Michel, G. *Eur. J. Biochem.* **1984**, *142*, 99-103.
33. Blaschek, W.; Schütz, M., Kraus, J., Franz, G. *Food Hydrocoll.* **1987**, *1*, 377-38034. Demleitner, S.; Kraus, J., Franz, G. *Carbohydr. Res.* **1992**, *226*, 247-252.
35. Bauer, R.; Remiger, S., Jurcic, K., Wagner, H. *Z. Phytother.* **1989**, *10*, 43-48.
36. Kraus, J.; Franz, G. *Dtsch. Apotheker Ztg.* **1986**, *124*, 2045-2049.
37. Bauer, R.; Wagner, H. *Economic and Medicinal Plant Res.* **1991**, *5*, 254-320.
38. Sazuki, T.; Sakurai, T., Hashimoto, K., Oikawa, S., Masuda, A., Ohsawa, M., Yadomae, T. *Chem. Pharm. Bull.* **1991**, 1606-1608.
39. Kraus, J. *Phytochem.* **1991**, *30*, 3017-3020.

Chapter 8

Secondary Metabolites from Plants as Antiretroviral Agents: Promising Lead Structures for Anti-HIV Drugs of the Future

Eckart Eich

Institute für Pharmazie II, Pharmazeutische Biologie, Freie Universität Berlin, Königin-Luise-Str. 2+4, D–14195 Berlin, Germany

Retroviral integrase, a virally encoded enzyme (like reverse transcriptase and protease) could be another interesting target for the therapy of HIV infections. Herbal constituents and their derivatives belonging to different structural classes, e.g. lignans, flavonoids, and curcuminoids, turned out to be first inhibitors of this enzyme. Studies on structure/activity relationships showed that the most potent of these compounds may be described in general as consisting of two aryl units containing the 1,2-dihydroxy pattern each, thus forming bis-catechols separated by an appropriate linker segment. On the basis of these specific integrase inhibiting leads which, however, lack antiviral activity, the development of HIV-inhibiting compounds is in progress as recent results with a bis-coumarin derivative show.

Herbal extracts, as well as pure plant constituents, with elucidated chemical structures have been screened from species of numerous plant families with respect to antiviral properties. This has been so, particularly during the past two decades. Indeed a large number of compounds from almost all classes of secondary metabolites, show confirmed activity in special *in vitro* screens of many different pathogenic viruses. Nevertheless, the contribution of phytomedicines to the treatment of viral infections has been rather limited to date. A cream containing 1 % of a specially prepared dried extract from lemon balm leaves (*Melissa officinalis* L., Labiatae), has been introduced to the German market for local therapy of herpes simplex of the skin. The effect of this melissa cream in the topical treatment, is statistically significant as proven by clinical studies *(1, 2)*. On the other hand, constituents of plants could serve as useful leads for developing the antiviral drugs of the future. This review focuses on such leads from well-known medicinal plants in the field of human immuno-deficiency viruses (HIV), with emphasis on a novel target, retroviral integrase.

Arctigenin, the First Integration Inhibiting Agent in HIV Infected Cells

(–)-Arctigenin **1**, a lignan isolated from the pantropical climbing shrub *Ipomoea cairica* (L.) SWEET belonging to the morning glory family (Convolvulaceae) *(3)*, proved to be a potent inhibitor of HIV-1 replication in different human cell systems *(4,5)*. *I. cairica* is a medicinal plant of various ethnomedicines in West Africa *(6)*. On the other hand, **1** is also a constituent of certain species of the daisy family (Compositae), *e.g.* *Arctium lappa* L., the greater burdock *(7)*, which is used in Chinese traditional medicine (Niu bang) *(8)*. This lignanolide turned out to be an efficient inhibitor of the nuclear matrix associated DNA topoisomerase II activity.

R = CH₃: (–)-Arctigenin **(1)**
R = H: 3-O-Demethyl-
              arctigenin **(3)**

Thus, initially it was assumed that it showed anti-HIV activity due to prevention of the increase of activity of this mammalian enzyme involved in virus replication. However, using the polymerase chain reaction (PCR) technique to determine the degree of integration of proviral DNA into the cellular DNA genome after 6 days of incubation, a strong suppression of the HIV-1 integration could be observed *(9,10)*. From these results it was concluded that the anti-HIV effect of arctigenin **1** is primarily caused by inhibition of the HIV integration reaction rather than by inhibition of DNA topoisomerase II. Until this discovery, no compounds with HIV integration inhibiting properties had been found.

HIV Integrase *in vitro* Assay

In principal there is the possibility for the development of multiple classes of antiviral agents that could act at each respective stage of HIV replication, *e.g.* attachment, penetration, uncoating, and reverse transcription. These antiviral agents could act at an early stage of the HIV life cycle, or at different later stages, *e.g.* as protease inhibitors in the maturation stage. To date, only inhibitors of reverse transcriptase and protease have been found to be of therapeutic relevance. For many years, pharmacological antiretroviral research, with virally encoded enzymes as targets, has focused principally on agents that inhibit these two enzymes.

The protein catalyzing the integration observed in the studies with arctigenin **1** is a target which had been vastly neglected until the early 1990s. This protein is called retroviral or HIV integrase and is contained within the virus particle. It integrates a double stranded DNA copy of the viral RNA genome, synthesized by reverse transcriptase, into a host chromosome *(11, 12)*. This represents the last, and somehow decisive, step of the viral infection: the point of no return.

The development of an *in vitro* assay for HIV-1 integrase function, in the early 1990s, permitted testing of potential inhibitors. The recombinant integrase enzyme produced *via* an *Escherichia coli* expression vector uses a blunt-ended 21-mer duplex oligonucleotide as both the specific donor and nonspecific target substrates. The initial step in the enzymatic reaction involves nucleolytic cleavage (= 3'-processing). It liberates the 3'-terminal dinucleotide GT, producing a 19-mer oligonucleotide from the ^{32}P-labeled 21-mer duplex substrate. The resultant 3'-termi-nal dA hydroxyl is the donor substrate for the second step, the subsequent joining to a 5'-phosphate of a second 21-mer duplex oligonucleotide. The nucleophilic attack on the phosphodiester bond of the target substrate leads to the DNA strand transfer or integration reaction. Thus, this reaction is a transesterification cutting the target DNA (*in vivo*: the host DNA) and binding its 5'-end to the viral DNA's just shortened 3'-end. It results in the insertion of one 3'-processed oligonucleotide into another nucleotide, yielding higher molecular weight species with slower migration in electrophoretic separation procedures than the 21-mer substrate (for further details see *13-16*).

Plant Constituents as Leads for HIV Integrase Inhibitors

Arctigenin and CAPE.   The results in this enzyme assay for arctigenin **1**, however, were not up to the expectations: the compound was almost inactive *(13-16)*. A structural comparison of **1** with another recently discovered HIV-1 integrase inhibitor, caffeic acid phenethylester (CAPE) **2** *(17)*, a secondary metabolite from the buds of poplar trees (*Populus* spp., Salicaceae) as well as a constituent of bee propolis, shows surprising similarities. On the other hand, an important difference is the presence (**2**) or lack (**1**) of a catechol partial structure (two adjacent hydroxyl groups). This led to the

Caffeic acid phenethylester (CAPE) **(2)**

assumption that the corresponding 3-*O*-demethylated derivative **3** ought to be active. Indeed, this was confirmed *(13-16)*. Compound **3** exhibited remarkable activities in both integrase assays: the cleavage or 3'-processing step was inhibited by 57 % at a 100 μM concentration of the compound, and the integration or strand transfer step was inhibited by 52 % (Figure 1). Thus, the observed integration inhibiting effect of **1**, in the HIV-infected human cell systems *(9, 10),* could be due to metabolization of **1** to the 3-*O*-demethylated derivative **3** during the incubation. Therefore, arctigenin **1** probably acts as a "prodrug" which may easily enter the cell due to its more lipophilic character compared with **3**. Such a metabolization was confirmed at least in V79

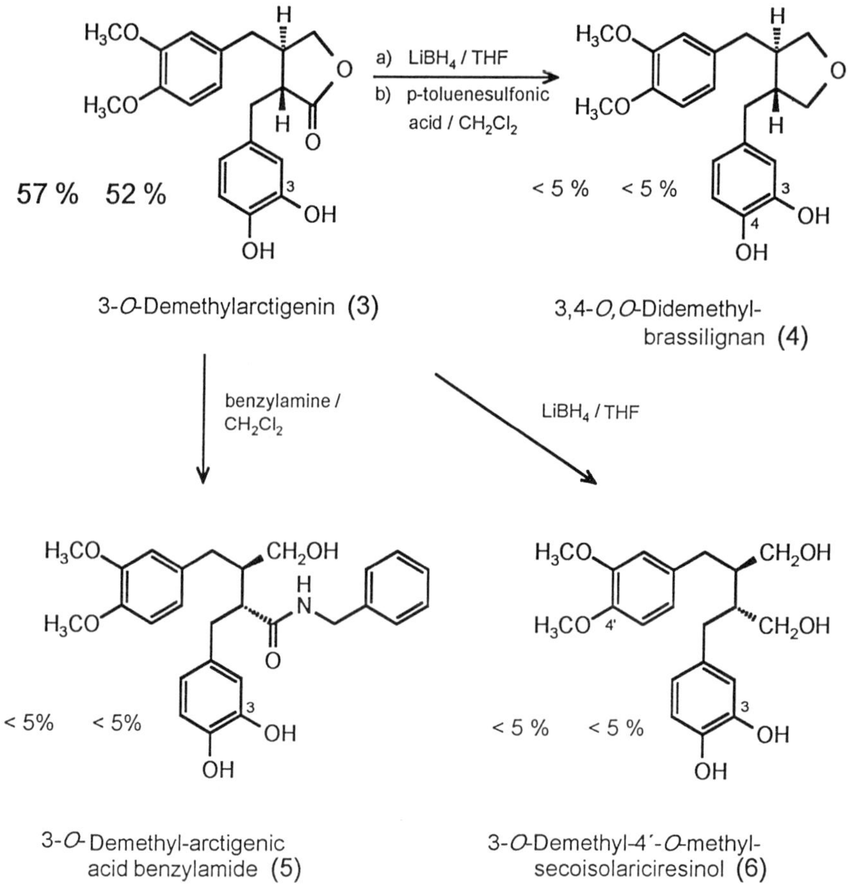

Figure 1: Derivatisation of the monocatechol **3** demonstrating the importance of the lactone moiety for the inhibition of the HIV-1 integrase reaction steps. First values represent the inhibiting activity for the cleavage (= 3'-processing) step, second values for the integration (= strand transfer) step at a concentration of 100 μM.

Chinese hamster cells genetically engineered for stable expression of the rat single cytochrome P450 isoform 2B1 *(18)*.

The activity of **3** led to detailed studies on structure-activity relationships with natural lignans of different types, as well as with certain semisynthetic, and also completely synthetic, analogs *(13-16)*. The importance of the lactone moiety is clearly demonstrated by the fact that there is an almost complete loss of activity, in spite of a catechol moiety, if the carbonyl oxygen is lacking (**4**), if a corresponding amide is used (**5**), or if the lignanolide is reduced to (**6**) (Figure 1). On the other hand, the synthetic racemic *trans*-cyclopentanone lignan analog **7** exhibits differential

(±)-*trans*-2-(3,4-Dihydroxybenzyl)-
3-(3,4-dimethoxybenzyl)cyclopentanone (**7**)

results in the two assays. The integration step is inhibited to a certain degree (38%; 100 µM), in contrast to the cleavage step (<5%). Retaining the lactone moiety, the influence of the number of phenolic hydroxy groups, and also the influence of the arrangement of these groups is decisive: if the catechol moiety is situated in the "upper" part of the lignanolide structure (Figure 2), the corresponding compound **8** shows less, and differential, activity compared with **3**, the catechol substructure of which belongs to the region "below." The synthetic racemic *trans*-lignanolide **9** with three vicinal hydroxy groups (pyrogallol moiety) is inactive in contrast to the corresponding catechol **3**. The bis-catechol congener **10**, however, turned out to be the most active compound for both assay functions. Total inhibition of the enzyme could be observed at 100 µM concentrations. The corresponding IC_{50} values (indicating those concentrations for which a 50 % inhibition is exhibited) confirm the potency of this bis-catechol **10**. Since 1993, **10** has been used as a reference compound in the research program of the French Rhône-Poulenc-Rorer company for the development of HIV integrase inhibitors *(19)*.

α-Conidendrin. In 1995, an NIH-conference on retroviral integrase, the first and currently the only meeting on this enzyme, took place in Bethesda, MD. During this meeting there was a report on a closely related bis-catechol **11**, α-conidendrol, which was ten times more active than **10** (Figures 3 and 4) *(20, 21)*. It had been the most active of 8000 natural and synthetic compounds in a random screening using a novel rapid assay for the strand-transfer reaction *(22)*. A comparison of the two structures

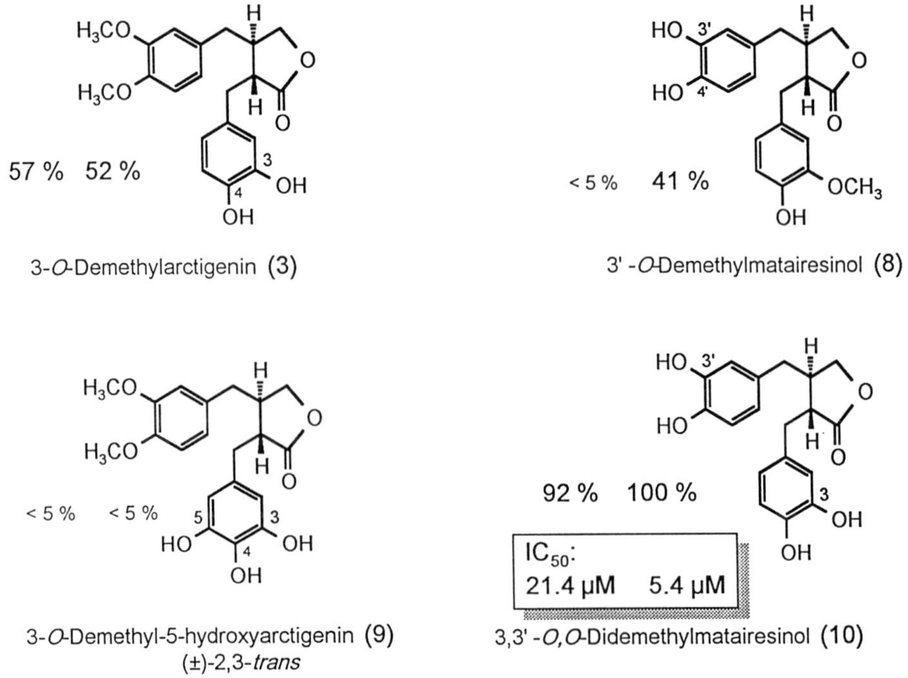

57 %   52 %

3-O-Demethylarctigenin (3)

< 5 %   41 %

3'-O-Demethylmatairesinol (8)

< 5 %   < 5 %

3-O-Demethyl-5-hydroxyarctigenin (9)
(±)-2,3-trans

92 %   100 %

IC₅₀:
21.4 µM   5.4 µM

3,3'-O,O-Didemethylmatairesinol (10)

Figure 2: Influence of the number and the positions of phenolic hydroxy groups on the inhibition of the HIV-1 integrase reaction steps. First values represent the inhibition activity for the cleavage (= 3'-processing) step, second values for the integration (= strand transfer) step at a concentration of 100 µM. In the case of 10 the data below show the corresponding IC₅₀ values.

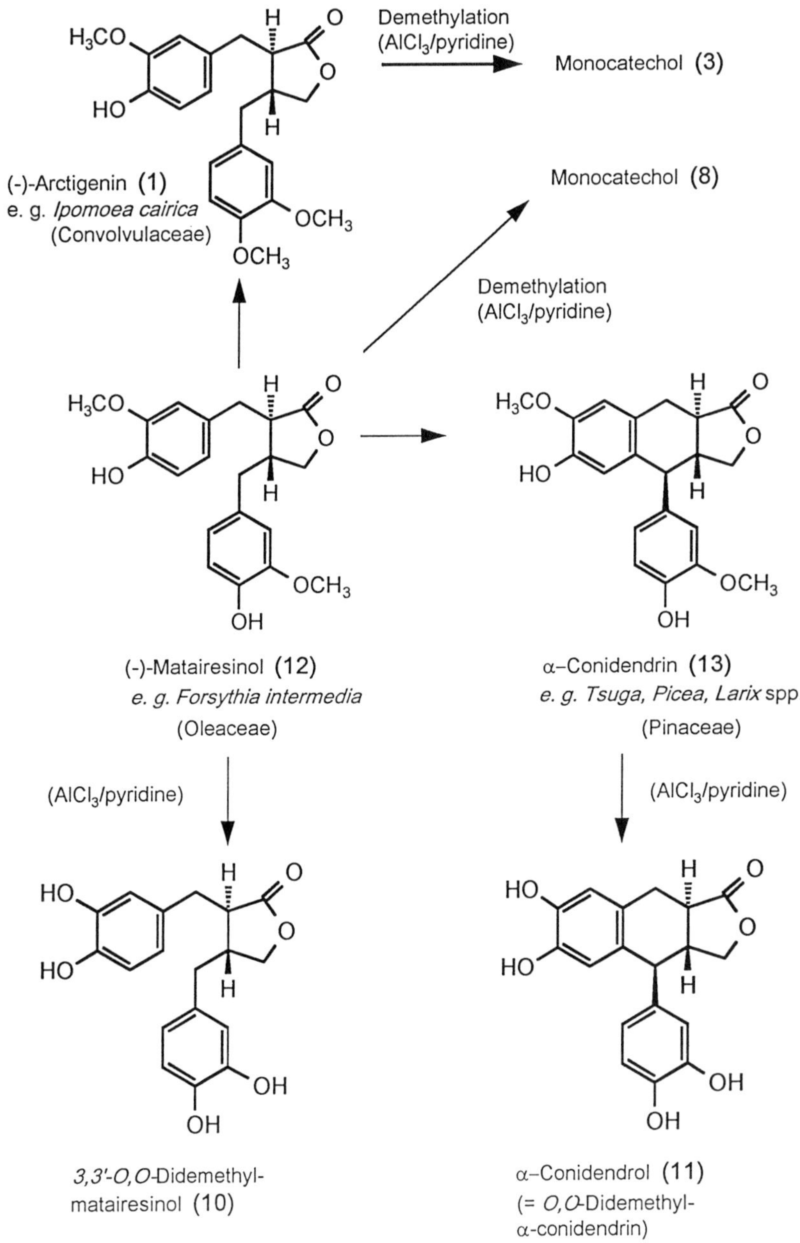

Figure 3: Biogenetic precursors of the mono- and bis-catechol lignanolides and plant sources for their semisynthetic preparation.

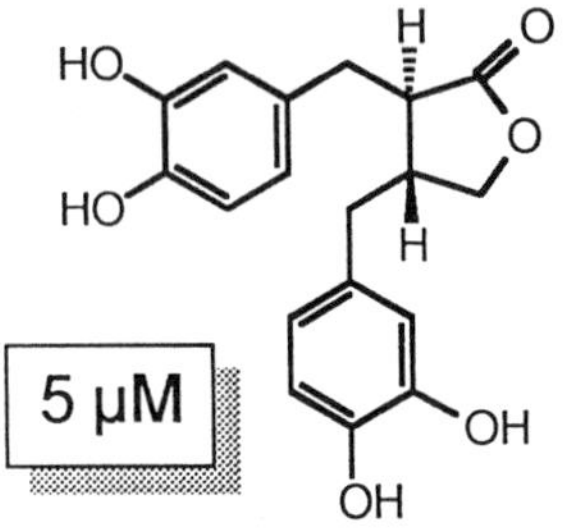

3,3'-*O*,*O*-Didemethylmatairesinol (10)
lead structure: (-)-arctigenin (1)

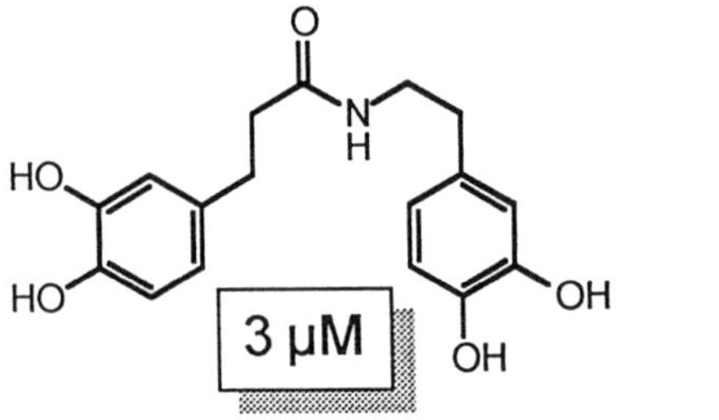

Dihydrocaffeic acid β-(3,4-dihydroxy-
phenylethyl)amide (18)
lead structure: CAPE (2)

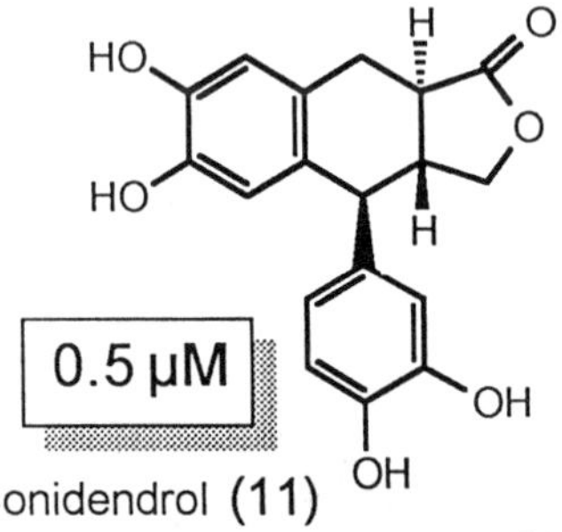

α-Conidendrol (11)
lead structure: α-conidendrin (13)

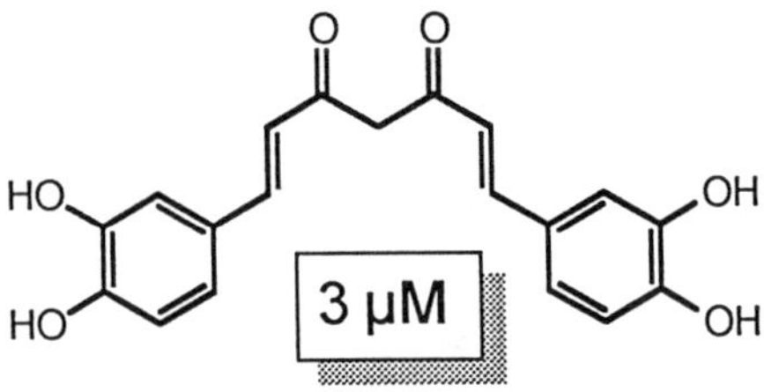

Dicaffeoylmethane (17)
lead structure: curcumin (15)

Figure 4: Semisynthetic bis-catechols as HIV-1 integrase inhibitors with their
IC$_{50}$ values for the integration (= strand transfer) step.

shows that the only difference between **10** and **11** is the ring closure to a so-called *retro*lignanolide, thus fixing the molecular structure (Figure 4). Whereas **10** is obtainable by demethylation of the inactive natural matairesinol **12**, from e.g. a *Forsythia* species (Oleaceae), **11** can be prepared from natural α-conidendrin **13**, which is a typical constituent of the wood of many Pinaceae species, where it is biosynthesized from **12** (Figure 3).

Specifity of Lignanoid Bis-Catechols. It can be convincingly demonstrated, that the inhibiting activity of these lignanoid bis-catechols is a specific one with respect to HIV integrase *(21)*. ß-Conidendrol **14,** which is the *cis*-isomer of **11** and is as active as the latter, did not inhibit a variety of nucleic acid-processing enzymes like EcoRI

α-Conidendrol **(11)**   β-Conidendrol **(14)**

(restriction enzyme), HIV-1 reverse transcriptase, HeLa RNA polymerase II, mammalian topoisomerase I, and mammalian topoisomerase II. The observation that **14** did not affect these assays suggests that this compound does not interact nonspecifically with nucleic acids.

Curcumin. Since ancient times turmeric rhizomes have been used as medicinal plants and as spices in South and South East Asia. *Curcuma longa (Zingiberaceae),* the long rooted turmeric, is a famous example as an ingredient of Indian curry. The German pharmacopoeia includes the rhizoma of *C. xanthorrhiza* from Java/Indonesia. A major constituent is curcumin **15**, the yellow colouring dye of turmerics, which exhibits a variety of pharmacological effects *(23)*. Compound **15** is also an inhibitor of HIV replication and is currently in clinical trials with AIDS patients *(24)*. Its mechanism of action against HIV is complex and not totally understood. In 1993 it was reported that **15** is a potent inhibitor of HIV-1 LTR-directed gene expression, p24 antigen production, and *Tat*-mediated transcription *(25)*. Furthermore, it has been demonstrated that **15** is a very interesting lead for HIV protease inhibitors. The moderate activity of this compound could be increased dramatically by dimerisation in a boron complex **16** (Table I) *(26)*. However, **15** exhibits only modest activity as an inhibitor of HIV-1 integrase, whereas, again, the bis-catechol character of its semisynthetic derivative dicaffeoylmethane **17** lead to a potent inhibitor *(27)*. There seems to be potential in combining the decisive partial structures in order to develop an inhibitor for attacking differential virus targets. Such dual, or even multiple inhibition, would be of therapeutical advantage. High-dose monotherapy with a drug possessing only one mechanism of action, exerts a high selective pressure for the generation of mutant viruses. Therefore, clinical benefits are only short term before resistance and decreased susceptibility arise.

Table I. Inhibition of HIV Proteases and Integrase by Curcuminoids (IC$_{50}$ conc.: µM)

	HIV-1 protease	HIV-2 protease	HIV-1 integrase: cleavage	integration
Curcumin **15**	100	250	150	140
Dicaffeoylmethane **17**	n.d.	n.d.	6	3
Boron complex **16**	6	5	n.d.	n.d.

n.d. = not determined

Curcumin-boron-complex chloride (**16**)

Common Structural Properties of Integrase Inhibitors of the First Generation

Figure 4 summarizes the semisynthetic bis-catechols obtained from different plant constituents, including the most active derivative **18** of the caffeic acid phenethylester (CAPE) lead *(28)*. All four compounds are characterized by certain common structural features. There is a strong correlation between potency and the number and arrangement of hydroxyl groups in all four leads. The main result of the structure-activity relationship study on arctigenin derivatives, namely that only the bis-catechol congener **10** is a really potent inhibitor *(13-16)*, has also been subsequently confirmed for the other three leads *(21, 27, 28)*. Compounds with one catechol moiety generally show only moderate inhibiting activities. The relative potency of the compounds in each structural set demonstrates that inhibitory activity is associated with the specific spatial disposition of the hydroxyl groups. The active pharmacophore is a bis-catechol, or more precisely: the most active compounds contain two pairs of adjacent hydroxyls on nonadjacent benzene rings. A certain fixed orientation of the bis-catechol moieties seems to be of advantage, as shown by the increased activity of the conidendrine derivatives **11** and **14** compared with the others. It is striking that in addition all four leads in Figure 4 possess at least one three membered side chain including a carbonyl function (**17, 18**) or an analogous partial structure integrated into a ring system (**10, 11**). Comparable results were obtained in structure-activity relationship studies on further natural compounds like flavonoids *(29)* and rosmarinic acid **19** which represents a natural derivative of CAPE *(27)*.

Rosmarinic acid (**19**)

Mechanisms of Action

All these active compounds are also potent inhibitors of the "disintegration" reaction, an apparent reversal of the natural integration step of the enzyme (for details see *16, 30*). This inhibitory effect is also exhibited on disintegration catalyzed by a deletion mutant integrase, the so-called truncated enzyme lacking the N-terminal zinc finger region and the C-terminal DNA binding region. Thus, it can be proven that the core domain of HIV-1 integrase represents the common site of action.

The prominence of catechol moieties in the structure-activity relationships could suggest oxidation-reduction and divalent metal chelation as factors in the mechanism of inhibition of HIV integrase. However, there is evidence that neither factor is likely. It was impossible to overcome the ß-conidendrol-(**14**)-mediated inhibition by adding extra zinc, manganese, or magnesium to the cleavage reaction *(21)*. However, it is proposed that coordination of the hydroxyl groups to the carboxylate groups present in the integrase active site, could inhibit binding of the DNA phosphodiester backbone, in such a way that an appropiately positioned transition state could not be generated. This hypothesis awaits further analysis *via* co-crystal structures of integrase bound to drug candidates *(31)*.

Developments Based on the Catechols

Unfortunately, none of these bis-catechols exhibit antiviral activity in cell cultures. The reasons may be trivial: lack of cell penetration or instability in the several days incubation. On the other hand, it is possible that there are significant differences between the *in vitro* and *in vivo* reactions catalyzed by the enzyme. In order to enable such an inhibitor to enter the cell, it could be applied to cell cultures as a "prodrug" which liberates the active drug by means of a mammalian enzyme inside the cell. To a moderate extent this seems to happen with arctigenin **1,** as already mentioned above.

Another possibility would be to introduce a lipophilic moiety into a highly active, hydrophilic bis-catechol. This could enable the inhibitor to pass through the cytoplasmic membrane while, hopefully, allowing the inhibiting activity to be conserved.

Very recently, a novel stage in the development of integrase inhibitors has been reached on the theoretical basis of the catechols. Again, an old medicinal plant is involved: the sweet clover, *Melilotus officinalis* (L.) Pall. (Fabaceae). The formation of dicumarol synthesized by certain fungi in improperly cured sweet clover hay (Figure 5), causes a well-known severe cattle disease. Dicumarol inhibits blood coagulation

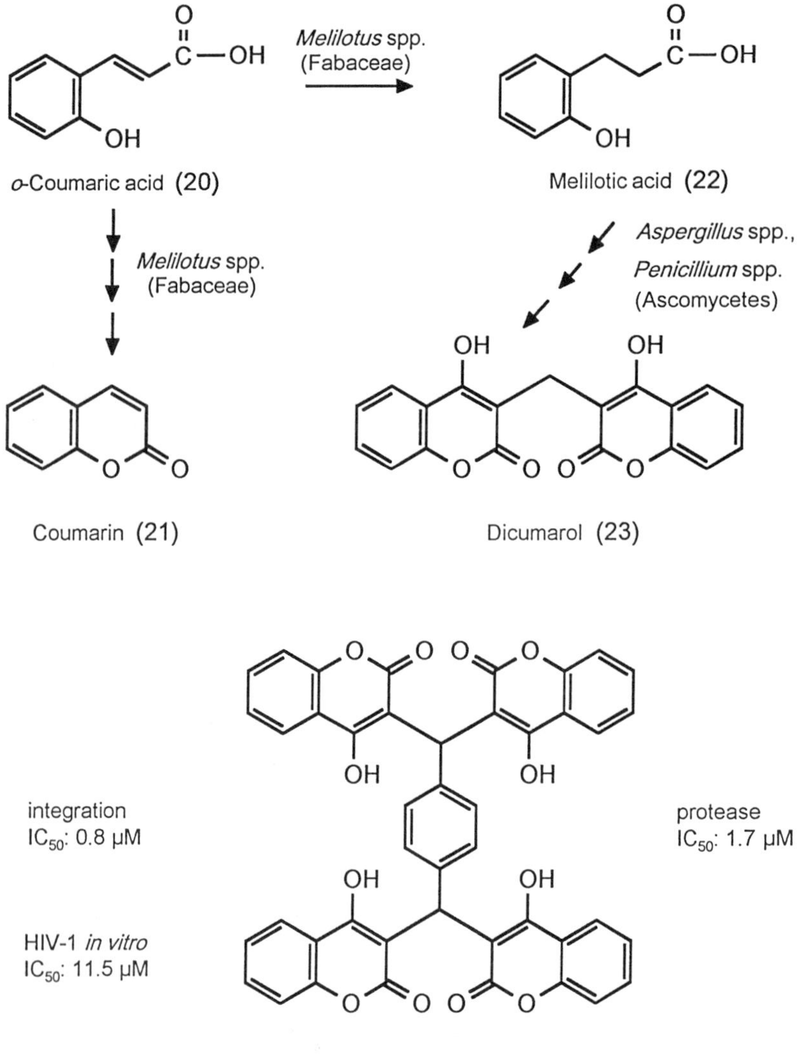

o-Coumaric acid (20) Melilotic acid (22)

Coumarin (21) Dicumarol (23)

integration
IC$_{50}$: 0.8 µM

protease
IC$_{50}$: 1.7 µM

HIV-1 *in vitro*
IC$_{50}$: 11.5 µM

NSC 158393 (24)

Figure 5: Antiviral, anti-integrase, and antiprotease activity (HIV-1) of the phenyl linked dimeric dicumarol conjugate NSC 158393 (24).

and is clinically used to prevent thromboses etc. An interesting dimeric compound, NSC 158393 (**24**), displayed complex activities *(30)*. It is a dual inhibitor, two enzymes are strongly inhibited. It even shows a remarkable antiviral activity. The two dicumarol pharmacophores are joined by a phenyl linker. Apparently this compound does not include a bis-catechol substructure. However, there are also four hydroxyls as well as carbonyl functions. It has been developed from a dicumarol derivative discovered through pharmacophore mapping of the known catechol integrase inhibitors, and a 3 D pharmacophore search of the NCI open 3 D database of 207,000 compounds.

Acknowledgments. The author is indebted to Dr. Jim Greatorex, Norges Landbrukshøgskole, Institutt for Tekniske Fag, Ås, Norway, for linguistic help in the preparation of this manuscript.

Literature Cited

1. Wölbling, R.H.; Leonhardt, K. *Phytomedicine* **1994**, *1*, 25-31.
2. Mohrig, A. *Dtsch. Apoth. Ztg.* **1996**, *136*, 4575-4580.
3. Trumm, S.; Eich, E. *Planta Med.* **1989**, *55*, 658-659.
4. Eich, E.; Schulz, J.; Trumm, S.; Sarin, J. S.; Maidhof, A.; Merz, H.; Schroeder, H. C.; Müller, W. E. G. *Planta Med.* **1990**, *56*, 506.
5. Schroeder, H. C.; Merz, H.; Steffen, R.; Mueller, W. E. G.; Sarin, P. S.; Trumm, S.; Schulz, J.; Eich, E. *Z. Naturforsch.* **1990**, *45c*, 1215-1221.
6. Kerharo, J.; Adam, J. G. *La pharmacopée sénégalaise traditionelle*; Editions Vigot frères: Paris, France, 1974; pp 367-368.
7. Haworth, R. D.; Kelly , W. *J. Chem. Soc. London* **1936**, 998-1003.
8. Paulus, E.; Yu-he, D. *Handbuch der traditionellen chinesischen Heilpflanzen*; Karl F. Haug Verlag: Heidelberg, Germany, 1987; pp 187-189.
9. Eich, E.; Schulz, J.; Kaloga, M.; Merz, H.; Schroeder, H. C.; Mueller, W. E. G. *Planta Med.* **1991**, *57* (Supp. 2), A7-A8.
10. Pfeifer, K.; Merz, H.; Steffen, R.; Mueller, W. E. G.; Trumm, S.; Schulz, J.; Eich, E.; Schroeder, H. C. *J. Pharm. Med.* **1992**, *2*, 75-97.
11. Katz, R. A.; Skalka, A. M. *Annu. Rev. Biochem.* **1994**, *63*, 133-173.
12. Rice, P.; Craigie, R.; Davies, D. R. *Curr. Opin. Struct. Biol.* **1996**, *6*, 76-83.
13. Eich, E.; Pertz, H.; Schulz, J.; Fesen, M. R.; Mazumder, A.; Pommier, Y. *PSE-Symposium on Phytochemistry of Plants Used in Traditional Medicine*, Lausanne/ Switzerland **1993**, Abstract Book, p. 97.
14. Eich, E.; Pertz, H.; Kaloga, M.; Schulz, J.; Fesen, M. R.; Mazumder, A.; Pommier, Y. *Eur. J. Pharm. Sci.* **1994**, *2*, 101.
15. Eich, E.; Pertz, H.; Kaloga, M.; Schulz, J.; Fesen, M. R.; Mazumder, A.; Pommier, Y. *NIH-Conference on Retroviral Integrase*, Bethesda/MD **1995**, Abstract Book, p. 42
16. Eich, E.; Pertz, H.; Kaloga, M.; Schulz, J.; Fesen, M. R.; Mazumder, A.; Pommier, Y. *J. Med. Chem.* **1996**, 39, 86-95.
17. Fesen, M. R.; Kohn, K. W.; Leteutre, F.; Pommier, Y. *Proc. Natl. Acad. Sci. USA* **1993**, *90*, 2399-2403.
18. Kasper, R.; Gansser, D.; Doehmer, J. *Planta Med.* **1994**, *60*, 441-444.
19. Gueguen, J. C., Rhone-Poulenc-Rorer, Vitry sur Seine, France, personal communication.

96

20. Hazuda, D. J.; Felock, P.; Hastings, J.; Uncapher, C.; Wolfe, A. *NIH-Conference on Retroviral Integrase*, Bethesda/MD **1995**, Abstract Book, p. 37

21. LaFemina, R. L.; Graham, P. L.; LeGrow, K.; Hastings, J.; Wolfe, A.; Young, S. D.; Emini, E. A.; Hazuda, D. J. *Antimicrob. Agents Chemother.* **1995**, *39*, 320-324.

22. Hazuda, D. J.; Hastings, J.; Wolfe, A.; Emini, E. A. *Nucleic Acids Res.* **1994**, *22*, 1121-1122.

23. Ammon, H. P. T.; Wahl, M. A. *Planta Med.* **1991**, *57*, 1-7.

24. Mazumder, A.; Raghavan, K.; Weinstein, J.; Kohn, K. W.; Pommier, Y. *Biochem. Pharmacol.* **1995**, *49*, 1165-1170.

25. Li C. J.; Zhang, L. J.; Dezube, B. J.; Crumpacker, C. S.; Pardee, A. B.; *Proc. Natl. Acad. Sci. USA* **1993**, *90*, 1839-1842.

26. Sui, Z.; Salto, R.; Li, J.; Craik, C.; Ortiz de Montellano, P. R. *Biorg. Med. Chem.* **1993**, *1*, 415-422.

27. Mazumder, A.; Neamati, N.; Sunder, S.; Schulz, J.; Pertz, H.; Eich, E.; Pommier, Y. *J. Med. Chem.*, in press.

28. Burke, T. R.; Fesen, M. R.; Mazumder, A.; Wang, J.; Carothers, A. M.;Grunberger, D.; Driscoll, J.; Kohn, K. W.; Pommier, Y. *J. Med. Chem.* **1995**, *38*, 4171-4178.

29. Fesen, M. R.; Pommier, Y.; Leteutre, F.; Hiroguchi, S.; Yung, J.; Kohn, K. W. *Biochem. Pharmacol.* **1994**, *48*, 595-608.

30. Mazumder A.; Wang, S.; Neamati, N.; Nicklaus, M.; Sunder, S.; Chen, J.; Milne, G.; Rice, W.; Burke, T. R.; Pommier, Y. *J. Med. Chem.* **1996**, *39*, 2472-2481.

31. Mazumder, A.; Gazit, A.; Levitzki, A.; Nicklaus, M.; Yung, J.; Kohlhagen, G.; Pommier, Y. *Biochemistry* **1995**, 34, 15111-15122.

Chapter 9

Herbal Laxatives: Influence of Anthrones–Anthraquinones on Energy Metabolism and Ion Transport in a Model System

H. W. Rauwald

Department of Pharmacy, Pharmaceutical Biology, Universitiy of Leipzig,
D-04103 Leipzig, Germany

There is good evidence that anthrones/anthraquinones, known as active metabolites of emodin-type O- and C-glycosyl compounds which are found in many plant laxatives, influence the ion transport across colon cells, although the target transport systems have not yet been elucidated. There are various contradictory results used to explain the laxative action of these drugs. To elucidate this problem we tested 25 different anthrone/anthraquinone metabolites of plant drugs for their influence on different ion transport systems in Ehrlich cells as a model system. Comparing the laxative potency of these substances with their influence on the different ion transport systems involved in trans-epithelial ion transport makes it possible to exclude some transport processes as primary targets of the drugs. The following results were found: (1) Na^+-K^+-$2Cl^-$ cotransport was not inhibited by any of the substances tested. (2) Na^+/K^+-ATPase (pump) was inhibited by those 1,8-dihydroxyanthrones/anthraquinones that bear an additional phenolic hydroxyl group. This inhibition is indirect by interference with oxidative ATP production. However, there is no correlation to laxative action. (3) Cation channels were not influenced by these drugs. (4) Cl^--channels were inhibited significantly by those drugs that also show a laxative action. It is not clear, whether this effect is due to a direct interference with the channels or indirectly due to influences on the signal cascade triggering the transport. These results make it very likely that inhibition of Cl^--channels is the primary action responsible for the laxative effect. Interference with oxidative ATP production as an additional effect may explain the known synergistic action described for the combination of different anthrones/anthraquinones or anthranoid drugs, respectively.

Part 1. Influence on Energy Metabolism: An Approach for Interpretation of Known Synergistic Effects in Laxative Action?

Plant drugs containing emodin-type anthranoids are world-wide the main contingent of phytopharmaceuticals (such as aloe, senna, rhubarb, frangula or buckthorn, and cascara) used against constipation. Although these drugs are widely used, their pharmacological action, like constipation itself, is nevertheless not yet fully understood. In recent years progress has been made in understanding gastrointestinal motility. The specific physiological influence of individual anthranoids, mainly the senna drug-derived metabolites rhein/rheinanthrone, on colon motility has gained increasing interest (*1,2*). However, attempts to explain the mechanism of their laxative action on the cellular level have not yet been successful and have been almost solely confined to senna drug-derived metabolites. Several influences on metabolic reaction are described that may help understanding the laxative action of the compounds. For some of these substances, influences on prostaglandin E synthesis (rhein/rheinanthrone) or on inhibition of the sodium potassium pump were shown (for recent reviews on senna see *3, 4*). Some anthraquinones or anthrones, respectively, known as active metabolites of emodin-type anthranoids (*5, 6*) are able to inhibit the Na^+-K^+-ATPase activity of animal cells. The mechanism of this effect. however, is a matter of some controversy: it may be direct, inhibiting the enzyme, or indirect, interfering with the ATP supply for the enzymatic reaction by reducing ATP generation. Some authors describe an inhibition of the respiratory chain (e.g. *7*), while others propose an uncoupling of oxidative phosphorylation (e.g. *8*). Both effects would reduce cellular ATP content and therefore indirectly influence Na^+-K^+-ATPase activity. To investigate which anthranoids inhibit cation transport and whether this inhibition is direct or indirect, studies on Ehrlich cells were performed to test the influence of fifteen different anthraquinones, ten corresponding anthrones, as well as three glycosylated characteristic genuine plant constituents on cation transport (Na^+-K^+-2Cl cotransport, Na^+-K -ATPase activity) and ATP content of the cells.

The results obtained by testing this broad spectrum of both anthrone and anthraquinone forms may show possible differences in activities between these oxidation stages or their variations in molecular structure and may help to elucidate structure-activity relationships yet unknown, apart from the known essential 1,8-dihydroxylation. Finally, the inhibition found has to be compared with the pharmacological action of these anthranoids, especially with the known synergistic effects described for the combination of different anthranoid drugs (e.g. *9-11*) or individual anthraquinones/anthrones, as recently demonstrated (*12*). For example, combinations of aloe and frangula or senna and frangula were shown to be more potent than extracts of the individual drugs (*10,11,13*).

Studies on Ehrlich ascites tumor cells were performed to test the influence of various free anthraquinones and anthrones (10^{-5} M) on the following parameters: inhibition of Na^+-K^+pump; inhibition of Na^+-K^+-2Cl cotransport; reduction of ATP content; and influence on mitochondrial energy metabolism (for methods for EAT cells see refs. *18, 29, 41;* for methods for Na^+-K^+-2Cl /Na^+-K^+-pump see *40, 43, 44;* for comparison and transferability to colon cells see *19, 40, 42*) The results obtained on testing the fifteen different anthraquinones and ten corresponding anthrones listed in Table I showed a structureactivity relationship which is summarized as follows: 1,8-

dihydroxy-anthraquinones and -anthrones with an additional phenolic hydroxyl or carboxyl group, for example frangulaemodin and its anthrone, inhibit drastically the Na^+-K^+ pump, while compounds without an additional phenolic hydroxyl group, for instance aloeemodin or its anthrone, despite being effective laxatives showed no influence on Na^+-K^+ pump activity (Figure 1). The graph illustrates this behavior for five representatives of the first group and seven representatives of the second group. The filled bars show the rate of K^+ transport via the Na^+-K^+ pump, as characterized by the ouabain-sensitive component of the uptake of $^{86}Rb^+$. The crossed bars show the K^+ transport via the $Na^+/K^+/2Cl^-$ cotransporter, as characterized by the piretanide-sensitive component of the uptake of $^{86}Rb^+$. The pump flux was drastically inhibited by substances of the first group represented by frangulaemodin and its anthrone, alaternin (2-hydroxyfrangulaemodin), and by rhein and its anthrone, while no or minor influences were seen by substances of the second group represented by aloeemodin, its anthrone, some other anthraquinones without an additional phenolic hydroxyl group, and by alizarin-type anthraquinones. This group also showed no influence on the piretanide-sensitive uptake. The moderate reduction - in piretanide sensitive uptake observed for representatives of the first group is the indirect result of the pump inhibition, since the furosemide-sensitive transport is a secondary active process depending on the ion gradients build up by the pump. Therefore, the $Na^+-K^+-2Cl^-$ cotransporter is an unlikely target for the laxatives, since no direct inhibition of this transport system could be shown.

By these experiments it cannot be decided whether the inhibition of the pump flux is due to a direct influence on the transport protein or an indirect one due to reduction in energy supply by reduction of cellular ATP-level. Therefore, the influence of frangualemodin and its anthrone on the pump flux was studied in the presence and absence of glucose as additional energy supply (Figure 2). Glucose alone stimulates the transport indicating that the endogenous energy support is not sufficient for optimal transport. While in the absence of glucose, transport inhibition is clearly seen, the effect is less pronounced in the presence of glucose. This observation can be explained by assuming that frangulaemodin does not influence the pump directly, but indirectly by interfering with the energy supply of the pump by inhibiting the ATP production by inhibiting oxidative phosphorylation. In addition, one can speculate that aloeemodin and other compounds that do not reduce the pump flux may have no influence on cellular ATP content. To test this assumption, the cellular ATP content was measured as a function of time after addition of the relevant anthranoids to the incubation medium (Figure 3). One sees a pronounced difference between the two emodin types—while frangulaemodin and its anthrone reduce the ATP level with a time constant of approx. 1 minute, aloeemodin shows a less pronounced influence. Thus, the anthranoids which inhibit pump flux also reduce the ATP content of the cells. This inhibitory effect of frangulaemodin on ATP content is not observed in presence of glucose.

There are two different possibilities to explain how a substance may inhibit oxidative energy production. One class of substances, the uncouplers, prevents ATP production without inhibiting oxidation. The other class inhibits oxidation by inhibiting the respiratory chain or enzymes of the citrate cycle. We tested this alternative by measuring the rate of oxidative decarboxylation of pyruvate by the pyruvate dehydrogenase complex (Figure 4). The graph shows the amount of

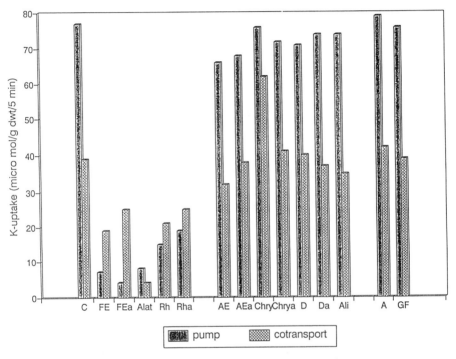

Figure 1. Influence of anthraquinones/anthrones on Na⁺/K⁺-pump and Na⁺/K⁺-2Cl⁻-cotransport. Ehrlich cells were preincubated in Krebs-Ringer-phosphate buffer for 10 minutes with various anthraquinones and anthrones (10^{-4} M). K^+- (^{86} Rb) uptake was then measured for 5 min. **C**, control; **FE**, frangulaemodin; **FEa**, frangulaemodinanthrone; **Alat**, alaternin; **Rh**, rhein; **Rha**, rheinanthrone; **AE**, aloeemodin; **Aea**, aloeemodinanthrone; **Chry**, chrysophanol; **Chrya**, chrysophanol-anthrone; **D**, danthron; **Da**, dithranol; **Ali**, alizarin; **A**, aloin A/B; **GF**, glucofrangulin A/B).

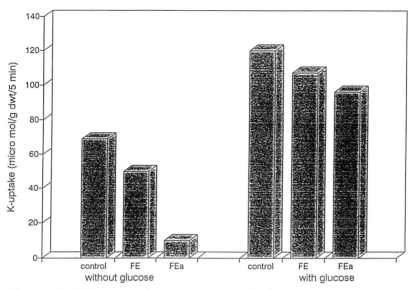

Figure 2. Influence of glucose on Na⁺/K⁺- pump inhibition by frangulaemodin/anthrone. Ehrlich cells were preincubated in Krebs-Ringer-phosphate buffer for 10 minutes with 10^{-4} M frangulaemodin (**FE**) or frangulaemodinanthrone (**FEa**) with or without 10 mM glucose. K^+-(^{86}Rb) uptake was then measured for 5 min.

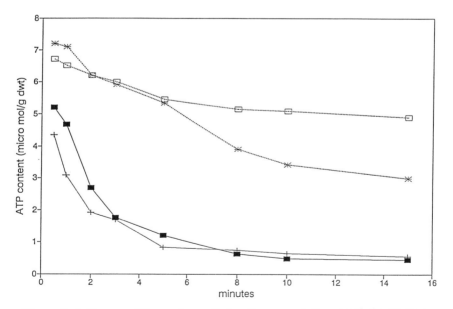

Figure 3. Influence of frangulaemodin/-anthrone and aloeemodin/-anthrone on ATP content. Ehrlich cells were preincubated in Krebs-Ringer-phosphate buffer for different times with 5×10^{-5} M frangulaemodin (■), frangula-emodinanthrone (+), aloe-emodin (*), or aloeemodinanthrone (□).

radioactive labeled CO_2 generated from 1-C 14-labeled pyruvate. One sees that the uncoupler FCCP stimulates this reaction by increasing the oxidation of NADH formed during the reaction, while antimycin, which prevents oxidation of NADH, drastically inhibits the oxidative decarboxylation of pyruvate. Since frangulaemodin behaves like antimycin and not like FCCP, it behaves as an inhibitor and not as an uncoupler under this condition. This type of experiment cannot elucidate which step in the citrate cycle or respiratory chain is inhibited. The reduction in oxidative decarboxylation of pyruvate is not observed by substances of the aloeemodin-group, lacking an additional hydroxyl group. For frangulaemodin, which shows structure similarity to ubiquinone (CoQ), we have some preliminary results that it inhibits electron transfer from complex I and/or II to complex III of the respiratory chain. Frangulaemodinanthrone may have an additional target at cytochrome c. It may also compete with cytochrome c_1 for reduction of cytochrome c, thus preventing electron flow from cytochrome c_1 to cytochrome c. To test this possibility, a solution of oxidized cytochrome c was incubated with Frangulaemodinanthrone. Figure 5 shows the increase in reduced cytochrome c as a function of time after addition of Frangulaemodinanthrone and the well-known antipsoriatic dithranol. One sees that Frangulaemodinanthrone reduces cytochrome c, while dithranol, a representative anthrone of the second group, has almost no effect.

Our results show that free 1,8-dihydroxyanthrones and -anthraquinones with an additional phenolic hydroxyl group or an carboxyl group inhibit Na^+-K^+-ATPase activation by reducing ATP level resulting from an interference with oxidative ATP production, while their corresponding O- and C-glycosyls—like aloins or glucofrangulins—have no influence (Table I). This is in agreement with the results of other authors who have demonstrated inhibition of Na^+-K^+-ATPase by several anthranoids (14, 15). The question still remains open whether this inhibition is sufficient to explain the observed laxative effects. There is no doubt that Na^+-K^+-ATPase is an essential part of the ion pumping system of the intestinal mucosa. However, it is not known whether the activity of this enzyme is rate-limiting under in vivo conditions and, therefore, whether or not partial inhibition of Na^+-K^+-ATPase by anthranoid laxatives is a determining factor for the inhibition or even the reversal of the sodium and fluid net transfer in the gut.

Furthermore, the results show that the inhibitory action of anthrones/anthraquinones on ion fluxes via the Na^+-K^+ pump results indirectly from reduction in the cellular ATP level due to interference with oxidative energy metabolism. Anthrones/anthraquinones with an additional phenolic hydroxyl group or a carboxyl group may act as uncouples or inhibitors of oxidative phosphorylation, as was postulated by Verhaeren (15). On the contrary, our results measuring pyruvate dehydrogenase reaction show that the compounds inhibit oxidative metabolism and are in agreement with those of Ubbink-Kok et al. (7) who showed that some free anthrones/anthraquinones interfere with electron transfer in the respiratory chain. It is likely that frangulaemodin influences electron flow from complex I (NADH : ubiquinone oxidoreductase) and/or complex II (succinate dehydrogenase) to complex III (ubiquinone : cytochrome c oxidoreductase). We have shown an additional effect of frangula-emodinanthrone, that competes with cytochrome c_1 for reduction of cytochrome c, thus preventing electron flow from cytochrome c_1 to cytochrome c. Since cytochrome c is reoxidized via complex IV, oxygen consumption may increase

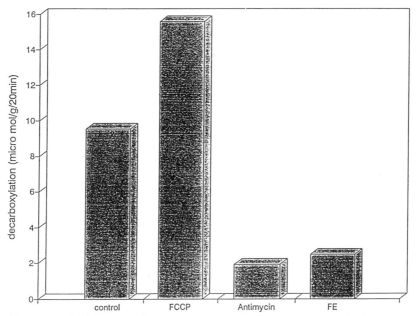

Figure 4. Influence of frangulaemodin on the pyruvate dehydrogenase reaction. Ehrlich cells were preincubated in Krebs-Ringer-phosphate buffer for 10 minutes with 5×10^{-5} M frangulaemodin (FE), 10^{-4} M antimycin, or 10^{-6} M FCCP. Decarboxylation of 1 mM $1\text{-}^{14}\text{C}$-pyruvate was then measured for 20 min.

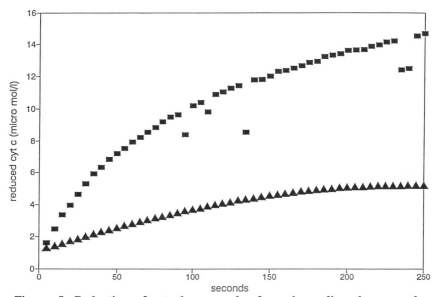

Figure 5. Reduction of cytochrome c by frangulaemodinanthrone and dithranol. Oxidized cytochrome c (50 µM) was incubated for 250 seconds with 150 µM frangulaemodinanthrone (■) or dithranol (▲). The reduction of cytochrome c was measured photometrically.

Table I. Influence of anthraquinones and corresponding anthrones on Na^+/K^+-pump and oxidative energy metabolism.

Compound	R_1	R_2	R_3	R_4	R_5	R_6	P.I.	ATP(-)	R.I.
Frangulaemodin*	H	CH$_3$	H	H	OH	H	+	+	+
Rhein*	H	COOH	H	H	H	H	+	+	+
7-Hydroxy-aloeemodin	H	CH$_2$OH	H	H	H	OH	+	+	+
5-Hydroxy-aloeemodin*	H	CH$_2$OH	H	OH	H	H	+	+	+
Alaternin*	OH	CH$_3$	H	H	H	H	+	+	+
Islandicin	H	CH$_3$	OH	H	H	H	+	+	+
Digitopurpone	CH$_3$	H	OH	H	H	H	+	+	+
Aloeemodin*	H	CH$_2$OH	H	H	H	H	-	-	-
Chrysophanol*	H	CH$_3$	H	H	H	H	-	-	-
Physcion*	H	CH$_3$	H	H	OCH$_3$	H	-	-	-
Danthron*	H	H	H	H	H	H	-	-	-
Nataloeemodin-8-methylether	H	CH$_3$ (8-0-CH$_3$)	H	H	H	OH	-	-	-
Alizarin* (1,2-Dihydroxy-AQ)							-	-	-
Quinizarin* (1,4-Dihydroxy-AQ)							-	-	-
Quinalizarin (1,2,5-Trihydroxy-AQ)							-	-	-
Aloin A (Aloe)							-	-	-
Glucofrangulin A (Frangula)							-	-	-
Sennoside A (Senna)							-	-	-

P.I., inhibition of Na^+/K^+-pump; ATP(-), reduction in ATP level, R.I., inhibition of mitochondrial respiration.
*Also tested as anthrones, where 10-C=O is replaced by 10-CH$_2$.

as in uncoupling. This might be an explanation for the uncoupling effect seen by measuring oxygen consumption (*15*). It seems possible that both redact states play a role in the inhibition of electron flow by competing with ubiquinone for its binding site at complex I and/or complex II, thus preventing electron transfer from these complexes to complex III and anthrones by competing with cytochrome c_1 for cytochrome c.

Summarizing, one can say that 1,8-dihydroxyantraquinones with an additional phenolic hydroxyl group and their anthrones. inhibit oxidative energy metabolism, presumably by competing with ubiquinone at the ubiquinone binding site at complex I and/or complex II. As a consequence, the ATP-level drops and ATP-dependent ion transports decrease. Representatives of this group show only a relatively weak laxative action, while the more potent laxative aloeemodin and others have no influence on energy metabolism. Therefore, inhibition of oxidative ATP production and inhibition of pump activity cannot be the main reason for the laxative potency of the drugs. However, pump inhibition may influence the laxative action of a combination of drugs, if other components influence other transport processes (e.g. Cl^- channels) (*16*). For instance, inhibition of the Na^+/K^+-ATPase may increase secretion in the upper parts of the intestine. Since the resorbing system for electrolytes in the colon normally operates far below its maximal capacity, stimulation of secretion is fully compensated by an increase in resorption. But, if resorption in the colon is partially inhibited, an increase in secretion can no longer be compensated for. Stimulation of secretion in the upper part of the intestine and inhibition of resorption in the colon may therefore act synergistically and may also give an explanation for the synergistic effect of a combination of drugs from both groups.

Part 2. Inhibition of Cl^--Channels as a Possible Basis of Laxative Action

There is good evidence that anthranoids, as laxatives, influence the net ion and fluid transport across colon cells, the water content, and the volume of the feces as the result of fluid and electrolyte secretion in the upper parts of the intestine and as the result of resorption of electrolytes and fluid in the colon. Since resorption usually operates below its maximal capacity, variations in secretion are often fully compensated by an increase in resorption. Only during severe stimulation of secretion, such as in cholera intoxication, is the maximal capacity of the resorption process insufficient to compensate secretion and an overflow diarrhea is observed. Therefore, it is likely that laxatives act by inhibiting resorption and not by stimulation of secretion.

In a resorbing colon cell, entrance of ions into the cells proceeds via the furosemidesensitive Na^+-K^+-2Cl^--cotransporter. At the contraluminal side sodium is extruded actively via the Na^+-K^+-pump. Potassium recirculates across luminal and contraluminal membranes while chloride leaves the cell across the contraluminal membrane via chloride channels. Inhibition of one of these transport systems should lead to inhibition of the overall process, reabsorption of ions followed by water. The target transport systems are not yet elucidated (*17*). As shown in *Part 1*, inhibition of the Na^+-K^+-pump is not sufficient to explain the laxative effects. The Na^+-K^+-2Cl^--cotransporter is an unlikely target for the laxatives, since no direct inhibition of this system could be shown. Therefore, we wonder whether the drugs would influence chloride and/or potassium permeabilities. The aim of this study is to test whether anthranoids influence chloride and/or potassium channels. If activation of chloride channels is observed it would indicate activation of secretion, while inhibition would indicate inhibition of resorption. If an effect is found, it will be compared with the laxative potency of the compounds in humans.

To study the influence of anthranoids on the various transport systems, Ehrlich ascites tumor cells were used as a model system since all relevant transport systems are well characterized for these cells. Chloride and potassium permeabilities are enhanced after osmotic swelling, resulting in extrusion of KCl from the cell followed by water. This process is called regulatory volume decrease (RVD) (18). Figure 6A shows the water content of Ehrlich ascites tumor cells at different times after reduction of osmolarity from 310 to 200 mOsm. To study the two different types of channels separately, specific inhibitors or activators of the two pathways are used. For instance the potassium conductance can be elevated by the potassium ionophore valinomycin accelerating the RVD. Under this condition Cl⁻-permeability is rate limiting. To test the influence of anthranoids on chloride and/or potassium channels, we chose aloeemodin which shows pronounced laxative effects but does not influence energy metabolism and sodium-potassium pump activity (see *Part 1*). The cells were swollen osmotically by reducing osmolarity. In the presence of aloeemodin shrinkage was retarded similar to using 130 B (compare Figures 6A and 6B). To determine the concentration dependence of the effect, the cell volume was observed eight minutes after osmotic challenge at concentrations above 10 µM, aloeemodin inhibits RVD in a dose dependent manner (for details see *19*). The influence of aloeemodin on RVD can be due to inhibition of chloride or potassium channels (Figure 6C). In the presence of the potassium ionophore valinomycin, potassium transport is no longer rate limiting for RVD. Therefore any inhibitory effect observed under valinomycin must be due to inhibition of chloride channels. This is clearly seen by this experiment for aloeemodin, demonstrating activity of chloride channels under the influence of the specific chloride channel blocker 130B and of aloeemodin. Since aloeemodin has an additional influence on RVD, it is likely that the two substances influence different classes of chloride channels. The results obtained by measuring the water content of cells were verified by direct measuring cellular volume by an electronic cell counter (for details see *19*).

The influence of anthranoids on membrane potential may give independent evidence whether K⁺- or Cl⁻-channels are inhibited by anthranoids. The membrane potential is mainly determined by the permeabilities of the cell membrane for chloride and potassium. If chloride conductance would be inhibited, the actual membrane potential would be shifted more closely to the potassium equilibrium potential, resulting in hyperpolarization. On the other hand, inhibition of potassium conductances would depolarize the cell towards the chloride equilibrium potential. In Figure 7 the accumulation of the lipophilic cation TPP⁺ (tetraphenylphosphonium) is plotted as a function of aloeemodin concentration. Increasing distribution ratio R indicates hyperpolarization. This experiment clearly shows aloeemodin reduces chloride and not potassium conductance.

Table II summarizes the results of this study and the parallel study on energy metabolism (*Part 1*). Besides aloeemodin fourteen other anthraquinones and ten anthrones were characterized in respect to their influence on sodium-potassium pump, cellular ATP-levels, and RVD as measures of chloride permeability. In contrast to results described by others, no significant differences were found between the anthrone and the anthraquinone form of each compound. Thus, we can arrange the substances into three different groups. The first group, frangulaemodin and some other compounds characterized by an additional phenolic or carboxylic group at the 1,8-dihydroxyanthrone chromophore, shows laxative action, reduction in ATP content, inhibition of ion transport via the sodium-potassium pump and some inhibition of RVD and of the permeability of chloride channels. The second group, aloeemodin and other compounds, despite being effective laxatives do not influence ATP content and sodium-potassium pump rate, but show pronounced inhibition of RVD and of chloride channels. The third group, alizarin-type quinones and other anthranoids without laxative action, lacking the essential 1,8-dihydroxyl substitution, e.g. the 8-O-methylated nataloeemodinmethylether, show no influence on the various parameters tested.

Table II. Inhibition of Cl⁻-channels by anthraquinones and their anthrones

Compounds	P.I.	I.Cl.	R
Aloeemodin*	-	+	1.21
Chrysophanol*	-	+	*1.05*
Physcion*	-	+	0.98
Danthron*	-	+	1.00
Frangulaemodin*	+	+	1.06
Rhein*	+	+	1.08
Alatemin	+	+	1.08
Alizarin	-	-	0.22
Quinizarin*	-	-	0.17
Glucofrangulin A	-	-	0.09

P.I., inhibition of the Na^+/K^+-pump; **I.Cl.**, inhibition of Cl--channels; **R**, inhibition of Cl⁻-channels relative to that by substance 130B (1.00).
*Also tested as anthrones, where 10-C=O is replaced by 10-CH2.

The first attempts to explain increased secretion after application of anthranoids centered especially on net potassium transport. However, the potassium secretion and the resulting osmotic effect is too small to explain the large amount of fluid in the colon. Leng-Peschlow (20) could demonstrate a non-stimulation of potassium and mucus secretion after pretreatment with sennosides. After having excluded the Na^+-K^+ pump and Na^+-K^+-$2Cl^-$ cotransport, we studied the influence on K^+ and Cl⁻ channels by an indirect method. In our test model, Ehrlich ascites tumor cells, chloride and potassium permeabilities are enhanced after osmotic swelling, resulting in extrusion of KCl from the cell followed by water (18). All emodin-type compounds tested show a very long delay with respect to RVD. This delay is due to inhibition of the permeability for Cl⁻ and not for K^+ as could be verified by two

108

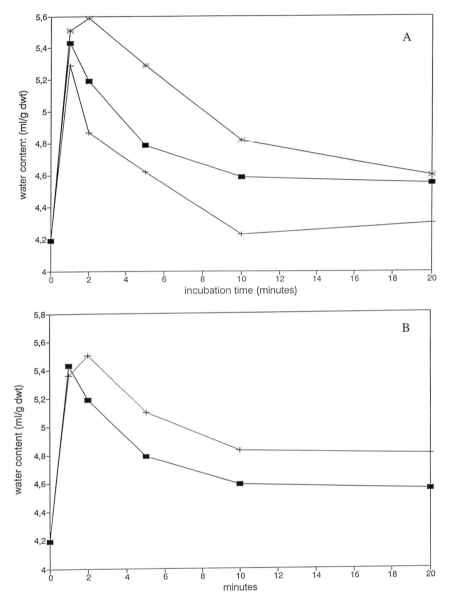

Figure 6. Influence on regulatory volume decrease (RVD) by (A) valinomycin ($2x10^{-6}$ M)(+) and the chloride channel blocker, 130B (10^{-6} M)(∗) (B) aloe-emodin ($5x10^{-5}$ M)(+) and (C) 130B plus valinomycin (+) or 130B plus aloe-emodin (∗). Cells were preincubated for 5 min in Krebs-Ringer-phosphate buffer at pH 7.4. At zero time, the compounds were added and the osmolarity of the solution was reduced to 200 mOsm by addition of distilled water. The cellular water content was then determined at different times. Control values, (■).

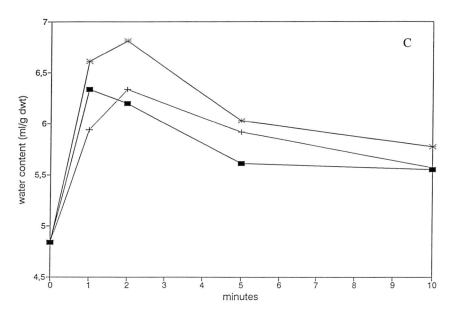

Figure 6. *Continued.*

independent methods: RVD under the influence of additional inhibitors or stimulators of ion permeabilities and by the influence of the anthranoids on membrane potential. Our results clearly show that anthranoids inhibit Cl^--transport as would be expected for inhibition of resorption. If stimulation of secretion (21) is the mode of action of the anthranoids, one would observe an increase in Cl^--permeability as in cholera intoxication, which acts via cAMP-mediated increased Cl^--permeability. Our results cannot decide whether the inhibition of Cl^--permeability is direct due to binding of the anthraquinone or anthrone to the channel or indirect due to interference with regulating properties. Several regulatory metabolites are shown to influence Cl^--permeability. For example, the role of cAMP in the secretary action of laxatives is also discussed, but the significance of these findings is still under debate. It may merely represent another non-specific membrane interaction, occurring simultaneously with effects on membrane structure and permeability, since cAMP would stimulate resorption by activation Cl^--channels (22). The cAMP increase may be indirect and not correlated to laxative action. For instance, reduced channel permeability may create a signal that stimulates cAMP production to reopen the channels.

Beubler et al. (23), Capasso et al. (24), Mascolo et al. (25) and Dharmsathaphorn et al. (26) have discussed histamine or 5-hydroxytryptamine as the modulator of the Cl^--secretion, but the exact mode of action is not known. Other results make it unlikely that they play a role under in vivo conditions (27). On the other hand, Prostaglandins are candidate autocoids which may mediate the effects of laxatives on motor and secretary activities of the gastrointestinal tract. Laxative agents such as aloe stimulate the formation of Prostaglandins E and F from arachidonic acid (28). While there is evidence linking enhanced release and mucosal content of Prostaglandins to the actions of laxatives, the effect on fluid transport cannot be explained entirely by Prostaglandins. Hypotonic swelling results in stimulation of the leukotriene synthesis and a concomitant reduction in the prostaglandin synthesis. PGE_2 significantly inhibits regulatory volume decrease following hypotonic swelling (29). The signals regulating prostaglandin and leukotriene synthesis are still unknown. The aim of this study was to identify the transport process which is directly or indirectly influenced by anthranoids. Our results are not in contrast to those on prostaglandin synthesis. In combination with the results of Lambert et al. (29) our findings would be explained by an increase in prostaglandin level which would inhibit RVD.

Part 3. Conclusions with Regard to Phytomedicinal Aspects

These results with a cellular model system demonstrate a correlation between laxative response and inhibition of chloride channels while the influence on energy metabolism may have some additional side effects. These differences in the target for the drugs may be an explanation for synergistic effects between different laxatives known by several pharmacological experiments. For example, combinations of frangula, senna and aloe were shown to be more potent laxatives than were extracts of the individual drugs [e.g. extractum frangulae 25 (ED 50/mg; mice), extractum aloes 55, and 18 for a 1:1 mixture of the two] (10,12,13). The synergistic action of aloeemodinanthrone and rheinanthrone as active metabolites of sennoside C, a minor bianthrone-O-glucoside of senna or rhubarb, has recently been demonstrated for their purgative effects on mice *(12)* (intracaecal ED 50/μmol kg[-1]: aloeemodinanthrone 24.1, rheinanthrone 11.4, and 11.2 for an equimolar mixture), large intestine propulsion and water secretion *(30)*. Concerning the anthraquinone form, synergistic effects between individual cascara metabolites, namely aloeemodin, chrysophanol and frangulaemodin were described already in 1938 *(31)*. Finally, the results on different influences on energy metabolism and Cl-channels imply an approach for interpretation of different activities due to synergistic effects of anthranoids within single drugs, as shown for laxative activities of official Cape aloe *(35)*. Cape aloe contains both types of anthrone-C-glycosyls, with and without an additional hydroxyl group, namely the aloins/aloinosides (active metabolites: aloeemodin/-anthrone) and 5-hydroxyaloin A (active metabolites: 5-hydroxyaloeemodin/-anthrone), which was recently isolated and elucidated in our laboratory *(32,33)* and previously designated as "periodate positive substance" (for review see *34*). These drugs were significantly more active in humans (dose response curves) than Cape aloes from other origins, which contained no hydroxylated aloin, or than an equimolar amount of aloin itself *(35)*. Mapp et al. *(35)* have stated "Neither can the aloin content alone serve as a criterion for purgative activity, since the Alicedalc sample (with 19% aloin plus 5-OH-aloin) ... had a greater activity than aloin in humans if dosage is estimated on aloin conent only." Analogous behavior may be expected for Aloe barbadensis (syn. A. vera/Curaçao-Aloe) having free and cinnamoylated aloin- and 7-hydroxyaloin-type C-glycosyls (see review 34). The postulated metabolic relationship between these diastereomers is shown in Figure 8. This may also explain pharmacological tests with several plant extracts from aloe, cascara, frangula and senna with the strongest laxative action shown by aloe extract *(36)*, but also the more potent activity of dualistic acting rheinanthrone in comparison to aloeemodinanthrone *(37)*.

Furthermore, our results are in direct connection with the pharmakokinetics of anthranoids and confirm recent information concerning their metabolism and pharmakokinetic characteristics. These anthranoids are present in plant drugs like aloe, senna etc. as O- or C-glycosylated compounds which are prodrugs and converted by bacterial enzymes into the metabolites responsible for the laxative action only after it reaches the large intestine. The metabolites are poorly absorbed and their action therefore confined to the large intestine. While the pharmakokinetics of anthraquinone-O-glycosides (main type of frangula, rhubarb) have scarcely been forthcoming—especially intestinal reduction to the anthrone form is yet unclear—so far the best characterized compounds are sennosides (bianthrone-O-glycosides) and their aglycone rheinanthrone (reductase from Peptostreptococcus intermedius/O-hydrolase from Bifidobacterium sp. SEN) and its oxidized metabolite rhein (see reviews on senna *3,4*).

In recent years, studies with anthrone-C-glycosyls of aloin-type (aloe, cascara) have resulted in some data regarding the fate and metabolism of these compounds, which are summarized in Figures 8 and 9. Aloins and hydroxyaloins (5-OH-derivatives in Cape aloe and free and cinnamoylated 7-OH derivatives in Curaço aloe, as shown in Figure 8; for structural details see *32-34),* being prodrugs, pass quantitatively and unchanged—apart from incipient formation of polymerization products—into the large intestine where they are metabolized by bacterial enzymes (reductase from Eubacterium sp. BAR) into the laxatively active metabolites aloeemodinanthrone and its 5-/7-hydroxylderivatives, respectively, which may be spontaneously oxidized to

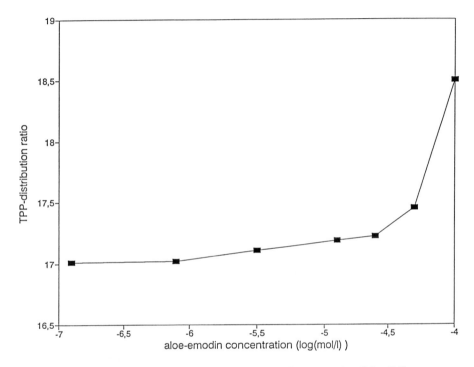

Figure 7. Influence of aloeemodin on membrane potential. Cells were preincubated for 5 min in Krebs-Ringer-phosphate buffer pH 7.4 in the presence of tetraphenylphosphoniume (TPP). The osmolarity of the medium was then reduced to 200 mOsm by addition of distilled water. At that time (0 min) aloeemodin was added. The cell/membrane distribution ratio of TPP was determined 5 min after osmotic challenge (control for cells in isotonic medium: 12).

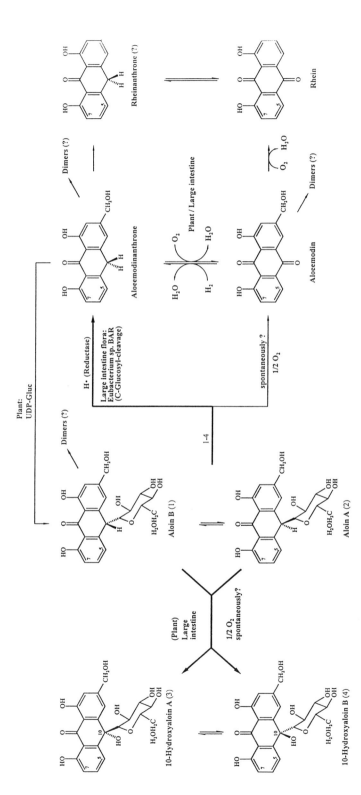

Figure 8. Postulated metabolic relationship of aloins A/B.
The analogous scheme may be assigned to hydroxyaloins bearing hydroxyl groups in position 7 (Aloe barbadensis) or position 5 (Aloe capensis). Both types of metabolites, without and with an additional OH, show different targets of action.

114

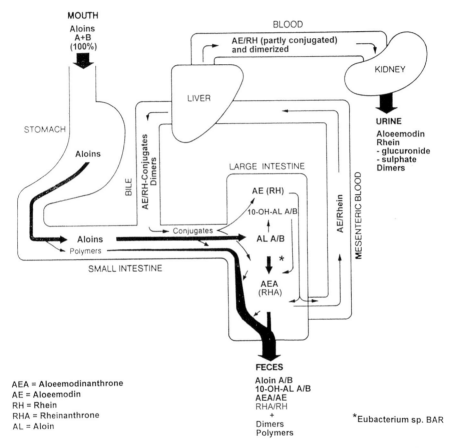

Figure 9. Hypothetical scheme of pharmakokinetics of the aloins A/B.

their corresponding anthraquinone form (aloeemodin, 5-/7-hydroxyaloeemodin). Recent experiments (*38*) have shown that Eubacterium sp. BAR is not only capable of transforming the diastereomeric aloins to aloeemodinanthrone, but simultaneously the aloins stimulate significantly the growth of this bacterium. Only a small portion of the aglycones is absorbed by the large intestine and passes in oxidized form into the blood stream. This process is attended with oxidation of the 3-hydroxymethyl side chain of the anthrone component to the 3-carboxyl yielding rhein/rheinanthrone as further metabolites. Furthermore, by simulating physiological conditions using artificial intestinal buffer medium, aloins are converted to the corresponding oxanthrone-C-glucosyls 10-hydroxyaloin A and B *(19, 39)*. The major part of the aloins or of the metabolites resulting from bacterial action is converted in the large intestine into polymers and excreted in the feces together with unchanged aloins, aloeemodinanthrone, aloeemodin and the 10-hydroxyaloins. The absorbed metabolites are conjugated to form sulphates and glucuronides, a process which takes place within the intestinal epithelium and in the liver, and are then excreted, partly via kidney and partly via the bile. The conjugates which re-enter the large intestine with the bile are split there to aloeemodinanthrone once again or converted into dimerization-polymerization products.

From these investigations of the metabolism and pharmakokinetics of aloins it can be concluded that (a) only a fraction of the aloins given by mouth becomes pharmacologically active, (b) their activity is confined exclusively to the large intestine, and (c) that any systemic loading attributable to absorbed metabolites is very small, which is of significance also for toxicological aspects.

Products containing anthranoid drugs are among the most widely used laxatives worldwide. Out of the products listed under "laxatives" in the German register of pharmaceutical preparations (Rote Liste 1991), 57% contained anthranoid drugs (38% senna; 15% aloe; 4% frangula, rhubarb, cascara, buckthorn), i.e. in more than half of all such products the action was based exclusively or largely on anthranoid components. In the Rote Liste of 1996 these were still 46%, indicating a significant decrease of anthranoid drugs. This decrease is due to trenchant administrative steps in the prescription of such natural products, which have a somewhat low reputation among the medical profession. As of 1996, this appears to be at least partly due to a poor knowledge of the pharmacodynamic, toxicological and pharmakokinetic properties of anthranoids and, above all, due to the strong, controversial discussion on their possible genotoxicity and cancerogenicity, an important topic which is beyond the scope of this chapter.

Literature Cited

1. Beubler, E.; Juan, H. *J. Pharm. Pharmacol.* **1979,** *31*, 681-685.
2. Donowitz, M.; Wicks, J.; Battisti, L. *Gastroenterology* **1984,** *87*, 503-512.
3. *Senna and its Rational Use;* Leng-Peschlow, E. *Pharmacology* **1992,** Vol..44, Suppl. 1.
4. *Second International Symposium on Senna;* Ewe, K.; Lemli, J.; Leng-Peschlow, E.; Sewing, K.-F., Eds.; *Pharmacology* **1993,** Vol. 47, Suppl. 1.
5. Lemmens, L. *Pharm. Weekbl.* **1979,** *114*, 178-185.
6. Che, Q.-M.; Akao, T.; Hattori, M.; Kobashi, K.; Namba, T. *Planta Med.* **1991,** *57,* 15-19.

116

7.   Ubbink-Kok, T.; Anderson, J.A.; Konnings, W.N. *Antimicrob. Agents and Chemoth.* **1986,** *30,* 147-151.
8.   Verhaeren, E.H.C. *Phytochemistry* **1980,** 19, 501-503.
9.   Auterhoff, H.; Knabe, J.; Höltje, H.-D. In *Lehrbuch der Pharmazeutischen Chemie;* Wissenschaftl. Verl. Ges., Stuttgart, 1991, Vol. 12; p 664.
10. Auterhoff, H. *Arzneim. Forsch.* **1953,** *3,* 23.
11. Vogel, G. *Arzneim. Forsch.* **1975,** *25,* 1356-1365.
12. Yamauchi, K.; Shinano, K.; Nakajima, K.; Yagi, T.; Kuwano, S. *J. Pharm.Pharmacol.* **1992,** *44,* 973 -976.
13. ·Auterhoff, H. *Arzneim. Forsch.* **1953,** *1,* 412-414.
14. Lemmens, L.; Borja, E. *J. Pharm. Pharmacol.* **1976,** *25,* 498-501.
15. Verhaeren, E. *J Pharm. Pharmacol.* **1980,** *29,* Suppl. *1,* 43-49.
16. Rauwald, H.W.; Hönig, J.; Geck, P. *Planta Med.* **1997,** submitted.
17. Wanitschke, R.; Karbach, U. *J Pharm. Pharmacol.* **1988,** *36,* Suppl. 1, 98-103.
18. Hoffmann, E.K.; Lambert, I.H.; Simonsen, L. 0. *J. Membrane Biol,* **1986,** *91,* 227-244.
19. Hönig, J. Dissertation; 1994, University of Frankfurt/Main.
20. Leng-Peschlow, E. *J Pharm. Pharmacol.* **1986,** *38,* Suppl. 1, 369-373.
21. Nell, G.; Rununel, W. In *Pharmacology ofintestinal Permeation* (Csaky, T., Ed.); Springer, Berlin/Heidelberg/New York, 1984.
22. Gulkinson, G.W.; Bass, P. in: see Reference 21, 1984.
23. Beubler, E.; Bukhave, K.; Rask-Madsen, J. *Gastroenterology* **1986,** *90,* 1972-1977.
24. Capasso, F.; Mascolo, G.; Romano, V. *J Pharm. Pharmacol.* **1986,** *38,* 627-629.
25. Mascolo, N.; Meli, R.; Autore, G.; Capasso, F. *J. Pharm. Pharmacol.* **1988,** *36,* Suppl. 1, 92-97.
26. Dharmsathaphom, K.; Wassermann, S.; Barret, K.E.; Huott, P.A.; Beuerlein, G.; Kagnoff, M.F. *Am. J. Physiol.* **1988,** *254,* C53-C62.
27. Liedtke, C.M. *Ann. Rev. Physiol.* **1989,** *51,* 143 -160.
28. Collier, H.O.J.; McDonald-Gibson, W.J.; Saeed, S.A. *Br. J .Pharmacol.* **1976,** *58,* 193-199.
29. Lambert, I.H.; Hoffmann, E.K.; Christensen, P. *J. Membrane Biol.* **1987,** *98,* 247-256.
30. ·Yagi, T.*J Pharm. PharmacoL* **1997,** 49, 22-25.
31. Greeen, M.W.; King, C. G.; Beal, G. D. *J. Am. Pharm. Assoc. Sci. Ed.* **1938,** *27,* 95-100.
32. Rauwald, H. W.- Beil, A. *J. Chromatogr.* **1993,** *639,* 359-362.
33. Rauwald, H. W.; Beil A. *Z. Naturforsch.* **1993,** *48c,* 1-4.
34. Beil, A.; Rauwald, H. W. In *'Aloe'/Hagers Handbuch der Pharmazeutischen Praxis;* Hänsel, R.; Keller, K.; Rimpler, H.; Schneider, G., Eds.; Springer: Berlin, 1992, Vol. 4, 209-232.
35. Mapp, R. K.; Mc Carthy, T.I. *Planta Med.* **1970,** *18,* 360-365.
36. Fairbairn, J. W.; Moss, M. *J Pharm. Pharmacol.* **1970,** *22,* 584-593.
37. Fairbairn, J. W. *J Pharm. Pharmacol* **1976,** *14,* 48-61.
38. Hattori, M.; Akao, T.; Kobashi, K.; Nambu, T. *Pharmacology* **1993,** *47,* Suppl.1, 125-133.
39. Rauwald, H.W.; Lohse, K. *Planta Med* **1992,** *58,* 259-262.
40. Greger, R. *Physiologie aktuell* **1986,** *2,* 47-49.
41. Hoffmann, L.K.; Simonsen, L.O.; Lambert, J.H. *J. Membrane Biol.* **1984,** *78,* 211-222.
42. Konturek, S.J.; Classen, M. In *Kolonabsorption/Gastrointestinale Physiologie,* G. Witzstrock, Köln, 1976, pp. 427-431.
43. Geck, P.; Heinz, E. *J. Membrane Biol.* **1986,** *91,* 97-105.
44. Geck, P.; Pfeiffer, B. *Ann. Rev. New York Acad. Sci.* **1985,** *456,* 166-182.

SPECIFIC PLANTS

Chapter 10

Arnica Flowers: Pharmacology, Toxicolgy, and Analysis of the Sesquiterpene Lactones—Their Main Active Substances

G. Willuhn

Institute für Pharmazeutische Biologie, Heinrich-Heine-Universität Düsseldorf, D–40225 Düsseldorf, Germany

Arnica flowers from the European Compositae *Arnica montana* have been used in traditional medicine to treat a large variety of different ailments. It has been shown that the known diverse effects of the flowers correlate closely with the pharmacological and toxicological profile of their constituents helenalin, 11α,13-dihydrohelenalin and their esters, which therefore must be considered as their main active compounds. Methods for their qualitative and quantitative determination have been established and the variability in flowers from different geographical origin investigated. *Arnica chamissonis* ssp. *foliosa* is used as alternative source for the official "Arnicae flos." In addition to the sesquiterpene lactones of *A. montana,* they contain 2,3-dihydro-2α-hydroxy- and 2,3-dihydro-2α-,4α-dihydroxy-helenanolide derivatives. The qualitative and quantitative variation is much larger than in the case of *A. montana* flowers, but populations exist equal to that of *A. montana*.

Arnica flowers from the European *Arnica montana* L. have been extensively used in traditional medicine to treat a large variety of different ailments. First written records exist from the 16th century when Arnica flowers were used externally and internally in injuries and accidents such as sprains, dislocations, haematomas and oedema associated with fractures. Many vernacular names for this medicinal plant from that time, such as "fall herb", "break herb" or "wound herb" point to this indication. In the following centuries the range of indications for Arnica flowers was enlarged to include treatment of heart and circulatory diseases, chronical diseases such as asthma and rheumatic pains, feverish conditions, menstruation complaints and many other diseases. The plant was named "Panacea lapsorum" after the hypothetical medicine "Panazee," of which the ancient physicians were always dreaming. This imagined universal remedy was named after Panakaia—meaning "universal healer"—the daughter of the old Greek healing god Asklepios (*1*).

But even at that time numerous adverse reactions were observed after internal use of the tincture or the infusion of Arnica flowers with high doses. Thus, already in the 18th century we find reports that the oral use of Arnica flowers must be carefully controlled because it is a very strong remedy, working even at very low doses, and that it should be used only by experienced physicians in especially severe cases (*1*).

Pharmacological and Toxicological Effects

From the 19th century to the present, a whole host of pharmacological effects have been proven for Arnica flowers by observing patients treated with simple preparations of the flowers (galenicals), by animal experiments, and by in vitro tests (*1-5*). The pharmacological profile includes antibacterial, antifungal, and antiinflammatory properties. They have analgesic effects for inflammation pains and support the absorption of internal bleeding and exsudates. They show antineuralgic effects, are cardioactive and vasoactive, and have respiratory analeptic and oxytocic (stimulates uterine contraction) activities. In addition, cholagogic (increases bile flow) and diuretic effects, as well as reflex-modulating effects on the central nervous system, have been described.

Following internal use of the tincture or the infusion of Arnica flowers with high doses, many adverse reactions have been observed, such as nausea, gastroenteritis, feeling of dizziness, strong vomiting, exaggerated pulse, heart and respiratory dysfunctions, tremor, mensis and abortions. Very high doses can even lead to death through cardiac arrest following labored breathing (*1-5*). External use can cause irritation and edematous contact dermatitis (*6, 7*).

This broad spectrum of pharmacological effects explains the mentioned wide range of indications, however it also shows a deficiency of selectivity.

Restricted Use of Arnica Flowers Today

The lack of selectivity together with the mentioned adverse effects upon oral use obliged the German Commission E to restrict the application of Arnica preparations to external use only with limited indications (*8*). These are related to injuries and accidents: hematomas, dislocations, sprains, bruising, edema associated with fractures, rheumatic muscle and joint complaints, inflammation of the mucous membranes of the mouth and throat, boils, inflamed insect bites, and surface phlebitis. The antiinflammatory, analgesic, and antiseptic effects are accepted by the Commission E to be well-documented.

Chemical Constituents of the Flowerheads

The examination of the pharmacological effects of Arnica has been accompanied by an investigation of its chemical constituents starting in the 19th century. More than 150 compounds have been identified in the flowerheads (*4, 5, 9*). They contain 0.23-0.35% essential oil (which consists of sesquiterpenes [9], thymol derivatives [6] and other monoterpenes [6]), triterpenes [4], carotenoids [14], 0.4-0.6% flavonoids [33], phenolic acids [7], caffeoylquinic acids [3], coumarins [2], polyacetylenes [4] and the nontoxic pyrrolizidines tussilagin acid and isotussilagin acid (brackets indicate number of identified compounds). It was not until the 1970s that the breakthrough in the search for the essential pharmacologically and toxicologically active compounds was achieved with the first isolation and identification of sesquiterpene lactones in *Arnica* (10, 11). The flowerheads of *Arnica montana* were found to contain the pseudoguaianolides helenalin, 11α,13-dihydrohelenalin, and predominantly their ester derivatives with short-chain fatty acids (12-14), as shown in Figure 1. As minor compounds, tetrahydrohelenalin isobutyrate (15) and mexicanin C isovalerate (16) were found.

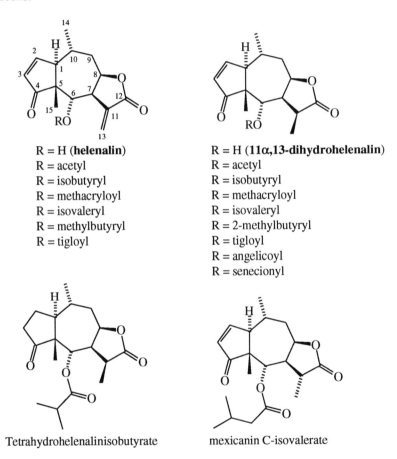

R = H (**helenalin**)
R = acetyl
R = isobutyryl
R = methacryloyl
R = isovaleryl
R = methylbutyryl
R = tigloyl

R = H (**11α,13-dihydrohelenalin**)
R = acetyl
R = isobutyryl
R = methacryloyl
R = isovaleryl
R = 2-methylbutyryl
R = tigloyl
R = angelicoyl
R = senecionyl

Tetrahydrohelenalinisobutyrate

mexicanin C-isovalerate

Figure 1. Sesquiterpene lactones of the flowerheads of *Arnica montana*.

Regardless of the fact that the flavonoids, the essential oil, as well as other components of the flowers, may take part in the diverse effects of the Arnica flowers, helenalin, 11α,13-dihydrohelenalin and their esters must be regarded as the main active compounds. In the following it will be shown that the pharmacological and toxicological profile of these potent compounds correlates closely with that of the Arnica flowers.

Biological Activities of Arnica Sesquiterpene Lactones

Sesquiterpene lactones have long been associated with medicinal properties. They contribute the "bitter principle" to plants because of their intensive bitter taste. There are more than 4000 known structures in this class of secondary plant metabolites. The majority occur in members of Compositae family. They have been given a multivalent role since many of them have been reported to possess cytotoxic and antitumoral properties together with a wide variety of other biological activities which have already been reviewed previously (17-20). The reported biological effects include antibacterial, antifungal, antiprotozoan, antiinflammatory, antiarthritic, antihyperlipidemic, antiulcerogenic, antiasthmatic, cholinergic, anticholinergic, hypotensive, analgesic, CNS stimulant, cardiotoxic, genotoxic and mutagenic activities.

Primary Targets of Sesquiterpene Lactones. The fundamental mechanism of this wide range of activities—established by chemical studies and by numerous in vitro and in vivo tests—involves the functional α,β-unsaturated carbonyl structures, commonly part of the α-methylene-γ-lactone, but also present in an α,β-unsubstituted cyclopentenone or conjugated ester structure. These α,β-unsaturated structure elements act as Michael acceptors to biological nucleophiles, especially sulfhydryl groups of cellular proteins, often present as cysteine residues (see Figure 2), thus causing the observed multitude of responses (e.g. 21, 22).

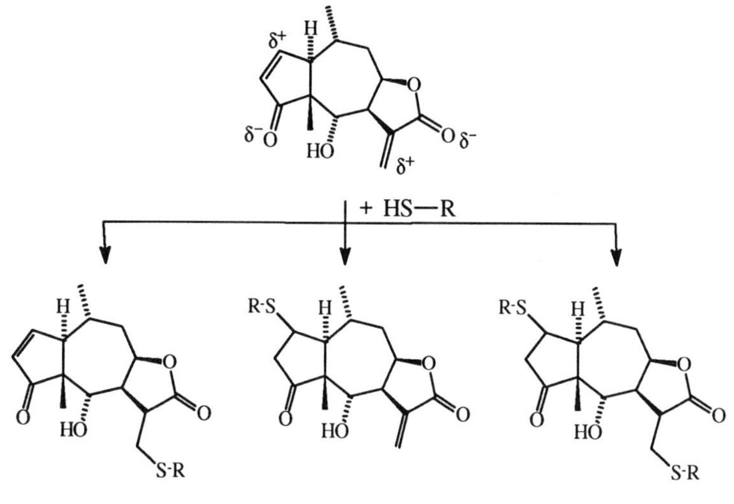

Figure 2. Michael addition of the α,β-unsaturated carbonyl structures of helenalin to sulfhydryl groups of biological molecules.

Thus, thiol-bearing enzymes or metabolically important proteins with sulfhydryl groups in exposed position appear to be the primary targets of sesquiterpene lactones. Among the tested sesquiterpene lactones helenalin has been shown to have a comparatively high activity. This has been explained by the fact that it possesses two alkylating structure elements (*e.g.* 23-27). However, bioactivity is not always correlated with the number of potential alkylating centers (28). Thus, other factors, such as the lipophilicity and especially the molecular geometry, for example the conformation of the basic carbocyclic skeleton, have been suggested to be important in the bioactivity and perhaps selectivity of sesquiterpene lactones (12, 28, 30).

In a more recent study it was shown that helenalin and its ester derivatives exist in solution in two different twist chair conformations with a very fast exchange at room temperature (31). This, too, might result in a higher and more widespread reactivity towards biological target molecules compared with more rigid structures.

Moreover, it has been demonstrated in vitro that the two alkylating structure elements of helenalin exhibit a different behaviour in the Michael-addition reaction towards glutathione or cysteine (Schmidt, T.J. *Bioorg. Med. Chem.*, in press.). It was shown by intensive NMR-studies that glutathione is bound preferentially by the cyclopentenone, while free cysteine reacts very quickly with the exomethylene group of the γ-lactone (see Figure 2). The importance of this regioselectivity of the reaction in dependence upon the chemical environment of the target molecule will certainly increase for reactions in more sterically demanding situations, for example on the surface or even in the pockets of polypeptides.

Concentrations Necessary to Obtain Biological Effects. The concentrations for helenalin, 11α,13-dihydrohelenalin, and their esters necessary to obtain biological effects in the test models used usually lie in the micromolar range or even below.

Antineoplastic Activity. Helenalin, dihydrohelenalin, and their ester derivatives have been shown to possess potent cytotoxic and antineoplastic activities against a battery of murine and human tumors. A very large number of studies deals with these effects of helenalin, its derivatives, and even of semisynthetic modifications (*e.g.* 32-37). The main modes of action are the inhibition of DNA, RNA and protein synthesis. Among others, the specific metabolic targets are the thiol-bearing enzymes IMP dehydrogenase, DNA polymerase α, ribonucleoside reductase (33, 36, 38), eIF-2 kinase, and the elongation factor eIF-3 (29, 30). Furthermore, the compounds lower the cellular glutathione level (39, 40).

Antibacterial and Antifungal Activity. Helenalin, dihydrohelenalin and their esters inhibit the growth of gram-positive and gram-negative bacteria. The minimum inhibition concentration (MIC) lies in the range of 7-100 µg/mL (41-43). Helenalin also has been reported to possess a strong antifungal activity (44). Thus, the sesquiterpene lactones contribute mainly to the known antiseptic effects of the Arnica flowers.

Anti-Inflammatory Activity. Of central importance for the medicinal use of Arnica flowers is their antiinflammatory activity. It was not until the 1970s that sesquiterpene lactones were examined directly for antiinflammatory activity in animal (45) and biochemical models, such as the inhibition of enzymes or the release of mediators involved in the inflammatory process (46). Among the tested sesquiterpene lactones, helenalin was proven to be the most active (45, 46).

Helenalin and 11α,13-dihydrohelenalin were tested against carrageenan-induced edema in the rat. The drugs were administered i.p.. At 2.5 mg/kg helenalin effected an inhibition of edema of 77%. Indometacin at 10 mg/kg was 78% active. The exomethylene-γ-lactone group was found to be mainly responsible for this inhibition, because dihydrohelenalin, which contains only the unsubstituted cyclopentenone, was less active.

Helenalin was also tested in the rat adjuvant arthritis model. At 2.5 mg/kg administered i.p. daily for three weeks, helenalin was again the most active, with a value of 77% inhibition compared to untreated controls. Indomethacin administered at 10 mg/kg per day for three weeks was 45% active. Thus, the compound is not only an antiinflammatory drug that relieves the symptoms of inflammation processes, but also an antiarthritic drug which may actually slow down or stop the degenerative process in arthritis.

Helenalin, as well as 11α,13-dihydrohelenalin, were also screened for analgesic activity using the mouse writhing reflex. Drugs were administered i.p. at 20 mg/kg. Helenalin was most active with 93% reflex reduction. In addition, the compounds also suppressed induced pleurisy, delayed hypersensitivity and passive cutaneous anaphylaxis and were mild immunostimulants of immunoglobulins (45). Thus, the sesquiterpene lactones of the Arnica flowers possess potent antiinflammatory and antiarthritic activity in rodents at a relatively low dose of 2.5 mg/kg. At the doses used, no severe toxic effects were observed in rodents.

Mode of Action. As for the mode of action, it has been shown that helenalin and its derivatives act at multiple sites. This is not uncommon, as many non-steroidal-antiinflammatory druges (NSAID's) undoubtedly are multivalent in that they effect a number of inflammatory parameters rather than act on a single one. Currently, no model system is completely satisfactory, as we still have much to learn about the etiology of inflammatory diseases.

Helenalin has been shown in vitro to effect a number of biochemical parameters that influence the inflammation process (46). It is a potent inhibitor of neutrophil migration and chemotaxis. It definitely uncouples the in vitro oxidative phosphory-lation respiration of neutrophils of humans, which may be associated with the lack of an efficient energy-generating system necessary for leucocyte chemotaxis, microtubular contraction and migration to the inflammation site. The compound has been shown to inhibit the release of neutrophil lysosomal enzymes, as well as their catalytic activity (46). Inhibition of lysosomal release and catalytic activity will reduce the spread of inflammation and tissue damage. Furthermore, cathepsin D enzymatic activity has been implicated in the inflammation process as causing the release of vasoamine peptides and chemotactic factors, which are chemical mediators that can attract phagocytic cells to the site of inflammation (47).

Helenalin has been shown in vivo to elevate the cyclic adenosine monophosphate (cAMP) level in rat and mouse liver (46). Elevated levels of cAMP have been correlated with the stabilization of lysosomal membranes (48). In addition, high cAMP levels inhibited rabbit and human polymorphonuclear neutrophile chemotaxis and migration and were associated with the formation of migration inhibitory factor, activation of phagocytic cell function, and release of histamine, slow-reacting substances and prostaglandins [cited according to (46)]. Furthermore, induced chronic arthritis was related to decreased cAMP levels, increased lysosomal enzymatic activity and release in the paw (49).

11α,13-Dihydrohelenalin isobutyrate and senecionate have been shown to inhibit the release of histamine from peritoneal mast cells of the rat and to possess antiallergic activity in vivo in a passive cutaneous anaphylaxis test (50). On the other hand, helenalin was found to be a potent stimulator of mast cell degranulation (51).

Helenalin and 11α,13-dihydrohelenalin inhibit the secretion of serotonin from human blood platelets in vitro (52). Both compounds inhibit the aggregation of human platelets in response to external chemical stimuli (collagen, arachidonic acid) (52), as well as thrombin-induced rat platelet aggregation (IC_{50} for helenalin: 2.6 x 10^{-6} M; Herbert, J.M., Lane, A., Willuhn, G. unpublished) in a concentration-dependent manner. Both compounds were shown to inhibit the prostaglandin biosynthesis in human platelets, probably not by directly targeting cyclooxygenase, but rather by indirectly interfering with phospholipase A_2 activity (52). Prostaglandin synthetase has been shown to be inhibited by helenalin in vitro only at a concentration of 10^{-3} M (46), which is not physiological. Whether helenalin is able to inhibit the 5-lipoxygenase, as has been demonstrated already for other sesquiterpene lactones (53, 54), has not yet been proven.

Effects on Heart and Circulation. In various in vivo and in vitro test systems, positive inotropic actions and, at higher doses, cardiotoxic effects have been documented for helenalin, 11α,13-dihydrohelenalin and their esters (12, 55-62).

In guinea-pig left atria, helenalin exerts a concentration-dependent positive inotropic action with an EC_{50}-value of 1.4 x 10^{-5} M, which is strongly diminished in reserpine-pretreated animals or after addition of the ß-blocking agent, bupranolol (60, 61). The same results were found using atrial strips from the left atrium of guinea-pigs (59).

In the test model of the left atria, the presence of 5 x 10^{-6} M phosphodiesterase inhibitor IMBX (1-methyl-3-isobutylxanthine) shifted the concentration-response curve of helenalin to the left and diminished the IC_{50}-value to 6.4 x 10^{-6} M, indicating a potentiating effect (60, 61). In contrast to this result, in the model of the isolated papillary muscle of guinea-pigs, helenalin (3 x 10^{-4} M) did not cause any positive inotropic effect in the presence of IBMX (10^{-3} M), but instead produced a slight negative inotropic effect (62).

In vitro, helenalin exhibited only a rather weak and concentration-independent inhibition of phosphodiesterase (20-30%) (60, 61). However, a highly significant helenalin-induced increase in cAMP content of atria could be detected in combination of helenalin and IMBX, supporting the sequential synergism of both compounds (60, 61).

At 10^{-4} M, toxic effects of helenalin on the atrium were observed, while the positive inotropic action was diminished. Helenalin prolonged relaxation time in isolated cat papillary muscles at concentrations exceeding 10^{-5} M, decreased maximum relaxation velocity and thus irreversibly disturbed relaxation (60, 61).

The results of these investigations led to the conclusion that the cardiotoxic effects of helenalin are based on a slow down of the restitution kinetics for activator calcium, whereas the positive inotropic action resulted from an indirect sympathomimetic effect (59-62). Recently, helenalin stimulated cellular $[Ca^+]$ responses have been reported in connection with its antitumor activity (63). The indirect sympathomimetic action differs from that of tyramine because no tachyphylaxia and no inhibition by imipramine could be observed (60, 61).

Blood pressure experiments on rabbits have shown that helenalin (55), its acetate and 11α,13-dihydrohelenalin acetate (12) adminstered intravenously (0.33-7 mg/kg) cause a sudden rise of 10 to 50 mm Hg, followed by a gradual fall to the same degree below normal. This variation in the blood pressure was accounted for as a secondary action due to the effect on the heart muscle, and on the respiratory center. More recently the sesquiterpene lactone parthenolide has been demonstrated to interfere markedly with both contractile and relaxant mechanisms in vascular smooth muscle (64, 65).

Respiratory Analeptic Activity. Helenalin acetate and 11α,13-dihydrohelenalin acetate have strong respiratory analeptic properties. Intravenous administration of the low dose of 0.33 mg/kg to rabbits caused an increase of respiratory rates of 50-85% and 35-50%, respectively, and an increase of tidal (breath) volume/min. of 40-50% and 20-30%, respectively (12).

Oxytocic Activity. 11α,13-Dihydrohelenalin acetate has been shown to induce contraction of the isolated uterus of the rat (66). The helenanolide arnifolin (2,3-dihydro-2α-hydroxyhelenalin tiglinate) isolated from *Arnica chamissonis* ssp. *foliosa* (see Figure 3), has been demonstrated in situ on rabbits to elevate the tonicity of the uterus and to increase its periodical contractions (67).

Antihyperlipidemic Activity. Helenalin and 11α,13-dihydrohelenalin significantly lowered, after intraperitoneal dosing for 2 weeks at 6 mg/kg/day and 20 mg/kg/day, serum cholesterol levels in mice by 37% and 15%, respectively. At the same time helenalin reduced serum triglyceride levels by 24% (68). The mode of action could be linked to the inhibition of thiol-bearing regulatory enzymes of the cholesterol and triglyceride syntheses, such as HMG-CoA reductase and citrate lyase, which was demonstrated in vitro (68).

Acute Toxicity and Contact Allergy. Helenalin-containing species of the genus *Helenium* are well known to be acutely toxic to livestock (69-72). Acute toxicity by helenalin has been related to hepatic glutathione levels, which are rapidly depleted by very low concentrations of helenalin (39). The oral LD_{50} values of helenalin for mammals varies form 85 mg/kg for hamster to 150 mg/kg for mouse (73). The intraperitoneal LD_{50} for mouse is 10 mg/kg (74). The acute toxicity of the more lipophilic helenanolide esters of the Arnica flowers was estimated with a purified fraction of these compounds in the mouse (57). The intraperitoneal and oral DL_{50} values for this fraction were 31 mg/kg and 123 mg/kg, respectively.

An initial study on the pharmacokinetics and tissue distribution of [^{3}H]-11α,13-dihydrohelenalin in mice following intravenous, intraperitoneal and oral administration showed a mean serum radioactivities of 34.4%, 8.5%, and 5.2%, respectively, of the total at 15 min (75). A two-compartment pharmacokinetic model predicted that the maximum terminal (beta) half-life of the compound was 57 hours. Urinary excretion accounted for 40.3% to 64.4% of the administered radioactivity with about one third contributed to unmetabolised 11α,13-dihydrohelenalin. Fecal excretion accounted for 9.3% (i.v. application) to 39.7% (oral application), indicating that the compound was secreted in the bile. Radioactivity in the major organs (brain, heart, lung, spleen, liver, kidney, stomach, intestine) paralleled the serum radioactivity, showing that these organs did not sequester [^{3}H]-11α,13-dihydrohelenalin. In contrast, significant radioactivity was found in muscle, bone, fat, and skin even after 24 days.

Contact Allergy. A well-known adverse reaction of Arnica flowers is their ability to cause allergic contact dermatitis. With longer and more frequent external application, edematous dermatitis may occur, with formation of small vesicles. The responsible contact sensitizers have been shown to be helenalin and its ester derivatives (6, 7, 76-78). Today this is known to be a common feature of many α-methylene butyrolactones, an effect which is probably due to the familiar Michael addition mechanism.

Conclusion. The selected examples of the proven pharmacological and toxicological activities of helenalin, 11α,13-dihydrohelenalin, and their ester derivatives clearly show that these activities correlate closely with the known effects of Arnica flowers, and, therefore, must be considered their main active constituents. Their broad spectrum of biological effects and the lack of predictable selectivity, as well as the multitude of unwanted side effects, makes it difficult to utilize them as therapeutic agents, especially when applied systemically. That is the reason why the German Commission E has restricted the application of Arnica preparations to external use only. However, here the contact allergenic potential must also be taken into account.

Flowers of Arnica chamissonis

Arnica montana is a protected species in Germany. Hence the German Pharmacopoeia permits an alternative source for the official "Arnicae flos": *Arnica chamissonis* Less. ssp. *foliosa* (Nutt.) Maguire (North American meadow arnica) which can be cultivated more easily than *A. montana*. This *Arnica* species (including both the *foliosa* and *chamissonis* subspecies) has also been used in folk medicine (79), and has already been accepted as a substitute in the pharmacopoeia of the Soviet Union and the former East Germany, because extracts from its flowers have been proven to possess the same activity as those from *A. montana* (80-82). It was later shown that extracts from the flowers of *A. chamissonis* ssp. *foliosa*, *A. chamissonis* ssp.

chamissonis, and from plants of different geographical regions differ in their antibacterial activity and acute toxicity (*83*). This suggested the existence of several chemical races.

Sesquiterpene Lactones. The sesquiterpene lactone pattern of the flowerheads of *A. chamissonis* has been found to be much more complex than that of *A. montana* (83-87). In addition to helenalin and 11α,13-dihydrohelenalin esters, the flowers of the subspecies *foliosa* contain their 2,3-dihydro-2α-hydroxy derivatives ("arni-folins" and "dihydroarnifolins," respectively) as well as 2,3-dihydro-2α,4α-dihydroxy or di-O-acyl derivatives ("chamissonolides") shown in Figure 3. Altogether, 29 helenanolides have been isolated from the flowerheads of the official subspecies *foliosa* (83-85). Flowers of the subspecies *chamissonis* contain almost only chamissonolides, but traces of the eudesmanolide ivalin are also present (86, 87).

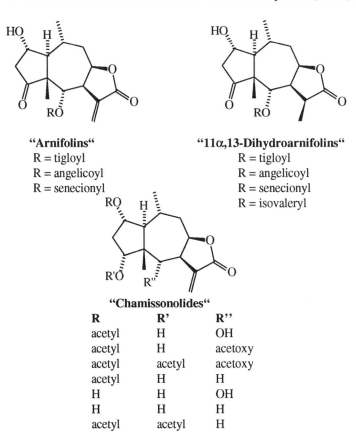

"Arnifolins"
R = tigloyl
R = angelicoyl
R = senecionyl

"11α,13-Dihydroarnifolins"
R = tigloyl
R = angelicoyl
R = senecionyl
R = isovaleryl

"Chamissonolides"

R	R'	R''
acetyl	H	OH
acetyl	H	acetoxy
acetyl	acetyl	acetoxy
acetyl	H	H
H	H	OH
H	H	H
acetyl	acetyl	H

Figure 3. Helenanolides occurring in flowerheads of *Arnica chamissonis* subsp. foliosa in addition to helenalin and 11α,13-dihydrohelenalin esters.

It has been shown that in chamissonolides the seven membered ring adopts a twisted boat conformation (*88, 89*), whereas arnifolins and dihydroarnifolins like 11α, 13-dihydrohelenalin and helenalin adopt a twisted chair conformation (*89*). Some chamissonolides, arnifolins and dihydroarnifolins have also been shown to possess antibacterial, antiinflammatory, and cytotoxic activities, which, however, are lower than that of helenalin and its ester derivatives (*83, 89*).

Analysis of the Sesquiterpene Lactones

The helenanolides of the Arnica flowers with cyclopenten-4-one and 2-hydroxycyclopentan-4-one structures, but not those with the 2α,4α-dihydroxycyclopentane structure (chamissonolides), react with m-dinitrobenzene in alkaline solution to red coloured reaction products (Zimmermann reaction). This reaction is used in the German pharmacopoeia as an identity test, by which e.g. flowerheads of *A. chamissonis* ssp. *chamissonis* can be excluded. A more convenient proof of identity is TLC (*90, 91*).

The Zimmermann reaction has also been used for photometric quantitative determinations (*92,93*). Another method for photometric determination is based on the hydroxamate reaction with hydroxylamine/iron (*93*). Moreover, high-performance liquid chromatographic (HPLC) and gas chromatographic (GC) methods have been developed. Hypersil-ODS columns with water-methanol elution gradient and UV detection at 225 nm in the HPLC system and OV-01-CB capillary column with flame ionization detection in the GC system have been used (*93, 94*), with eudesmanolide santonin as the internal standard for both methods.

Quantitative and Qualitative Variation of the Sesquiterpene Lactones in Arnica Flowers. As part of an effort to set quality control guidelines for the official Arnica flowers, the quantitative and qualitative variation of the sesquiterpene lactones were studied by HPLC and GC analysis (*91*).

In flowerheads of *A. montana* from 39 different origins the quantitative content ranges from 0.31 to 0.91% with an average of 0.5%. For standardization of the official flowers, a minimum content of 0.4% has been suggested.

Concerning to the qualitative composition of the sesquiterpene lactones, a Central European type and a Spanish type of *A. montana* have been found (*91*). Central European flowers, showing no differences in the qualitative sesquiterpene lactone pattern, contain principally helenalin esters, contributing 60-90% of the sesquiterpene lactone content. Spanish flowers contain nearly exclusively 11α, 13-dihydrohelenalin esters with the methacrylate, tiglate and isobutyrate as predominant compounds.

Compared with *A. montana*, the flowers of *A. chamissonis* ssp. *foliosa* show a much larger quantitative and qualitative variation in their sesquiterpene lactone content, depending on the origin of the plant material. In flowers, originating from plants grown under identical conditions from seeds of 11 different geographical regions, the quantitative content was found to vary from 0.07% to 1.4% (*91*).

With regard to the qualitative composition, there are populations with only four compounds, as well as those with up to 18 compounds, out of a total of 29 sesquiterpene lactones identified in the flowers of this subspecies (*91*). There are sources in which the dihydrohelenalin esters (80-100%) or helenalin and 11α,13-dihydrohelenalin esters (> 70%) dominate, and are thus similar to *A. montana* flowers. There are also sources in which the "dihydrohelenalins" and "dihydroarnifolins" are predominant in more or less equal proportions or in which the "arnifolins" or "chamissonolides" predominate.

The cultivation of *A. chamissonis* ssp. *foliosa*, which has already been started, needs to select those populations which are most similar to *A. montana*. Now, as in the past, commercially available "Arnicae flos" is derived from *A. montana*.

Content of Sesquiterpene Lactones in Arnica Tinctures and Infusions. Arnica flowers are used externally as infusions (2%) or as official tinctures and as ointments containing tincture or "Arnica oil" (extract from 1 part drug and 5 parts vegetable oil) as crude active substances. According to the German Commission E monograph (*8*) for poultices the tincture has to be diluted 3-10 times, and as a mouth wash, 10 times. Ointments are limited to have a maximum of 20-25% of the tincture or a maximum of 15% "arnica oil." Infusions for poultices have to be prepared from a ratio of 2 g of drug to 100 mL of water.

When making the officinal tincture according to the German pharmacopoeia or the infusion according to the German Standard Licence [see (*5*)], about 90% and 75%, respectively, of the sesquiterpene lactones are extracted (95). Applied topically, the resulting concentrations (450 µg/mL and 75 µg/mL, respectively) should be high enough to achieve the desired effects on the site of application even if the absorption will be incomplete.

Conclusions

In conclusion, the therapeutic effects of the Arnica flowers reported in traditional medicine for its simple preparations (galenicals) are plausible and can be explained on the basis of the sesquiterpene lactones as their main active constituents. Nevertheless, clinical tests with standardized preparations have not yet been published. Such tests under controlled conditions are an essential requirement for the use of this drug in the future.

Literature Cited

1. Willuhn, G. *Pharm. Ztg.* **1991**, *135*, 2453-2468.
2. Faber, K. *Pharmazie* **1953**, *8*, 179-187, 286-298, 340-346.
3. List, P.H.; Hörhammer, L., Eds.; *Hagers Handbuch der Pharmazeutischen Praxis*, 4. Edition; Springer- Verlag: Berlin, Heidelberg 1972; Vol. 3, pp. 214-233.
4. Merfort, I. *In Hagers Handbuch der Pharmazeutischen Praxis*; Hänsel, R.; Keller, K.; Rimpler, H.; Schneider, G., Eds.; 5. Edition; Springer-Verlag: Berlin, Heidelberg, New York 1990, Vol. 4; pp 342-357.
5. Bisset, N.G., Ed., *Herbal Drugs and Phytopharmaceuticals, a Handbook for Practice on Scientific Basis*; ed. and transl. from the 2. German edition, Wichtl, M., Ed.; medpharm Scientific Publishers: Stuttgart 1994, CRC Press, pp. 83-87.
6. Hausen, B.M. *Hautarzt* **1980**, *31*, 10-17.

7. Willuhn, G. *Dtsch. Apoth. Ztg.* **1986**, *126*, 2038-2044.

8. Bundesanzeiger Nr. 228 vom 05.12.1984.

9. Willuhn, G.; Merfort, I.; Paßreiter, C.M.; Schmidt, T.J. In *Advances in Compositae Systematics*; Hind, D.J.N.; Jeffrey, C.; Pope, G.V., Eds.; Royal Botanic Gardens, Kew 1995; pp 167-195.

10. Poplawski, J.; Holub, M.; Samek, Z.; Herout, V. *Collect. Czech. Chem. Commun.* **1971**, *36*, 2189-2199.

11. Evstratova, R.I.; Ban`kowski, A.J.; Sheichenko, V.I.; Rybalko, K.S. *Khim. Prir. Soedin* **1971**, 270-272.

12. List, P.H.; Friebel, B. *Arzneim.-Forsch.* **1974**, *24*, 148-151.

13. Willuhn, G.; Röttger, P-M.; Matthiesen, U. *Planta Med.* **1983**, *49*, 226-231.

14. Berner, M., PhD thesis; 1995; University of Düsseldorf, Germany.

15. Willuhn, G.; Röttger, P.M.; Wendisch, D. *Planta Med.* **1984**, *50*, 35-37.

16. Kresken, J.; Willuhn, G. *Pharm. Ztg.* **1983**, *128*, 1610.

17. Willuhn, G. *Dtsch. Apoth. Ztg.* **1987**, *127*, 2511-2517.

18. Picman, A.K. *Biochem. Syst. & Ecol.* **1986**, *14*, 255-281.

19. Ivie, G.W.; Witzel, D.A. In *Handbook of Natural Toxins* Vol. 1: Plant and Fungal Toxins. Keeler, R.F., Tu, A.T., Eds.; Marcel Dekker, Inc., New York 1983, pp 543-584.

20. Rodriguez, E.; Towers, G.H.N.; Mitchell, J.C. *Phytochemistry* **1976**, *15*, 1573-1580.

21. Kupchan, S.M.; Fessler, D.C.; Eakin, M.A.; Giacobba, T.J. *Science* **1970**, *168*, 376-378.

22. Lee, K.H.; Hall, I.H.; Marr, E.C.; Starnes, C.O.; Eigeбally, S.A.; Waddel, T.G.; Hadgraft, R.I.; Ruffner, C.G.; Weidner, I. *Science* **1977**, *196*, 533-536.

23. Kupchan, S.M.; Eakin, M.A.; Thomas, A.M. *J.Med. Chem.* **1971**, *14*, 1147-1152.

24. Lee, K.H.; Meck, R.; Piantadosi, C.; Huang, E.S. *J. Med. Chem.* **1973**, *16*, 299-301.

25. Picman, A.K.; Rodriguez, E.; Towers, G.H.N. *Chem. Biol. Interact.* **1979**, *28*, 83-89.

26. Hall, I.H.,;Lee, K.H.; Okano, M.; Sims, D.; Ikuba, T.; Liou, Y.F.; Imakura, Y.J. *Pharm. Sci.* **1981**, *70*, 1147-1150.

27. Hall, I.H.; Lee, K.H.; Starnes, C.O.; Eigebaly, S.A.; Ikuba, T.; Wu, Y.S.; Kimura, T.; Haruna, M. *J. Pharm. Sci.* **1978**, *67*, 1235-1239.

28. Hartwell, J.L.; Abbot, B.J. *Adv. Pharmacol. Chemother.* **1969**, *7*, 117-209.

29. Williams, W.L.; Chaney, S.G.; Willigham, W.; Considine, R.T.; Hall, I.H.; Lee, K.H. *Biochem. Biophys. Acta* **1983**, *740*, 152-162.

30. Williams, W.L.; Chaney, S.G.; Hall, I.H.; Lee, K.H. *Biochemistry* **1984**, *23*, 5637-5644.

31. Schmidt, T.J. *J. Mol. Struct.* **1996**, *385*, 99-112.

32. Page, J.D.; Chaney, S.G.; Hall, I.H.; Lee, K.H.; Holbrook, D.J. *Biochem. Biophys. Acta* **1987**, *926*, 186-194.

33. Hall, I.H.; Williams, W.L.; Grippo, A.A.; Lee, K.H.; Holbrook, D.J.; Chaney, S.G. *Anticancer Drugs* **1988**, *8*, 33-42.

34. Grippo, A.A.; Hall, I.H.; Kiyokawa, H.; Muraoka, O.; Shen, Y.-C.; Lee, K.-H. *Drug Des. Discov.* **1992**, *8*, 191-206.

35. Hall, I.H.; Lee, K.H.; Mar, E.C.; Starnes, C.O.; Waddell, T.G. *J. Med. Chem.* **1977**, 333-337.

36. Hall, I.H.; Lee, K.H.; Grippo, A.A.; Chaney, S.G.; Holbrook, D.J. *Pharm. Res.* **1987**, *4*, 509-514.

37. Woerdenbag, H.J.; Merfort, I.; Paßreiter, C.M.; Schmidt, T.J.; Willuhn, G.; van Uden, W.; Pras, N.; Kampinga, H.H.; Konings, W.T. *Planta Medica* **1994**, *60*, 434-437.

38. Williams, W.L.; Hall, I.H.; Grippo, A.A.; Oswald, C.B.; Lee, K.H.; Holbrook, D.J.; Chaney, S.G. *J. Pharm. Sci.* **1988**, *77*, 178-184.

39. Merrill, J.C.; Kim, H.L.; Safe, S.; Murray, C.A.; Hayes, M.A. *J. Toxicol. Environ. Health* **1988**, *23*, 159-169.
40. Hall, I.H.; Grippo, A.A.; Holbrook, D.J.; Roberts, G.; Hang-Ching Liu; Kim, H.L. *Planta Med.* **1989**, *55*, 513-517.
41. Lee, K.-H.; Ikuba, T.; Wu, R.Y.; Geisman, T.A. *Phytochemistry* **1977**, *16*, 1177-1781.
42. Willuhn, G.; Röttger, P.-M.; Quack, W. *Pharm. Ztg.* **1982**, *127*, 2183-2185.
43. Picman, A.K.,;Towers, G.H.N. *Biochem. Syst. Ecol.* **1983**, *11*, 321-327.
44. Picman, A.K. *Biochem. Syst. Ecol.* **1984**, *12*, 13-18.
45. Hall, I.H.; Lee, K.H.; Starnes, C.O.; Sumidi, Y.; Wu, R.Y.; Waddell T.G.; Cochran, J.W.; Gerhart, K.G. *J. Pharm. Sci.* **1979**, *68*, 537-542.
46. Hall, I.H.; Starnes, C.O.; Lee, K.H.; Waddell, T.G. *J. Pharm. Sci.* **1980**, *69*, 537-543.
47. Rojas-Espinosa, O.; Dannenberg, A.M. In *Inflammation and Anti-inflammatory Therapy*. Katona, G., Blengio, J.R., Eds.; Spectrum, New York, N.Y. 1974, p. 31.
48. Arrigoni-Martelli, E. *Inflammation and Antiinflammatories*. Spectrum, New York, N.Y. 1977, pp. 85-98.
49. Parnheim, M.J.; Bonta, I.L.; Adolfs, M.J.P. In *Perspectives in Inflammation*. Willoughby, D.A., Giroud, J.P., Velo, G.P. Eds.; University Park Press, Baltimore, Md. 1977, pp. 279-287.
50. Wu, J.B.; Chun, Y.T.; Ebizuka, Y.; Sankawa, U. *Chem. Pharm. Bull.* **1985**, *33*, 4091-4094.
51. Elissade, M.H.; Ivie, G.W.; Row, L.D.; Elissade, G.S. *Am. J. Vet. Res.* **1983**, *44*, 1894-1899.
52. Schröder, H.; Lösche, W.; Strobach, H.; Leven, W.; Willuhn, G.; Till, U.; Schrör, U. *Thromb. Res.* **1990**, *57*, 839-845.
53. Ysreal, M.C.; Croft, K.D. *Planta Med.* **1990**, *56*, 268-270.
54. Summer, H.; Salen, U.; Knight, D.W.; Hoult, J.R.S. *Biochem. Pharmacol.* **1992**, *43*, 2313-2320.
55. Lamson, P.D. *J. Pharmacol. Exp. Ther.* **1913**, *4*, 471-489.
56. Szabuniewicz, M.; Kim, H.L. *Southwest. Vet.* **1972**, 305-311.
57. Röttger, P.-M. PhD thesis; 1981; University of Düsseldorf, Germany. Willuhn, G.; Röttger, P.-W. *Planta Med.* **1982**, *45*, 131.
58. Grevel, J. PhD thesis; 1982, University of München, Germany.
59. Takeya, K.; Itoigawa, M.; Furukawa, H., *Chem. Pharm. Bull.* **1983**, *31*, 1719-1725.
60. Kondor, A.; Bayer, R.; Mannhold, R.; Noack, E.; Willuhn, G. *Pharm. Tydschr. Belg.* **1984**, *61*, 240.
61. Eggert, A. PhD thesis 1989; University of Düsseldorf, Germany.
62. Itoigawa, M.; Takeya, K.; Furokawa, H.; Ito, K. J. *Cardiovas. Pharmacol.* **1987**, *9*, 193-201.
63. Powis, G.; Gallagos, A.; Abraham, R.T.; Ashendal, C.L.; Zalkon, L.H.; Grindey, G.B.; Bonjou-Klian, R. *Cancer Chemother. Pharmacol.* **1994**, *34*, 344-350.
64. Barsly, R.W.J.; Salan, K..; Knight, D.W.; Hoult, J.R.S. *Planta Med.* **1993**, *59*, 20 -25.
65. Hay, A.J.B.; Hamburger, M.; Hostettmann, K.; Hoult, J.R.S. *Br. J. Pharmacol.* **1994**, *112*, 9 - 14.
66. Herrmann, D., PhD thesis; 1978, University of Frankfurt/Main, Germany.
67. Rybalko, K.S.; Trutnew, E.A.; Kibaltschitsch, B.N. *Aptetschn. Delo* **1965**, *14*, 32 - 33.
68 Hall, I.H.; Lee, K.H.; Starnes, C.O.; Muraoka, O.; Sumida, Y.; Waddell, T.G. *J. Pharm. Sci.* **1980**, *69*, 694-697.
69. Kingsbury, J.M. *Poisonous plants of the United States and Canada*. Prentice Hall, Englewood Cliffs 1964.

70. Dollahite, J.W.; Hardy, W.T.; Henson, J.B. *J. Amer. Vet. Med. Assoc.* **1964**, *145*, 694-696.

71. Dollahite, J.W.; Rowe, L.D.; Kim, L.H.; Camp, B.J. *Southwest. Vet.* **1972**, *26*, 135-137.

72. Herz, W. In *Effects of poisonous plant of livestock.* Keeler, R.F., van Kampen, K.R., James, L.F. Eds.; Acad. Press. New York 1978, pp. 487-497.

73. Witzel, D.A.; Ivie, G.W.; Dollahite, J.W. *Am. J. Vet.* **1976**, *37*, 859-861.

74. Kim, H.L. *Res. Comm. Chem. Pathol. Pharmacol.* **1980**, *28*, 189-192.

75. Grippo, A.A.; Wyrick, S.D.; Lee, K.A.; Shrewsbury, R.P.; Hall, I.H. *Planta Med.* **1991**, *57*, 309-314.

76. Mitchell, J.C.; Dupois, G. *Br. J. Dermatol.* **1971**, *84*, 139-150.

77. Herrmann, H.-D.,;Willuhn, G.; Hausen, B.M. *Planta Med.* **1978**, *34*, 299-304.

78. Hausen, B.M.; Herrmann, H.-D.; Willuhn, G. *Contact Dermatitis* **1978**, *4*, 3-10.

79. Tschirsch, A. *Handbuch der Pharmakognosie*, Vol. 2, Publishing House Chr. Herrmann Tauchnitz, Leipzig 1917, pp. 1174-1184.

80. Brunzell, A.; Wester, S. *Svensk Farm. Tidskr.* **1947**, *51*, 645-651, 693-703.

81. Akselrow, D.M.; Bereschinskaja, W.W. *Med. Ind. UdSSR* **1958**, *12*, 19-22.

82. Luckner, M.; Bessler, O; Luckner R. *Pharmazie* **1965**, *20*, 681-685.

83. Kresken, J. PhD thesis; 1984; University of Düssesdorf, Germany.

84. Willuhn, G.; Kresken, J.; Wendisch, D. *Planta Med.* **1983**, *47*, 157-160.

85. Willuhn, G.; Kresken, J.; Leven, W. *Planta Med.* **1990**, *56*, 111-114.

86. Willuhn, G.; Pretzsch, G.; Wendisch, D. *Tetrahedron* **1980**, *37*, 773-776.

87. Willuhn, G.; Junior, I.; Kresken, J.; Pretzsch, G.; Wendisch, D. *Planta Med.* **1985**, *51*, 398-401.

88. Wiebcke, G.; Kresken, J.; Mootz, D.; Willuhn, G. *Tetrahedron* **1982**, *35*, 2709-2714.

89. Schmidt, T.J.; Fronczek, F.R.; Liu, Y.-H. *J. Mol. Struct.* **1996**, *385*, 113-121.

90. Willuhn, G.; Kresken, J.; Merfort, I. *Dtsch. Apoth. Ztg.* **1983**, *123*, 2431-2434.

91. Willuhn, G.; Leven, W.; Luley, C. *Dtsch. Apoth. Ztg.* **1994**, *134*, 4077-4085.

92. Leven, W.; Willuhn, G. *Planta Med.* **1986**, *52*, 537-538.

93. Willuhn, G.; Leven, W. *Pharm. Ztg. Wiss.* **1991**, *4/139*, 32-39.

94. Leven, W.; Willuhn, G. *J. Chromatogr.* **1987**, *410*, 329-342.

95. Willuhn, G.; Leven, W. *Dtsch. Apoth. Ztg.* **1995**, *135*, 1939-1942.

Chapter 11

Curcuma (Tumeric): Biological Activity and Active Compounds

M. Wichtl

Institute of Pharmaceutical Biology, Department of Pharmacy, Philipps-University, D-35032 Marburg, Germany

Extracts of *Curcuma domestica* and *Curcuma xanthorrhiza* rhizomes (turmeric, Javanese turmeric resp.) are used in Europe as cholagogues. In Eastern Asia only *C. xanthorrhiza* is in use for the treatment of biliary disorders. The active compounds, diarylheptanes, esp. curcuminoids, and constituents of the essential oils, mainly sesquiterpenes, have been—and are still—objects of pharmacological and clinical studies. Many interesting effects, such as antiinflammatory, lipid reducing, and sedative actions, have been reported. Nevertheless, more research is necessary to better understand the mechanism of action of Javanese turmeric.

Biliary disorders and their therapy are an important part of the practice of family doctors and specialists for internal diseases. In addition to several synthetic drugs, some phytopharmaceuticals are therapeutically used in Europe. The most effective are those prepared from the rhizomes of two *Curcuma* species, turmeric and Javanese turmeric.

Remedies to cure biliary disorders are (less exactly) named cholagogues. They are differentiated into choleretics (raise the production of bile) and cholekinetics (stimulate the contraction and emptying of the gall bladder). Both remedies are used in the therapy of cholecystitis and cholangitis, or generally to stimulate production of bile, but they are contraindicated for gallstones or hepatitis.

Curcuma xanthorrhiza (Javanese turmeric) and *Curcuma domestica* (syn. *C. longa*, turmeric) are perennial herbs, belonging to the Zingiberaceae family and are cultivated in Indonesia and South India. The very large, oblong lanceolate leaves show a more or less parallel nervature. The zygomorphic flowers appear at ground level and have three fused sepals and three large yellow petals, with only one fertile stamen The rhizomes of both species are more or less tuberousovoid, in case of *C. domestica* also cylindrical or fingershaped. When the rhizomes of *C.domestica* are scalded with boiling water in toto, the starch becomes largely gelatinized and the surface turns an intensive yellow. The rhizome of *Curcuma xanthorrhiza* is purified superficially and then cut into small slices of about 3 mm thickness. The rhizomes of Javanese turmeric have been used in Indonesia medicinally for centuries under the name „Temu lavak" as a cholagogue. This remedy was made known in Europe by the Dutch, the former colonizers of Indonesia, and was (and is still) official in the Nederlandse Pharmacopoeia. Since 1978, Curcumae xanthorrhizae rhizoma has been official in the German Pharmacopoeia as well. In contrast to the Javanese turmeric, *Curcuma domestica* has been used in tropical Asia only as a spice (it is one of the component parts of Curry powder), not as a remedy. In Germany this *Curcuma* species is also used as a cholagogue, since it has very similar constituents.

Constituents

The rhizomes of both *Curcuma* species contain (a) diarylheptane-derivatives, esp. the "curcuminoids," dicinnamoylmethane-derivatives, and (b) essential oil, comprising mainly sesquiterpenes.

Turmeric, Curcumae domesticae rhizoma, contains 3-5% yellow pigments that are not volatile in steam; diarylheptanoids, consisting mainly of three compounds, curcumin (diferuloylmethane), monodesmethoxycurcumin (feruloyl-p-hydroxycinnamoylmethane) and bisdesmethoxycurcumin (*1*); and 2-7% essential oil, comprising sesquiterpenes of the germacrane-, bisabolane- and guajane-types. The main compounds of the volatile oil are turmerone, ar-turmerone, zingiberene and curlone. The high content of bisabolane derivatives is characteristic for turmeric and cannot be observed in other *Curcuma* species (*2*). Currently, more than 100 constituents have been isolated and their structures elucidated by spectral methods. Turmeric also contains small amounts of a complex acidic arabinogalactan, ukonan A (*3*) and starch, largely gelatinized.

Javanese turmeric, Curcumae xanthorrhizae rhizoma, contains only 1-2% diarylheptanoids (curcuminoids), mainly curcumin and monodesmethoxycurcumin (*1, 4*), but the content of essential oil, 3-12%, is higher than that of *C. domestica*. The essential oil comprises mainly sesquiterpenes, such as β-curcumene, ar-curcumene, germacrone, and a phenolic sesquiterpene xanthorrhizol, a characteristic constituent of this drug. p-Tolylmethylcarbinol, earlier described as a compound of the essential oil, is no doubt an artefact, which arises during the distillation of the oil.

Pharmacological and Clinical Investigations

The use of both plant drugs for the treatment of biliary disorders, such as chronic forms of cholangitis and cholecystitis, has been confirmed by some pharmacological and clinical investigations (*5-8*). Therefore, in Germany many preparations, prepared from the rhizomes of both *Curcuma* species, are licensed by the Institute for Remedies (the government health agency). The German Commission E has published monographs for both plant drugs, with the indication "dyspeptic complaints." For the efficacy of the extracts, prepared from these *Curcuma* species, both the curcuminoids and the essential oils are responsible. The curcuminoids act mainly as cholekinetics, but they also stimulate the choleresis. Extracts of *C. xanthorrhiza* are distinctly more effective than extracts of *C. domestica*, in accordance with the empirical medicinal use

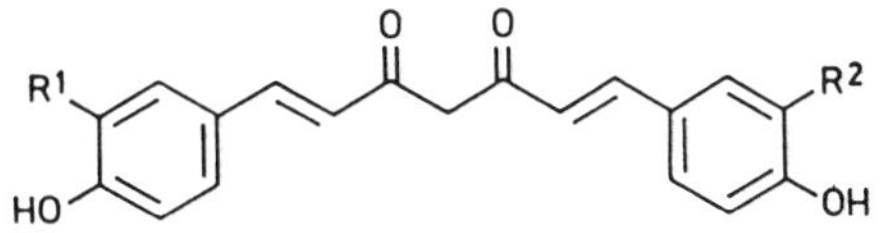

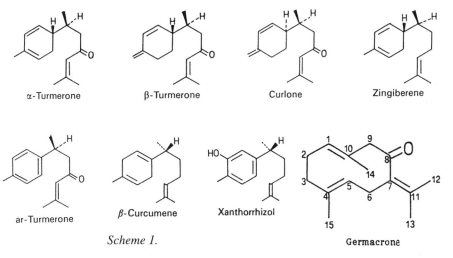

	R¹	R²
Curcumin	OCH₃	OCH₃
Desmethoxycurcumin	OCH₃	H
Bisdesmethoxycurcumin	H	H

α-Turmerone β-Turmerone Curlone Zingiberene

ar-Turmerone β-Curcumene Xanthorrhizol Germacrone

Scheme 1.

Scheme 2.

in Indonesia. Maybe the choleresis-inhibiting property, described for bis-desmethoxycurcumin (present only in *C. domestica*), is responsible for this phenomenon (*1*).

Between 1992 and 1996 some interesting pharmacological effects of isolated, pure constituents were reported. Curcumin, the main constitutent of both *Curcuma*-species, parenterally administered, acts strongly as an antiphlogistic and inhibits the formation of leukotriene B_4, a mediator-substance of inflammation (*9*). The in vitro inhibition of prostaglandin biosynthesis by curcumin had been observed earlier by H.Wagner and coworkers (*10*). Curcumin demonstrates in vitro effects on the biosynthesis of prostacyclin and on platelet aggregation in rats and in monkey plasma (*11*). Evaluation of nonsteroidal anti-inflammatory drugs in patients with postoperative inflammation produced an anti-inflammatory response for curcumin at the same level as phenylbutazone (*12*). Experiments on rats to find out the oral resorption of curcumin showed the major biliary metabolites were glucuronides of tetrahydro- and hexahydrocurcumin (*13*).

Antiphlogistic effects have also been recently detected for some the phenolic and nonphenolic diarylheptanoids, as well as for trans-trans-1,7-diphenyl-1,3-heptadien-5-one (alnustone), trans-1,7-diphenyl-1-hepten-5-ol, and trans-trans-1,7-diphenyl-1,3-heptadien-5-ol (*14*).

An ethanolic extract of turmeric, as well as an ointment of curcumin (0.5% curcumin), were found to produce remarkable symptomatic relief in patients with external cancerous lesions after application of the ointment three times daily for four weeks; in many patients the effect continued for several months (*15*). Also, in studies on experimentally induced skin tumors in mouse (induction by 12-O-tetradecanonylphorbol-13-acetat) curcumin was found to markedly inhibit tumor promotion: topical application of 1, 3 or 10 μmol of curcumin inhibited the induced tumors by 39, 77 or 98% respectively (*16*). The antitumor activity of curcumin is probably related to the inhibition of a phospholipid-dependent protein kinase C (PKC), because some other inhibitors of PKC can also inhibit tumor promotion caused by phorbol esters (*17, 18*). Recent investigations demonstrate the activity of curcumin at a molecular level. Treatment of human myeloid ML-1a cells with tumor necrosis factor rapidly activated the transcription factor NF-$_κ$B, and this activation was inhibited by curcumin (*19, 20*). In an other experiment, it could be demonstrated that curcumin inhibits the phorbolester-induced expression of c-fos, c-jun, and c-myc proto-oncogenes messenger RNA (*21*). Curcumin, at 5 μM, inhibited in vitro lipopolysaccharide-induced production of TNF (tumor necrosis factor α) and interleukin-1β by a human monocytic macrophage cell line, Mono Mac. 6 (*22*).

For some diarylheptanoids a hypolipidemic action, evaluated in Triton-induced hyperlipidemic hamsters by inhibiting hepatic triglyceride secretion, has been reported (*23*). Besides curcumin, the most effective compounds were 5-hydroxy-7(4-hydroxyphenyl)-1-phenyl-(1E)-1-heptene and 7-(3,4-dihydroxyphenyl)-5-hydroxy-1-phenyl-(1E)-1-heptene, the latter being more potent than the former, but there is up to now no practical use of this observation.

A very interesting discovery was made in 1995 with evidence of anti-HIV-activity for curcumin. On purified human HIV-1 integrase, curcumin was shown to

trans-trans-1,7-diphenyl-1,3-heptadien-4-one

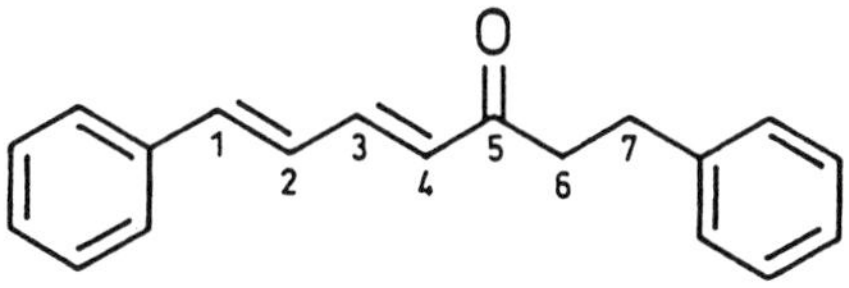

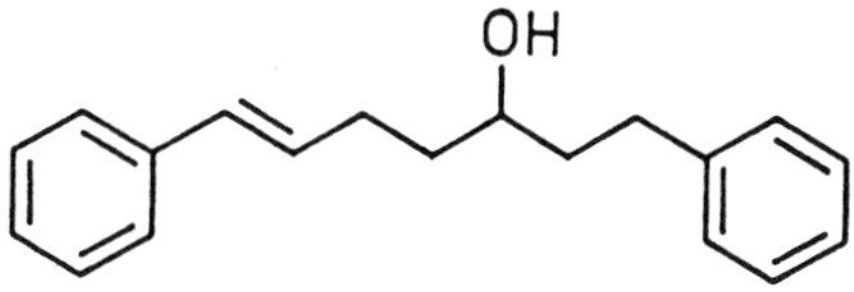

trans-1,7-diphenyl-1-hepten-5-ol

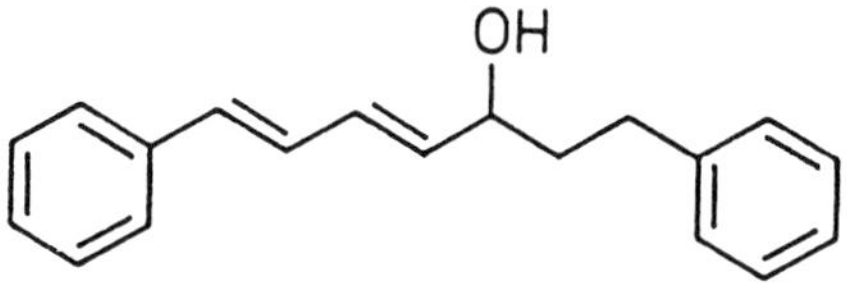

trans,trans-1,7-diphenyl-1,3-heptadien-5-ol

Scheme 3.

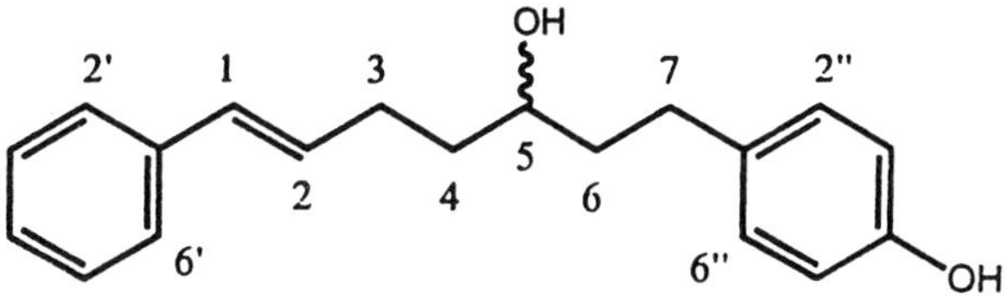

5-hydroxy-7-(4-hydroxyphenyl)-1-phenyl-(1*E*)-1-heptene

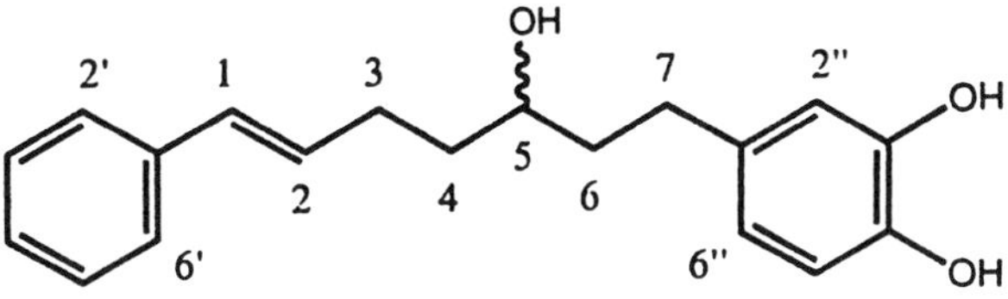

7-(3,4-dihydroxyphenyl)-5-hydroxy-1-phenyl-(1*E*)-1-heptene

Scheme 4.

have an inhibitory concentration (I_{50}) for strand transfer of 40 μM, probably interacting with the integrase catalytic core (24).

Also, for some constituents of the essential oils, remarkable pharmacological effects have been demonstrated. Xanthorrhizol, the species-specific compound of Javanese turmeric, acts as a sedative: oral administration to mice at 50 mg/kg, xanthorrhizol prolongated the pentobarbital-induced sleeping time (328%), and its activity was almost the same as that of chlorpromazine at 5 mg/kg (25). Germacrone, a sesquiterpene-ketone of *C.xanthorrhiza*, upon oral administration to mice at 100-200 mg/kg, showed a significant hypothermic effect (ΔT ca 2-2.8° C) (26). At least nine sesquiterpenes of the essential oil of *Curcuma xanthorrhiza*, (e.g. xanthorrhizol, furanodienone, and others) are very effective contact-insecticides (27).

Many other pharmacodynamic studies have been published in a review by H.P.T.Ammon et al. (28). Most of the demonstrated activities have not (or not yet) been clinically demonstrated. Nevertheless a European patent was granted for a fixed combination of *Curcuma*-constituents as an antiphlogistic (29), and a Japanese patent was granted for germacrone (30) as a component part of an CNS-depressant remedy.

Besides their use as cholagogues, both *Curcuma* species should demand an increasing interest as sources of medicinally useful substances.

Literature cited

1.   Jentzsch, K.; Gonda, Th.; Hoeller, H. *Pharm.Acta Helv.* **1959**, *34*,181-188.
2.   Ohshiro, M.; Kuroyanagi, M..; Ueno, A. *Phytochemistry* **1990**, *29*, 2201-2205.
3.   Tomoda, M.; Gonda, R.; Shimizu, N.; Kanari, M.; Kimura, M. *Phytochemistry* **1990**, *29*, 1083-1086.
4.   Jentzsch, K.; Spiegl, P.; Kamitz, R. *Sci.Pharm.* **1968**, *36*, 251-256.
5.   Pietschmann, P., Doctoral thesis, University of Marburg, **1989**.
6.   Harnischfeger, G.; Stolze, H., *notabene medici* **1982**, *12*, 562-573.
7.   Baumann, J.C. *Med.Monatsschr.* **1975**, *29*, 173-175.
8.   Kalk, H.; Nissen, K. *Dtsch.Med.Wochenschr.* **1932**, *58*, 1718-1721; **1931**, *57*, 1613-1617.
9.   Ammon, H.P.T.; Anazodo, M.I.; Safayhi, H.; Dhawan, B.N.; Srimal, R.C. *Planta Med.* **1992**, *58*, 226.
10.   Wagner, H.; Wierer, M.; Bauer, R. *Planta Med.* **1986,** *52*, 184-187.
11.   Srivastava, R.; Puri, V.; Srimal, R.C.; Dhawan, B.N. *Arzneim.-Forsch.* **1986**,   *36*, 715-717.
12.   Satoskar, R.R.; Shah, S.J.; Shenoy, S.G., *J.Clin.Pharmacol.Ther.Toxicol.* **1986**, *24*, 651-654.
13.   Holder, G.M.; Plummer, J.L.; Ryan, A.J. *Xenobiotica* **1978**, *8*, 761-768.
14.   Claeson, P.; Pongprayoon, U.; Sematong, T.; Tuchinda, P.; Reutrakul, V.; Soontornsaratune, P.; Taylor, W.C. *Planta Med.* **1996**, *62*, 236-240.
15.   Kuttan, R.; Sudheeran, P.C.; Josph, C.D. *Tumori* **1987**, *73*, 29-31.
16.   Huang, M.T.; Smart, R.C.; Wong, C.Q.; Conney, A.H. *Cancer Res.* **1988**, *48*, 5941-5946..
17.   Huang, M.T.; Wie, M.; Lu, Y.P.; Chang, R.L.; Fisher, C.; Manchang, P.S.; Newmark, H.L.; Conney, A.H., *Carcinogenesis* **1995**, *16*, 2493-2497.
18.   Hasmeda M.; Polya, G.M. *Phytochemistry* **1996,** *42*, 599-605.

19. Singh, S.; Aggarwal, B.B. *J.Biol.Chem.* **1995**, *270*, 24995-24500.
20. Chan, M.M.Y. *Biochem.Pharmacol.* **1995**, *49*, 1551-1556.
21. Kakar, S.S.; Roy, D. *Cancer Lett.* **1994**, *87*, 85-89.
22. Chan, M.M.-Y., *Biochem.Pharmacol.* **1995**, *49*, 1551-1556.
23. Suksamrarn, A.; Eiamong, S.; Piyachaturawat, P.; Charoenpiboonsin, J. *Phytochemistry* **1994**, *36*, 1505-1508.
24. Mazumder, A.; Raghavan, K.; Weinstein, J.; Kohn, K.W.; Pommier, Y. *Biochem.Pharmacol.* **1995**, *49*, 1165-1170.
25. Yamazaki, M.; Maebayashi, Y.; Iwase, N.; Kaneko, T. *Chem.Pharm.Bull* **1988**, *36*, 2070-2074.
26. Yamazaki, M.; Maebayashi, Y.;Iwase, N.; Kaneko, T. *Chem.Pharm.Bull* **1988**, *36*, 2075-2078.
27. Pandji, Ch.; Grimm, C.; Wray, V.; Witte, L.; Proksch, P. *Phytochemistry* **1993**, *34*, 415-419.
28. Ammon, H.P.T.; Wahl, M.A., *Planta Med.* **1991**, *57*, 1-7.
29. Europ. Pat., Appl. EP440,855 (Cl. A61K35/78), 14 Aug. 1991; Chem. Abstr. **1992**, *116*, 46284.
30. Chem. Abstr. **1990**, *112*, 681.

Chapter 12

Echinacea: Biological Effects and Active Principles

R. Bauer

Institute für Pharmazeutische Biologie, Heinrich-Heine-Universität,
Universitätsstr. 1, D–40225 Düsseldorf, Germany

In recent years, Echinacea preparations have developed into the best selling herbal immunostimulants. Most of the products are derived from either the aerial or the underground parts of *Echinacea purpurea*, and from the roots of *E. angustifolia* or *E. pallida*. The common products in Europe mainly represent expressed juice preparations or alcoholic tinctures. The different species and preparations can be clearly distinguished phytochemically by their typical constituents. The most relevant compounds for standardization are caffeic acid derivatives (cichoric acid, echinacoside), alkamides, polyacetylenes, and glyco-proteins/polysaccharides. Pharmacological studies have shown that cichoric acid, alkamides, and glycoproteins/polysaccharides possess immunomodulatory activity. Therefore they might be considered as active principles and best suited for standardization purposes. Clinical effects have been demonstrated for the expressed juice of the aerial parts of *Echinacea purpurea* in the adjuvant therapy of relapsing infections of the respiratory and urinary tracts, as well as for alcoholic tinctures of *E. pallida* and *E. purpurea* roots as adjuvants in the therapy of common cold and flu.

Historical Use and Botanical Aspects of *Echinacea* Species

The genus *Echinacea* is endemic to North America, where it occurs in the Great Plains in between the Appalachian Mountains in the east and the Rocky Mountains in the west. Its use can be traced back to the American Indians, who regarded *Echinacea* as one of the most favored remedies to treat wounds, snake bites, headache and the common cold (*1*). The territories of the tribes that most frequently used Echinacea, show a close correspondence with the distribution range of *Echinacea angustifolia*. *Echinacea pallida* and *E. purpurea*, and possibly other *Echinacea* species have also been used, however.

In the second half of the 19th century, white settlers took began to use Echinacea. H.G.F. Meyer, a German "quack" doctor, distributed a tincture in Missouri made from *E. angustifolia* roots, called "Meyer´s Blood Purifier", which he recommended for rheumatism, neuralgia, headache, erysipelas, dyspepsia, tumors and boils, open wounds, vertigo, scrofula and bad eyes, as well as "poisoning by herbs", and rattlesnake bite (*2*). At the beginning of the 20th century, the use of Echinacea was introduced in Europe.

More than 800 Echinacea containing drugs, including homeopathic preparations, are currently on the market in Germany. Most of the preparations contain the expressed juice of *Echinacea purpurea* aerial parts, or hydroalcoholic tinctures of *E. pallida* or *E. purpurea* roots. They are mainly used for nonspecific stimulation of the immune system. In the U.S.A., Echinacea preparations have recently become very popular as well. There, mostly tinctures from the roots are used. But also freeze dried plants and encapsuled powders from roots and aerial parts are sold. According to a recent report, Echinacea products are the best selling herbal product in health food stores in the U.S.A. (*3*).

When analyzing the large number of products, it becomes obvious that completely different preparations are sold under the name "Echinacea". At least three different species are used medicinally: *Echinacea purpurea*, *E. angustifolia* and *E. pallida*. Some preparations are prepared from the roots (mainly from *E. angustifolia* and *E. pallida*, but also from *E. purpurea*), some from the aerial parts (*E. purpurea*), and some from the whole plant (homeopathic mother tinctures of *E. angustifolia* and *E. pallida*).

Another variable is the extraction mode: alcoholic tinctures, hydro-alcoholic extracts, tea preparations and, as a special form, the expressed juice of the aerial parts are on the market. Therefore, it is necessary to establish the botanical and chemical standardization of every preparation in order to be able to obtain reproducible pharmacological and therapeutic results.

Echinacea angustifolia was the plant originally used by H.C.F. Meyer and by the Lloyd-Brothers. *E. pallida* is much more abundant and a much taller species with bigger roots than *E. angustifolia*. When the monograph in the National Formulary of the U.S. was published in 1916, the roots of both *E. angustifolia* and *E. pallida* were made official, with the result that differentiation between these two species was neglected after 1916. It ended in the very confusing situation, in which most of the "Echinacea angustifolia" available in the market and in botanical gardens in Europe was in fact *E. pallida* (*4*). In the middle of this century, *E. purpurea* was introduced as a medicinal plant in Europe. The roots and the aerial parts are used separately for medicinal purposes.

Echinacea purpurea roots have been substituted for a long time with *Parthenium integrifolium*. The sesquiterpene esters, echinadiol-, epoxyechinadiol-, echinaxanthol- and dihydroxy-nardol-cinnamate (**1-4**), described as constituents of *Echinacea purpurea* roots (*5*), were in fact derived from the adulterant, *Parthenium integrifolium*, which was mistakenly processed at that time (Figure 1). Since both species contain different constituents, HPLC and TLC methods have been developed to distinguish them chemically (*6*). *Parthenium integrifolium* is characterized by the sesquiterpene esters, which cannot be found in *Echinacea purpurea*. Recently, adulterations of *E. angustifolia* roots with roots of *Parthenium integrifolium* have also been detected.

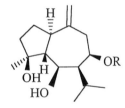

1 Cinnamoyl-echinadiol

2 Cinnymoyl-epoxyechinadiol

3 Cinnamoyl-echinaxanthol

4 Cinnamoyl-dihydroxynardol

R =

Figure 1. Constituents of the roots of *Parthenium integrifolium*, an adulterant of *Echinacea purpurea* roots.

Active Principles, Pharmacological Effects and Standardization

The constituents of *Echinacea*, as in any other plant, cover a wide range of polarity, ranging from the polar polysaccharides and glycoproteins, to the moderately polar caffeic acid derivatives and to the rather lipophilic polyacetylenes and alkamides. This makes it necessary to study the activity of different polar extracts of *Echinacea*, such as aqueous preparations, alcoholic tinctures, and hexane or chloroform extracts.

Systematic fractionation and subsequent pharmacological testing of the aqueous extracts of the aerial parts of *E. purpurea* led to the isolation of two polysaccharides (PS I and PS II) with immunostimulatory properties (*7,8*). They were shown to stimulate phagocytosis *in vitro* and *in vivo*, and to enhance the production of oxygen radicals by macrophages in a dose dependent way (*8,9*). Structural analysis showed PS I to be a 4-O-methyl-glucuronoarabinoxylan with an average MW of 35,000 D, while PS II was shown to be an acidic arabinorhamnogalactan of MW 45,000 D (*10*). A xyloglucan, MW 79,500 D, was isolated from the leaves and stems of *Echinacea purpurea*, and a pectin-like polysaccharide from the expressed juice (*11*).

Polysaccharides have also been obtained from cell cultures of *Echinacea purpurea*. From the growth medium of *E. purpurea* cell cultures, three homogeneous polysaccharides, i.e. two neutral fucogalactoxyloglucans with MW of 10,000 and 25,000 D, and an acidic arabinogalactan, MW 75,000 D, have been isolated (*12,13*). The structure of the tissue culture polysaccharides differed from those of the aerial parts, since cells in suspension culture possess exclusively primary cell wall components.

Luettig et al. (*14*) have shown that different concentrations of polysaccharides from *E. purpurea* could stimulate macrophages to release tumor necrosis factor α (TNFα) and interferon β_2 (= IL-6).

In an infection stress test, mice were treated with *E. purpurea* polysaccharides (10 mg/kg i.v.). Twenty four hours later, they were infected with *Candida albicans*. After 24 hours, the kidneys of the mice were examined for *Candida*. Compared to the untreated group, a significant reduction of *Candida* organisms was observed in the polysaccharide treated group. This means that *E. purpurea* polysaccharides can protect against *Candida* infections. In separate tests, it was shown that polysaccharides can enhance the survival rate of lethal *Candida* infections up to 100 %, and that they are also active against *Listeria* infections (*15,16,17*). Polysaccharides from *E. angustifolia* have also been found to possess antiinflammatory activity (*18,19*).

In a phase-I clinical trial, a polysaccharide fraction (EPO VIIa), isolated from *E. purpurea* tissue culture and injected at doses of 1 mg and 5 mg, caused an increase in the number of leukocytes, segmented granulocytes and TNFα (*20*).

Three glycoproteins, MW 17,000, 21,000 and 30,000 D, containing about 3% protein, have been isolated from *E. angustifolia* and *E. purpurea* roots. The dominant sugars were found to be arabinose (64-84 %), galactose (1.9-5.3 %) and glucosamines (6 %). The protein moiety contained high amounts of aspartate, glycine, glutamate and alanine (*21*). Purified extracts containing this glycoprotein-polysaccharide complex exhibited B-cell stimulating activity and induced the release of interleukin 1, TNFα and IFNα,β in macrophages, which could also be reproduced in mice (*22,23,24*).

An ELISA method has been developed for the detection and determination of glycoproteins in Echinacea preparations (*25*). It seems that *E. angustifolia* and *E.*

purpurea roots contain similar amounts of glycoproteins, while *E. pallida* contains less (*22*). One product containing an *E. purpurea* root extract is on the market which is standardized on glycoproteins (25 µg/ml) by this ELISA method.

Alcoholic tinctures of *Echinacea* aerial parts and roots contain caffeic acid derivatives (Figures 2-4) and lipophilic, polyacetylene-derived compounds. The different extracts can be distinguished by HPLC analysis of these low molecular weight constituents. The roots of *E. angustifolia* and *E. pallida* have been shown to contain 0.3-1.7 % echinacoside (**5**) (*22*). Both species can be discriminated by the occurrence of 1,3- and 1,5-O-dicaffeoyl-quinic acids (**11**, **12**), which are only present in the roots of *E. angustifolia*. Echinacoside has low antibacterial and antiviral activity, but does not show immunostimulatory effects (*23, 24*).

The roots of *E. purpurea* do not contain echinacoside, but cichoric acid (1*R*,3*R*-dicaffeoyl tartaric acid; **14**) and caftaric acid (monocaffeoyl tartaric acid; **13**) are the main constituents. Cichoric acid (**14**) is also a major constituent in the aerial parts of *Echinacea* species (*25, 26*). The aerial parts of *E. angustifolia* and *E. pallida* have also been shown to contain verbascoside (**7**), a structural analogue of echinacoside (**5**) (*30*). Cheminat *et al.* (*28*) have isolated des-rhamnosyl-verbascoside (**8**) and 6-O-caffeoyl-echinacoside (**6**) from *E. pallida*. The latter compound was present only in the roots.

From the leaves of *E. purpurea*, cichoric acid methyl ester (**15**), as well as 2-O-caffeoyl-3-O-feruloyl-tartaric acid (**18**) and 2,3-O-diferuloyl-tartaric acid were isolated (*29*). Later 2-O-feruloyl-tartaric acid (**16**) and 2-O-caffeoyl-3-O-cumaroyl-tartaric acid (**17**) were also found (*27*). From *E. pallida*, 2-O-caffeoyl-3-O--feruloyl-tartaric acid (**18**), 2,3-O-di-5-[α-carboxy-ß-(3,4-dihydroxy-phenyl)-ethyl-caffeoyl-tartaric acid (**19**), and 2-O-caffeoyl-3-O-{5-[α-carboxy-ß-(3,4-dihydroxy-phenyl)-ethyl]-caffeoyl}-tartaric acid (**20**) were isolated (*28*). Echinacoside (**5**) and cichoric acid (**14**) were also detected in tissue cultures of *E. purpurea* and *E. angustifolia* (*28*). In the leaves, flavonoids, such as rutoside, have also been identified (*29*).

Cichoric acid (**14**) has been shown to possess phagocytosis stimmulatory activity *in vitro* and *in vivo*, while echinacoside (**5**), verbascoside (**7**) and 2-caffeoyl-tartaric acid (**13**) do not exhibit this activity (*30*). Cichoric acid was also recently shown to inhibit hyaluronidase (Figure 5) (*31*) and to protect collagen type III from free radical induced degradation (*32*). Therefore, it could also play an important role in the activity of Echinacea extracts.

Cichoric acid is especially abundant in the flowers of all *Echinacea* species and in the roots of *E. purpurea* (1.2-3.1% and 0.6-2.1%, respectively). Much less is present in the leaves and stems. *E. angustifolia* contains the lowest amount of cichoric acid (*30*). The content, however, strongly depends on the season and the stage of development of the plant (*33*).

Cichoric acid undergoes degradation during preparation of alcoholic tinctures. During five days of ethanol 50 % maceration of *E. purpurea* tops, it was found that the content of cichoric acid decreased rapidly, although it was stable when the plant material was filtered off after the first day (Figure 6). Therefore, we assume enzymatic degradation occurred during the extraction process (*34*).

When analyzing six different products containing the same amount of *E. purpurea* expressed juice, we found dramatically different contents of cichoric acid (0.0-0.4 %), even within different batches of the same brand (Figure 7). This may be due to different

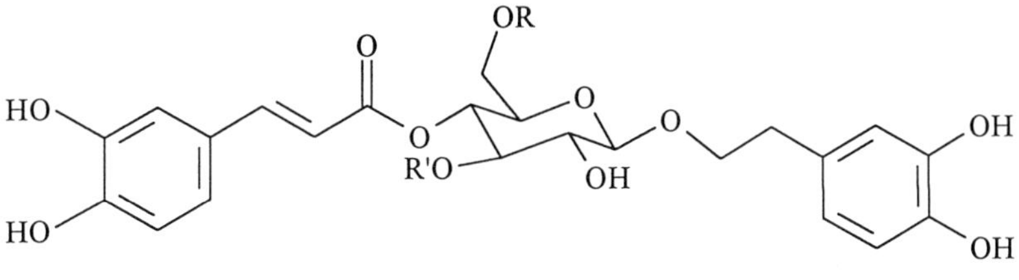

		R	R'
5	Echinacoside	Glucose (1,6-)	Rhamnose (1,3-)
6	6-O-Caffeoyl-echinacoside	6-O-Caffeoyl-glucose (1,6-)	Rhamnose (1,3-)
7	Verbascoside	H	Rhamnose (1,3-)
8	Desrhamnosyl-verbascoside	H	H

Figure 2. Phenylpropanoid glycosides found in *Echinacea* species.

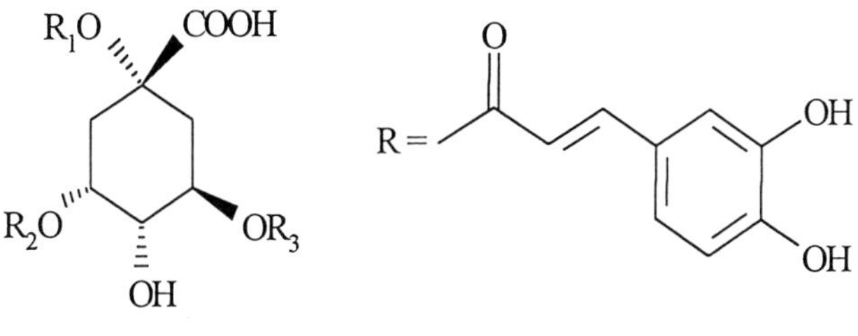

		R_1	R_2	R_3
9	Quinic acid	H	H	H
10	Chlorogenic acid	H	R	H
11	1,3-Dicaffeoyl-quinic acid (Cynarin)	R	R	H
12	1,5-Dicaffeoyl-quinic acid	R	H	R

Figure 3. Quinic acid derivatives from *Echinacea* species.

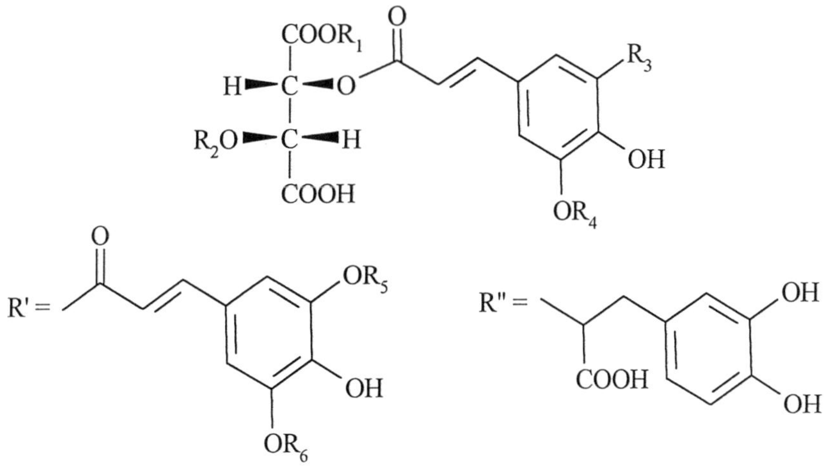

		R_1	R_2	R_3	R_4	R_5	R_6
13	2-O-Caffeoyl tartaric acid (Caftaric acid)	H	H	OH	H	-	-
14	2,3-O-Di-caffeoyl tartaric acid (Cichoric acid)	H	R'	OH	H	OH	H
15	2,3-O-Di-caffeoyl tartaric acid methyl ester	CH_3	R'	OH	H	OH	H
16	2-O-Feruloyl tartaric acid	H	H	OCH_3	H	-	-
17	2-O-Caffeoyl-3-O-cumaroyl tartaric acid	H	R'	H	H	H	H
18	2-O-Caffeoyl-3-O-feruloyl tartaric acid	H	R'	OH	H	OCH_3	H
19	2,3-O-Di-5-[α-carboxy-β-(3,4-dihydroxy-phenyl)-ethyl] tartaric acid	H	R'	OH	R"	OH	R"
20	2-O-Caffeoyl-3-O-{5-[α-carboxy-β-(3,4-dihydroxy-phenyl)-ethyl]-caffeoyl} tartaric acid	H	R'	OH	H	OH	R"

Figure 4. Tartaric acid derivatives from *Echinacea* species.

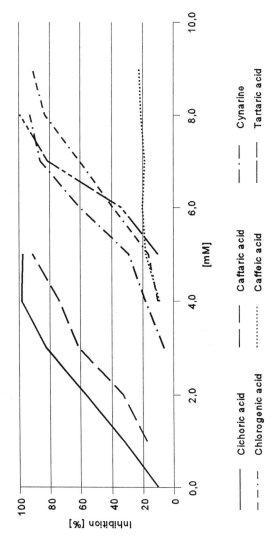

Figure 5. Inhibitory effects on hyaluronidase by caffeoyl esters, caffeic acid, and tartaric acid (35).

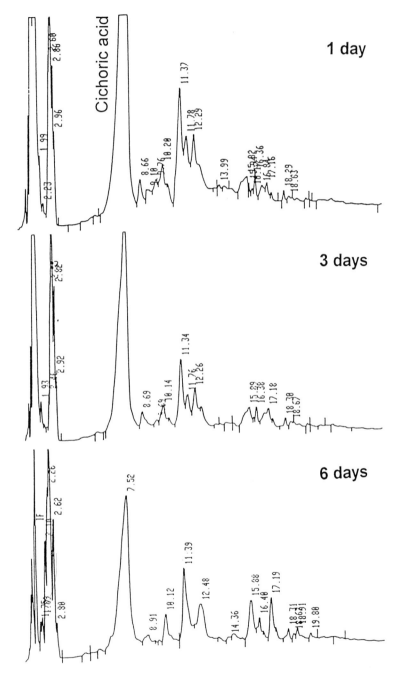

Figure 6. Degradation of cichoric acid during the preparation of alcoholic *E. purpurea* mother tincture, as shown by the HPLC fingerprints (*38*).

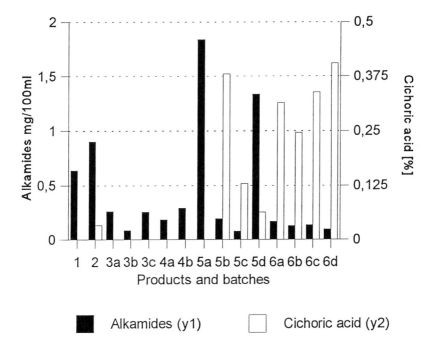

Figure 7. Content of cichoric acid and dodeca-2E,4E,8Z,10E,Z-tetraenoic acid-isobutylamide in different preparations containing *Echinacea purpurea* expressed juices (*39*).

manufacturing processes which allow or inhibit enzymatic activity. In products prepared by special procedure (Schoenenberger process) cichoric acid seemed to be preserved (*35*).

Standardization of the products seems to be necessary. Cichoric acid and other caffeoylics present in *Echinacea* can be determined by HPLC or by a recently developed capillary zone electrophoresis method. This method provides an excellent resolution and a very high sensitivity (*36*).

The lipophilic constituents of *Echinacea* roots can be analyzed by HPLC as well. Striking differences have been observed between the constituents of *E. angustifolia* (alkamides; Figure 8) and *E. pallida* (ketoalkenynes; Figure 9). *Echinacea purpurea* roots also contain alkamides, however with two double bonds in conjugation to the carbonyl group, while *E. angustifolia* mainly contains compounds with a monoene chromophore.

The major lipophilic constituents of *E. pallida* roots have been identified as ketoalkenes and ketoalkynes with a carbonyl group in the 2-position (*26, 37, 38*). The main constituents are tetradeca-8Z-ene-11,13-diyn-2-one (**21**), pentadeca-8Z-ene-11,13-diyn-2-one (**22**), pentadeca-8Z,13Z-diene-11-yn-2-one (**23**), pentadeca-8Z,11Z,13E-triene-2-one (**24**), pentadeca-8Z,11E,13Z-triene-2-one (**25**) and pentadeca-8Z,11Z-diene-2-one (**26**). They occur only in trace amounts in *E. angustifolia* and *E. purpurea* roots. Therefore, they are suitable as markers for the identification of *E. pallida* roots. It has been observed that these compounds undergo autoxidation when the roots are stored in powdered form (*42*). Then, mainly the hydroxylated artifacts, 8-hydroxy-tetradeca-9E-ene-11,13-diyn-2-one (**29**), 8-hydroxy-pentadeca-9E-ene-11,13-diyn-2-one (**30**) and 8-hydroxypentadeca-9E,13Z-diene-11-yn-2-one (**31**) can be found (Figure 10), often with only small residual quantities of the native compounds (*39*). Therefore, the roots are best stored in whole form.

About 15 alkamides have been identified as major lipophilic constituents of *E. angustifolia* roots (*40-42*). They are mainly derived from undeca- and dodecanoic acid, and differ in the degree of unsaturation and the configuration of the double bonds (Figure 8). The main structural type is a 2-monoene-8,10-diynoic acid isobutylamide, but also some 2'-methyl-butylamides have been found.

In *Echinacea purpurea* roots, 11 alkamides have been identified (*30, 45*). In contrast to *E. angustifolia*, most of these alkamides possess a 2,4-diene moiety with the isomeric mixture of dodeca-2E,4E,8Z,10E/Z-tetraenoic acid isobutylamides (**36, 37**) as the main compounds. Therefore, the roots of *E. purpurea* and *E. angustifolia* can clearly be discriminated also by HPLC of the lipophilic constituents (*41, 42*).

The aerial parts of all three Echinacea species contain alkamides of the type found in *E. purpurea* roots, also with dodeca-2E,4E,8Z,10E/Z-tetraenoic acid isobutylamides (**36, 37**) as the main constituents (*30, 43*).

When testing the alcoholic extracts obtained from the aerial parts and from the roots for phagocytosis stimulating activity, the lipophilic fractions showed the highest activity, indicating that the lipophilic constituents should also represent an active principle (*34, 44*). Ethanolic extracts of the aerial parts of *E. angustifolia* and *E. purpurea* showed immunomodulating activity on the phagocytic, metabolic and bactericidal activities of peritoneal macrophages in mice (*45-47*).

Purified alkamide fractions from *E. purpurea* and *E. angustifolia* roots were shown

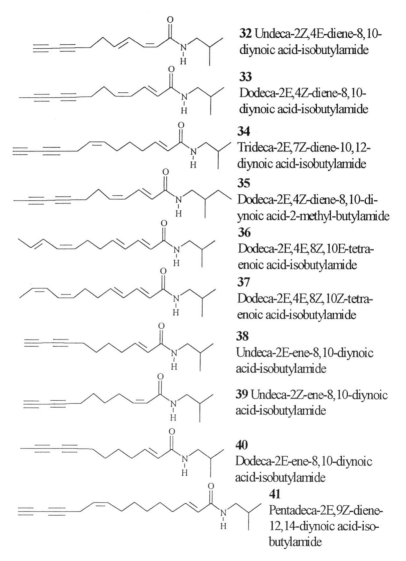

32 Undeca-2Z,4E-diene-8,10-diynoic acid-isobutylamide

33 Dodeca-2E,4Z-diene-8,10-diynoic acid-isobutylamide

34 Trideca-2E,7Z-diene-10,12-diynoic acid-isobutylamide

35 Dodeca-2E,4Z-diene-8,10-diynoic acid-2-methyl-butylamide

36 Dodeca-2E,4E,8Z,10E-tetraenoic acid-isobutylamide

37 Dodeca-2E,4E,8Z,10Z-tetraenoic acid-isobutylamide

38 Undeca-2E-ene-8,10-diynoic acid-isobutylamide

39 Undeca-2Z-ene-8,10-diynoic acid-isobutylamide

40 Dodeca-2E-ene-8,10-diynoic acid-isobutylamide

41 Pentadeca-2E,9Z-diene-12,14-diynoic acid-isobutylamide

Figure 8. Main alkamides found in *Echinacea* aerial parts and *E. purpurea* and *E. angustifolia* roots.

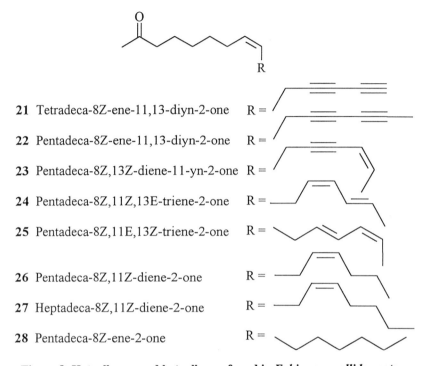

Figure 9. Ketoalkenes and ketoalkynes found in *Echinacea pallida* roots.

21 Tetradeca-8Z-ene-11,13-diyn-2-one   R =

22 Pentadeca-8Z-ene-11,13-diyn-2-one   R =

23 Pentadeca-8Z,13Z-diene-11-yn-2-one   R =

24 Pentadeca-8Z,11Z,13E-triene-2-one   R =

25 Pentadeca-8Z,11E,13Z-triene-2-one   R =

26 Pentadeca-8Z,11Z-diene-2-one   R =

27 Heptadeca-8Z,11Z-diene-2-one   R =

28 Pentadeca-8Z-ene-2-one   R =

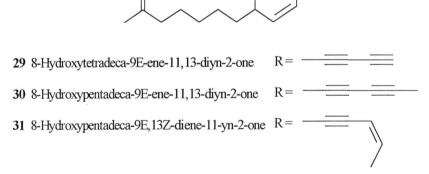

29 8-Hydroxytetradeca-9E-ene-11,13-diyn-2-one   R =

30 8-Hydroxypentadeca-9E-ene-11,13-diyn-2-one   R =

31 8-Hydroxypentadeca-9E,13Z-diene-11-yn-2-one   R =

Figure 10. 8-Hydroxy-ketoalkenynes formed via autoxidation in powdered *Echinacea pallida* roots.

to enhance phagocytosis in the Carbon-Clearance-Test by a factor of 1.5 to 1.7 (*34*). So, they also contribute to the immunostimulatory activity of Echinacea tinctures. Since the main constituent, dodecatetraenoic acid-isobutylamide, only exhibited weak activity, the most effective constituent remains to be found. Alkamides also displayed marked inhibitory activity in vitro in the 5-lipoxygenase (porcine leukocytes) and cyclooxygenase (microsomes from ram seminal vesicles) assays. Dodeca-2*E*,4*E*,8*Z*,10*E*/*Z*-tetraenoic acid isobutylamides (**36**, **37**) inhibited cyclooxygenase at a concentration of 50 µg/ml by 54.7 % and 5-lipoxygenase at a concentration of 50 µM by 62.2 % (*46*).

Determination of the contents of alkamides (dodeca-2*E*,4*E*,8*Z*,10*E*/*Z*-tetraenoic acid isobutylamides) by HPLC showed that they accumulate primarily in the roots and inflorescences, the highest content being found in *E. angustifolia*. *E. pallida* roots contain only trace amounts of alkamides, roots of *E. purpurea* 0.004 - 0.039%, and those of *E. angustifolia* 0.009 - 0.151%. The yield in the leaves is 0.001 - 0.03% (*48*). Small amounts of alkamides can also be found in the expressed juice of *E. purpurea*. They may, therefore, also contribute to its immunomodulatory activity. However, the content also varies considerably between the different products on the market (Figure 7) (*39*). Standardization seems to be necessary.

It is obvious that not a single, but several constituents, like the alkamides, cichoric acid, glycoproteins and polysaccharides, are responsible for the immunostimulatory activity of *Echinacea* extracts. Therefore, the application of extracts appears to be reasonable. However, standardization of these extracts is a must for generating reproducible products and reproducible activity.

Clinical Studies

Many case reports exist which report the effectiveness especially of *Echinacea purpurea* expressed juice [for reviews see (*4*) and (*47*)], but few controlled clinical studies have been performed to demonstrate the therapeutic value of Echinacea preparations (Table I).

In one comparative study, the effect of adjuvant application of *E. purpurea* expressed juice on the recurrence of vaginal *Candida* mycoses within six months was tested. Econazol nitrate was used as a local therapeutic agent over 6 days. The recurrence frequency in the group which was treated only with econazol nitrate was 60.5%. This decreased to between 5 and 16% (depending on the mode of application) after adjuvant application of *E. purpurea* expressed juice (*56*).

A retrospective study with 1280 children with bronchitis demonstrated that treatment with *E. purpurea* expressed juice caused a faster healing process than in the group treated with an antibiotic (*57*). This can be explained by the fact that bronchitis is mostly a viral infection in which antibiotics have no effects, while the Echinacea preparation stimulates the unspecific defense mechanisms in order to attack the virus infected cells.

In a placebo-controlled double-blind study on 108 patients with the efficacy of treatment with *E. purpurea* expressed juice (orally 2x4 mL daily over 8 weeks) was tested. There were less and less severe infections observed in the verum group. In patients which started the study with a T4/T8 ratio of < 1.5, a reduction of the duration of the disease was achieved (5.34 days instead of 7.54 days) (*58*).

154

Table I. Recent Clinical Studies with Echinacea Preparations

Aim of the study	Preparation	Effect	Ref.
Adjuvant therapy of recurrent vaginal *Candida* infections	*E. purpurea* expressed juice prep. (various amounts)	reduction of recurrence rate from 60.5% to 5-16%	*48*
Treatment of acute bronchitis in children	*E. purpurea* expressed juice prep. (1-3 ml/day i.v.)	faster healing process	*49*
Treatment of the common cold	*E. purpurea* expressed juice prep. (2x4 mL/day orally)	lower infection frequency; faster healing process	*50*
Treatment of the common cold	hydroalcoholic extract of *E. purpurea* roots (900 mg root/day)	significantly faster reduction of symptoms than in the placebo group	*51*
Treatment of infections of the upper respiratory tract	hydroalcoholic tincture (1:5) of *E. pallida* roots (900 mg root/day)	significantly faster recovery than in the placebo group	*52*
Treatment of common cold	vitamin C and *E. purpurea* root extract	length of the common cold was reduced	*53*

A placebo-controlled double-blind study has been performed to prove the therapeutic effectiveness of an alcoholic extract of *E. purpurea* roots on 180 patients with the common cold (*59*). They evaluated a score of symptoms, like irritated nose, frontal headache, lymphnode swelling, coated tongue, and rale, at the beginning of the study and after 3-4 and 8-10 days. The dose equivalent of 900 mg root/day (3 x 60 drops) caused a significantly faster reduction of symptoms compared to the placebo group and compared to a group treated with 450 mg/day (3 x 30 drops) only.

A similar placebo controlled double blind study with 160 patients has been performed with a hydroalcoholic tincture (1:5) of *E. pallida* roots (*60*). The group of patients treated with a dose of 90 drops (equivalent of 900 mg roots) *E. pallida* tincture recovered significantly faster from infections of the upper respiratory tract than the placebo group.

In a randomized, single-blind, and placebo-controlled study with 32 subjects (17 male and 15 female, ages 18 to 71 years), the efficacy of an anti-cold remedy including vitamin C (dose#) and *E. purpurea* root extract (dose#) in the treatment of the common cold was investigated. The evaluation parameter was the duration of the illness based on rhinorrea (nasal drip) and the number of paper tissues used daily by each subject. The length of the common cold was 3.37 days in the verum-treated patients and 4.37 days in the placebo-treated patients (p < 0.01). Also, the number of tissues used was significantly different between the two groups (882 treated, 1168 placebo). The preparation was found to be useful and safe for the treatment of the common cold (*61*).

A few studies have also been performed on the effectiveness of Echinacea preparations in the adjuvant treatment of cancer (*62-65*).

The clinical studies performed with Echinacea preparations have recently been reviewed by Melchart *et al.* (*62*). The authors concluded that the studies prove that Echinacea preparations can be effective for the enhancement of the body´s defense mechanisms, but that further investigations are necessary to find the best dosage and application form.

Adverse Effects and Toxicological Considerations

Little information exists on the side effects and toxicological risks of Echinacea treatment. Clinical reports provide indications of a good tolerability. For the parenteral application of Echinacea preparations, a rise in body temperature by 0.5 to 1.0°C has been reported. Shivering and other "influenza-like" symptoms have been occasionally observed. Brief fever can be the result of the secretion of interferon-α and interleukin 1 (endogenous pyrogen) from macrophages, i.e. it always occurs when macrophages are stimulated. Rarely, acute allergic reactions can occur (*63*). Adverse effects on oral administration of the expressed juice for up to 12 weeks are infrequent and consist mainly of unpleasant taste (*64*).

Because of general considerations, *Echinacea* preparations should not be used in case of progressive systemic diseases like tuberculosis, leukoses, collagenoses, multiple sclerosis, AIDS disease, HIV infection, and other autoimmune diseases (*65*). In case of diabetes, the metabolic status may become worse (*66, 67*). Mutagenicity studies did not show any tumour initiating properties (*68, 69*).

The acute toxicity of *E. purpurea* root extract and *E. purpurea* aerial parts expressed juice is extremely low (*72*). Also in long-term treatment, the expressed juice of *E. purpurea* was shown to be well-tolerated (*70*).

Conclusions

As a consequence of the research and experience with Echinacea plants, two preparations have been approved in Germany by the health authorities to be effective drugs: the expressed juice of *E. purpurea* aerial parts for the internal adjuvant therapy of relapsing infections of the respiratory and urinary tract, and externally for poorly healing superficial wounds; and tinctures of *E. pallida* roots for adjuvant therapy of common cold like infections. Extracts of *E. purpurea* roots have not yet been approved in Germany, but they have also shown effectiveness in the treatment of the common cold (*70, 71, 75*).

Literature Cited

1. Moerman, D.E. *Medicinal Plants of Native America*; Res. Rep. Ethnobotany, Contrib. 2, Technical Reports No.19, University of Michigan Museum of Anthropology, 1986.
2. Lloyd, J.U. *Pharm. Review* **1904,** *22,* 9-14.
3. Brevoort, P. *Herbalgram* **1996,** *36,* 49-57.
4. Bauer, R.; Wagner, H. *Echinacea - Ein Handbuch für Ärzte, Apotheker und andere Naturwissenschaftler*, Wissenschaftliche Verlagsgesellschaft, Stuttgart, 1990.

156

5. Bauer, R.; Khan, I.A.; Lotter, H.; Wagner, H.; Wray, V. *Helv. Chim. Acta* **1985**, *68*, 2355-2358.
6. Bauer, R.; Khan, I.A.; Wagner, H. *Dtsch. Apoth. Ztg.* **1987**, *127*, 1325-1330.
7. Wagner, H.; Proksch, A. Z. *angew. Phytother.* **1981**, *2*, 166-171.
8. Stimpel, M.; Proksch, A.; Wagner, H.; Lohmann-Matthes, M.-L. *Infect. Immun.* **1984**, *46*, 845-849.
9. Lohmann-Matthes, M.-L.; Wagner, H. Z. *Phytother.* **1989**, *10*, 52-59.
10. Proksch, A.; Wagner, H. *Phytochemistry* **1987**, *26*, 1989-1993.
11. Stuppner, H. *PhD thesis*, München, 1985.
12. Wagner, H.; Stuppner, H.; Puhlmann J.; Jurcic, K.; Zenk, M.A.; Lohmann-Matthes, M.-L. Z. *Phytother.* **1987**, *8*, 125-126.
13. Wagner, H; Stuppner, H.; Schäfer, W.; Zenk, M.A. *Phytochemistry* **1988**, *27*, 119-126.
14. Luettig, B.; Steinmüller, C.; Gifford, G.E.; Wagner, H.; Lohmann-Matthes, M.-L. *J. Nat. Cancer Inst.* **1989**, *81*, 669-675.
15. Roesler, J.; Steinmüller, C.; Kiderlen, A.; Emmendörfer, A.; Wagner, H.; Lohmann-Matthes, M.-L. *Int. J. Immunopharmacol.* **1991**, *13*, 27-37.
16. Steinmüller, C.; Roesler, J.; Gröttrup, E.; Franke, G.; Wagner, H.; Lohmann-Matthes, M.-L. *Int. J. Immunopharmacol.* **1993**, *15*, 605-614.
17. Roesler, J.; Emmendörfer, A.; Steinmüller, C.; Luettig, B.; Wagner, H.; Lohmann-Matthes, M.-L. *Int. J. Immunopharmacol.* **1991**, *13*, 931-941.
18. Tubaro, A.; Tragni, E.; Del Negro, P.; Galli, C.L.; Della Loggia, R. *J. Pharm. Pharmacol.* **1987**, *39*, 567-569.
19. Tragni, E.; Galli, C.L.; Tubaro, A.; Del Negro, P.; Della Logia, R. *Pharm. Res. Comm.* **1988**, *20*, Suppl. *V*, 87-90.
20. Melchart, D.; Worku, F.; Linde, K.; Flesche, C.; Eife, R.; Wagner, H. *Erfahrungsheilkunde* **1993**, 316-323.
21. Beuscher, N.; Kopanski, L.; Ernwein, C. *Adv. Biosci.* **1987**, *68*, 329-336.
22. Beuscher, N.; Bodinet, C.; Willigmann, I.; Egert, D. Z. *Phytotherapie* **1995**, *16*, 157-166.
23. Beuscher, N.; Beuscher, H.N.; Bodinet, C. *Planta Medica* **1989**, *55*, 660.
24. Bodinet, C.; Beuscher, N. *Planta Med.* **1991**, *57*, Suppl.2: A33-A34.
25. Egert, D.; Beuscher, N. *Planta Medica* **1992**, *58*, 163-165.
26. Bauer, R.; Khan, I.A.; Wagner, H. *Planta Medica* **1988**, *54*, 426-430.
27. Stoll, A.; Renz, J.; Brack, A. *Helv. Chim. Acta* **1950**, *33*, 1877-1893.
28. Cheminat, A.; Zawatzky, R.; Becker, H.; Brouillard, R. *Phytochemistry* **1988**, *27*, 2787-2794.
29. Becker, H.; Hsieh, W.-Ch. Z. *Naturforsch.* **1985**, *40c*, 585-587.
30. Bauer, R.; Remiger, P.; Wagner, H. *Dtsch. Apoth. Ztg.* **1988**, *128*, 174-180.
31. Soicke, H.; Al-Hassan, G.; Görler, K. *Planta Medica* **1988**, *54*, 175-176.
32. Sicha, J.; Hubik, J.; Dusek, J. *Cesk. Farm.* **1989**, *38*, 124-129.
33. Malonga-Makosi, J.-P. *PhD Thesis*, Universität Heidelberg, 1983.
34. Bauer, R.; Remiger, P.; Jurcic, K.; Wagner, H. Z. *Phytother.* **1989**, *10*, 43-48.
35. Maffei Facino, R.; Carini, M.; Aldini, C.; Marinello, C.; Arlandini, E.; Franzoi, L.; Colombo, M.; Pietta, P.; Mauri, P. *Il Farmaco* **1993**, *48*, 1447-1461.
36. Maffei Facino, R.; Carini, M.; Aldini, G.; Saibene, L.; Pietta, P.; Mauri, P. *Planta Medica* **1995**, *61*, 510-514.
37. Alhorn, R. *PhD Thesis*, Universität Marburg/Lahn, 1992.
38. Remiger, P. *PhD thesis*, Universität München, 1989.
39. Bauer, R. Z. *Phytother.*, in press.
40. Pietta, P.; Mauri, P.; Bauer, R., in preparation.

41. Schulte, K.E.; Rücker, G.; Perlick, J. *Arzneim.-Forsch.* **1967**, *17*, 825-829.
42. Khan, I.A. *PhD thesis*, Universität München, 1987.
43. Bauer, R.; Khan, I.A.; Wray, V.; Wagner, H. *Phytochemistry* **1987**, *26*, 1198-1200.
44. Jacobson, M. *J. Org. Chem.* **1967**, *32*, 1646-1647.
45. Bohlmann, F.; Grenz, M. *Chem. Ber.* **1966**, *99*, 3197-3200.
46. Bauer, R.; Remiger, P.; Wagner, H. *Phytochemistry* **1989**, *28*, 505-508.
47. Bauer, R.; Remiger, P.; Wagner, H. *Phytochemistry* **1988**, *27*, 2339-2342.
48. Bauer, R.; Remiger, P. *Planta Medica* **1989**, *55*, 367-371.
49. Bohlmann, F.; Hoffmann, H. *Phytochemistry* **1983**, *22*, 1173-1175.
50. Bauer, R.; Jurcic, K.; Puhlmann, J.; Wagner, H. *Arzneim.-Forsch.* **1988**, *38*, 276-281.
51. Bukovsky, M.; Vaverkova, S.; Kostalova, D.; Magnusova, R. *Cesk. Farm.* **1993**, *42*, 184-187.
52. Bukovsky, M.; Kostalova, D.; Magnusova, R.; Vaverkova, S. *Cesk-Farm.* **1993**, *42*, 228-31.
53. Bukovsky, M.; Vaverkova, S.; Kostalova, D. *Pol. J. Pharmacol.* **1995**, *47*, 175-177.
54. Müller-Jakic, B.; Breu, W.; Pröbstle, A.; Redl, K.; Greger, H.; Bauer, R. *Planta Medica* **1994**, *60*, 7-40.
55. Bauer, R. *Dtsch. Apoth. Ztg.* **1994**, *134*, 94-103.
56. Coeugniet, E.G. Kühnast, R. *Therapiewoche* **1986**, *36*, 3352-3358.
57. Baetgen, D. *TW Paediatrie* **1988**, *1*, 66-70.
58. Schöneberger, D. *Forum Immunologie* **1992**, *2*, 18-22.
59. Bräunig, B.; Dorn, M.; Knick, E. *Z. Phytother.* **1992**, *13*, 7-13 .
60. Bräunig, B.; Knick, E. *Naturheilpraxis* **1993**, *1*, 72-75.
61. Scaglione, F.; Lund, B. *Int. J. Immunother.* **1995**, *11*, 163-166.
62. Lersch, C.; Zeuner, M.; Bauer, A.; Siebenrock, K.; Hart, R.; Wagner, F.; Fink, U.; Dancygier, H.; Classen, M. *Arch. Geschwulstforsch.* **1990**, *60*, 379-383.
63. Lersch, C.; Zeuner, M.; Bauer, A.; Siemens, M.; Hart, R.; Drescher, M.; Fink, U.; Dancygier, H. and Classen, M. *Cancer-Invest.* **1992**, *10*, 343-348.
64. Lersch, C.; Gain, T.; Lorenz, R.; Classen, M. *Immun. Infekt.* **1994**, *22*, 58-59.
65. Nagel, G.A. *Arch. Gynecol. Obstet.* **1995**, *257*, 283-294.
66. Melchart, D.; Linde, K.; Worku, F.; Bauer, R.; Wagner H. *Phytomedicine* **1994**, *1*, 245-254.
67. Bauer, R.; Liersch, R. In: *Hagers Handbuch der Pharmazeutischen Praxis;* R. Hänsel; K. Keller; H. Rimpler; G. Schneider, Eds.; Springer Verlag, Berlin, Heidelberg, New York, 1993, Vol. 5; pp 1-34.
68. Parnham, M.J. *Phytomedicine* **1996**, *3*, 95-102.
69. Shohan, J. *TIPS* **1985**, *6*, 178-182.
70. Kommission E 1989, . Echinaceae purpureae herba. *BAnz.* 43, p. 1070, 02.03.1989.
71. Kommission E 1992, . Echinaceae pallidae radix. *BAnz.* 162, p. 7360, 29.08.1992.
72. Mengs, U.; Clare, C.B.; Poiley, J.A. *Arzneim.-Forsch.***1991**, *41*, 1076-1088.
73. Schimmer, O.; Abel, G.; Behninger, C. *Z. Phytother.* **1989**, *10*, 39-42.
74. Parnham, M.J. *Phytomedicine* **1996**, *3*, 95-102.
75. Kommission E 1992, . Echinaceae purpureae radix. *BAnz.* 162, p. 7360, 29.08.1992.

Chapter 13

Feverfew: A Review of Its History, Its Biological and Medicinal Properties, and the Status of Commercial Preparations of the Herb

S. Heptinstall[1] and D. V. C. Awang[2]

[1]Cardiovascular Medicine, University Hospital, Nottingham NG7 2UH, United Kingdom
[2]MediPlant Consulting Services, P.O. Box 8693, Station T, Ottawa K1G 3J1, Canada

Feverfew (*Tanacetum parthenium* (L.) Schultz Bip.) has been used since antiquity to reduce fever and pain and there are many anecdotal claims of efficacy. Placebo-controlled clinical trials have been performed in prophylaxis of migraine and treatment of rheumatoid arthritis. In migraine good evidence has been obtained for a reduction in frequency and severity of attacks and degree of nausea and vomiting following administration of feverfew leaf. No clear clinically demonstrable benefit has been achieved in rheumatoid arthritis. Extracts of feverfew tested *in vitro* have been shown to inhibit platelet aggregation and granule secretion (release of serotonin) and to inhibit some functions of white blood cells and vascular smooth muscle cells. Most of these inhibitory effects appear to be mediated by parthenolide, the principal sesquiterpene lactone present in the efficacious feverfew, probably through Michael addition of sulphydryl groups to its α–methylenebutyrolactone function. Commercial preparations of feverfew leaf in capsules or tablet form vary enormously in their parthenolide content, and if parthenolide is a clinically significant active constituent of feverfew some of these may be expected to offer poor therapeutic effectiveness. However, the parthenolide-based hypothesis is challenged by results of a recent trial of a parthenolide-rich ethanolic extract of feverfew in migraine prophylaxis.

There has been a resurgence of interest in feverfew in recent years particularly in the U.K., U.S.A. and Canada. This review is in four parts. In Part 1 we consider the nature of the plant and some of the anecdotal claims that have been made regarding its value as a medicinal agent. In Part 2 we report on the formal clinical trials that have been performed to assess its efficacy as a prophylactic for migraine and in treatment of rheumatoid arthritis. In Part 3 we consider the mechanisms through which feverfew may act as a medicinal agent, especially addressing the possibility that the main active constituent may be the sesquiterpene lactone parthenolide. In Part 4 we review the status of some of the commercial preparations of feverfew that are available and the prospects for regulation and control of feverfew products.

Part 1. What is Feverfew and Why is it of Interest?

Feverfew was originally assigned to the genus *Rudbeckia* then *Matricaria* by Linnaeus and eventually reassigned to *Tanacetum* by Schultz Bipontinus who is responsible for the designation *Tanacetum parthenium.* (L.) Schultz Bip. It has also been known as *Chrysanthemum parthenium* (L.) Bernh., currently the most frequently encountered synonym, as well as *Leucanthemum parthenium* (L.) Gren and Godron and *Pyrethrum parthenium* (L.) Sm. (*1*). It is a member of the aster/daisy family. In the old herbals it has been referred to as federfoy, featherfew, flirtwort, bachelor's buttons, nosebleed and midsummer daisy. In Germany it is known as *mutterkraut* (mother herb), in Wales as *feddygen fenyw* (woman's plant) and in Latin America as *Santa Maria*, the Blessed Virgin. Foster (*2*) claims that the generic name *Tanacetum* derives from an altered form of Athanasis (Greek, *athanatos*) meaning immortal, an allusion to the ever-lasting nature of the dried flowers. The reference to immortality could also derive from a classical legend in which a drink made from tansy (*Tanacetum vulgare*) was given to a beautiful young man, Ganymede, so that he could serve as Zeus' cup-bearer. The specific name *parthenium* (from the Greek *parthenos* meaning virgin) was originally applied to another plant by Hippocrates, referring to traditional use by women to help relieve menstrual cramps (*3*) .

A distinctive feature of feverfew is its pungent smell and bitter taste which is always evident irrespective of variety. Its leaves are chrysanthemum-like in appearance and the colour, dependent on the variety, can be dark to light green or yellow. The latter is known as golden feverfew or *Tanacetum parthenium aureus* (*4*). In summer the plant displays daisy-like flowers in clusters, usually with yellow centres and radiating white florets which can be present in single or double layers, spaced or overlapped. Some varieties are devoid of ray florets. The most common variety found growing in the wild in the U.K. is a dark green leafed variety with a single row of ray florets. At maturity it is about 50 cm tall. Most of the experimental and clinical studies that have been performed have involved the use of leaves taken from this particular variety.

Feverfew is a plant for which a great many medicinal properties have been claimed, and these can be divided into ancient and fairly recent uses.

Ancient Uses

The ancient uses of feverfew are derived from the old herbals and include agues, fever, headache, stomach ache, insect bites, toothache, vertigo, difficulties in labour, threatened miscarriage and regulation of menstruation. These uses are illustrated below by a series of quotations taken from the herbals themselves and from reviews written on the subject.

An early reference to "agues" can be found in Culpeper's Complete Herbal, originally produced in 1649 (*5*): "the decoction drank warm, and the herb bruised, with a few corns of bay-salt, and applied to the wrists before the coming of the ague fits, does take them away." The same source contains a reference to headache: "[Feverfew] is very effectual for all pains in the head coming of a cold cause, the herb being bruised and applied to the crown of the head." John Hill, in "The Family Herbal" published in 1772 affirmed: "In the worst headache [feverfew] exceeds whatever else is known." (6). According to Berry (7): "Feverfew has been called the aspirin of the 18th century."

There is an early reference in Bancke's herbal (1525) to the use of feverfew for comforting stomach ache and for soothing insect bites: "This is named Federfoy. His virtue is to comforte a mannus stomake. It is good to asswage the axes cotydyan, ye crampe, and to tempre it that cometh of colde stomackes. Also it is good to lay a soore that is byten with venymous beestes; it will hele it shortly on it be layed thereto". As regards toothache, according to Charles Estienne "Stampt and applied unto teeth or eare of the side that aketh, it wholy asswageth the paine of the teethe" (8). In 1597 Gerard advocated feverfew as being "very good for them that are giddie in the head, or which have the turning called Vertigo, that is, a swimming and turning in the head" (9).

Back to Culpeper for a statement on "female complaints": "Venus commands this herb, and has commended it to succour her sisters to be a general strengthener of their wombs, and to remedy such infirmities as a careless midwife has there caused; if they will be pleased to make use of her herb boiled in white wine, and drink the decoction, it cleanses the womb, expels the afterbirth, and does a woman all the good she can desire of an herb." (5)

More Recent Uses

More recently the main focus of the herb has been migraine and arthritis, for which there have been a large number of anecdotal claims of efficacy made in letters to newspapers, claims that have found their way into popular books on the subject. An early advocate was a certain Mrs. Ann Jenkins who, in 1973, at the age of 68 and having suffered migraine since the age of 16, was persuaded to try feverfew by a friend of her sister, whose father had found it effective in arthritis. It worked, and it also worked for her friends.

The story is recounted in detail in the book by Johnson (10) who also records other, even earlier, success stories in migraine: "A 60-year old woman from Berkshire who had suffered two or three migraine attacks a week since childhood began taking two leaves of feverfew daily in June 1971. Five months later she wrote 'Since then I have experienced only one very minor migraine attack. All through my school life and right until the beginning of July this year, at least one day a week was shear misery.' "

In a book by Hancock (11) there is an enthusiastic claim of effectiveness in arthritis: "The results are fantastic, after 20 years of suffering and so crippled I could barely walk, with advanced arthritis of both knees and feet. I am very grateful for the help I am getting with feverfew."

However, not all correspondents were quite so enthusiastic, and some adverse effects have also been reported. For example, the following two quotations appear in Johnson's book: "over four months the feverfew did not make any difference", and "she only took feverfew for about six weeks and then had a very intense allergic reaction to it - her mouth and lips swelled as though stung. She has not taken any since."

In his book Johnson records his own early work on feverfew which involved obtaining information via questionnaires from 300 people who had used the herb for headache and migraine. Over 70% of all respondents claimed that the herb produced benefit as a consequence of taking it over prolonged periods.

In general, reports of side-effects have been limited. Some people do experience mouth ulceration which can be serious enough for them to discontinue taking the herb. Contact-dermatitis is also quite common which can be a particular problem to those who grow feverfew (*12*).

One of the authors (SH) recently received a letter from the retired physiologist Professor Mary Pickford, who wrote as follows: "When my housekeeper had an attack of migraine I picked 5 leaves and told her to chew them. Within 20-30 minutes her pain had vanished." She concluded that ingredients extracted from the leaves into saliva provide a means of treating migraine as well feverfew acting as a preventative. Such symptomatic relief goes against the widely held view that feverfew is only of value in migraine as a prophylactic.

Of course, none of these claims of efficacy, either ancient or modern, can be taken to provide any proof whatsoever that feverfew has any medicinal properties. But what these claims of efficacy in migraine and arthritis did do was encourage prospective evaluation of the value of feverfew in these conditions, and encourage scientific investigation on possible mechanisms of action of the herb, and on the nature of the chemical constituents that may be responsible for any therapeutic effects.

Part 2. Have any of the Reported Benefits of Feverfew Been Confirmed in Prospective Clinical Trials?

Three prospective clinical trials have been performed involving dried whole feverfew leaf, two to assess the value of feverfew in migraine and one in rheumatoid arthritis. In short, the two dried leaf studies in migraine provided good evidence of benefit from feverfew in this condition. The one in rheumatoid arthritis did not provide good evidence of benefit. More recently a further study has been performed in migraine, this time using an ethanolic leaf extract. In contrast to the results with dried leaf, the ethanolic extract did not appear to provide any benefit.

The first trial of feverfew in migraine was conducted by Johnson et al (*13*). It was a double blind, placebo controlled trial of the herb that involved 17 migraineurs who had already been taking feverfew for at least three months, 8 of whom continued to take feverfew in the form of encapsulated dried leaf (25 mg feverfew leaf per capsule), and 9 who switched to placebo. The patients were instructed to take two capsules every morning with food for six periods of four weeks. In the 8 who continued there was no change in the frequency or severity of migraine. In the 9 who switched to placebo there was a significant increase in frequency and severity. For the 8 patients who continued on feverfew, mean values of monthly frequencies of headache were 1.63 and 1.69 attacks per month. The 9 patients who changed to placebo went from 1.22 attacks per month up to 3.13 attacks per month. Also nausea and vomiting was more common in the placebo than in the feverfew treated group.

The second of the migraine trials was performed by Murphy et al (*14*) at the University of Nottingham. This was also a double blind, placebo controlled trial of feverfew in migraine prevention, and the results were published in the Lancet in 1988. The trial involved 72 migraineurs who at the time of recruitment were not taking feverfew. All the vounteers received placebo for 1 month and then were randomly allocated to receive feverfew or placebo for four months, after which time they swapped over to receive the other treatment for a further four months. The feverfew was again in the form of dried feverfew leaf placed in gelatin capsules (mean weight 82 mg containing 2.19μmol [0.54 mg] parthenolide per capsule); the placebo was dried cabbage leaves in identical capsules. The effects of the treatment were assessed every two months. At the end of this long study full information was available for 59 of the volunteers. Treatment with feverfew was associated with a significant decrease in the frequency and severity of attacks, and in the degree of vomiting. The mean number of attacks in each two month period was reduced from 4.7 to 3.6, which was highly significant. Interestingly, there was no change in the mean duration of individual attacks. This trial was in line with the guidelines of the International Headache Society Committee on Clinical Trials in Migraine which recommends a minimum of 40 patients (*15*).

The third study performed again at the University of Nottingham involved patients with rheumatoid arthritis (*16*). This was a simple double blind, placebo controlled study that involved 40 female patients, 20 of whom took feverfew for 6 weeks and 20 who took placebo for 6 weeks. The feverfew (mean weight 76 mg) and placebo capsules were prepared in the same way as those used in the study by Murphy et al (*14*). Numerous measurements were performed before and after treatment; however the only significant change was a slight increase in grip strength in the feverfew treated group. In view of the number of different measurements performed it was considered possible that this improvement had occurred by chance. Although feverfew did not, then, prove to be of clear benefit in this particular trial, it should be noted that the patients studied were also unresponsive to other, more conventional therapies and that feverfew was being provided in addition to those therapies. It remains possible that feverfew would have proved more effective in people with less severe disease, and the authors noted that there may be possible benefit for osteoarthritis and soft tissue lesions for which self-treatment with feverfew is probably most common.

As part of all the trials, extensive data on possible adverse effects were collected. These included extensive haematological investigations and liver function tests. In short, nothing was found to suggest that feverfew may produce serious side effects at the doses that were used and for the durations of treatments that were investigated. Studies of possible genotoxic effects of taking feverfew have also been performed, with negative results (*17, 18*).

One outcome of these investigations was that feverfew found a place in Martindale's pharmacopeoia (30th Edition) (*19*), as well as the British Herbal Pharmacopeoia (*20*). In both cases reference is made to its use in migraine prophylaxis and its purported value in arthritis. Another outcome is that organisations such as the British Migraine Association have produced information leaflets on feverfew intended to be of benefit to the lay public (*21*).

Very recently the results of a fourth study involving feverfew have been published (*22*). This was another prospective study in migraine but differed from those conducted previously in that it compared an ethanolic extract of feverfew with placebo rather than utilizing a whole dried leaf preparation. The "active" capsules were filled with dried ethanolic extract of feverfew on microcrystalline cellulose and contained 0.5 mg parthenolide. It was a placebo-controlled cross-over study involving 50 patients in which "active" capsules or placebo were each administered for 4 months. The results of this more recent trial were disappointing. Although 44 patients completed the study there was no apparent difference between the two treatment regimes in terms of number of attacks or number of lost working days. Further comments on the results of this trial are provided below.

Part 3. Through what Mechanisms might Feverfew act as a Medicinal Agent and what is the Evidence that Parthenolide might be Involved?

In an attempt to gain some insight into the mechanisms through which feverfew might act in man, various aqueous extracts of feverfew leaf have been prepared and tested *in vitro* in a number of biological systems. The first such experiment was described by Clothier et al (*23*) in a short letter to the Lancet published in 1980. These authors determined the effect of an extract of feverfew on prostaglandin synthesis in a cell-free test system and reported inhibition of prostaglandin synthesis, but apparently not via an effect on the enzyme cyclo-oxygenase. Soon afterwards (in 1981/2) Makheja and Bailey showed that an extract of feverfew inhibited thromboxane synthesis in intact cells as well as phospholipase A_2 in a cell-free system (*24 ,25*). Phospholipase A_2 is an enzyme that liberates arachidonic acid from phospholipid for subsequent conversion into prostaglandins and thromboxanes. The cells used were blood platelets and the authors also demonstrated inhibition by the feverfew extract of platelet aggregation induced by adenosine diphosphate (ADP), collagen or thrombin, but not that induced by arachidonic acid. In another study a feverfew extract was also shown to be an inhibitor of smooth muscle phospholipase A_2 (*26*).

Heptinstall et al 1985 (*27*) confirmed that extracts of feverfew do inhibit prostaglandin and thromboxane synthesis in platelets, at least under some circumstances, and found inhibition of the platelet aggregation induced by adrenaline, collagen, ADP, U46619 (a stable thromboxane A_2 mimetic). Unlike Makheja and Bailey, they also found some inhibition of the aggregation induced by arachidonic acid. Their findings were consistent with an effect on phospholipase A_2 but they concluded that other effects of feverfew were also involved. Heptinstall et al also discovered that extracts of feverfew inhibit the platelet release reaction, the process through which the contents of intracellular storage granules, including 5-hydroxytryptamine (serotonin), are released from platelets. There was also inhibition of release of granule contents (measured as vitamin B12-binding protein) from another cell type, polymorphonuclear leukocytes. The inhibitory effects of feverfew on serotonin release from platelets were considered to be particularly interesting because it had already been speculated that platelet-derived serotonin may be involved in migraine (*28*). It was also considered that the inhibitory effects of feverfew on leukocytes could account for its claimed benefit in rheumatoid arthritis.

The aerial parts of feverfew are rich in sesquiterpene lactones (*29*), and in the feverfew that is grown in the U.K. and that was used in the various clinical trials, the sesquiterpene lactone that predominates is parthenolide. It is located in trichomes on the underside of the leaves as well as in the flowers and seeds and is thought to serve in a defensive capacity (*30*). The amounts present in leaves are normally well in excess of 2g/kg. Other sesquiterpene lactones are only present in mg/kg quantities (*31*). It is now known that, in comparison with the other sesquiterpene lactones present, parthenolide is unusual in that it is remarkably soluble in water as well as in organic solvents (amphiphilic) and large quantities are easily extracted into aqueous buffers (*31, 32*). Parthenolide (I), like many of the other sesquiterpene lactones found in the plant, contains an α–methylenebutyrolactone unit which, in this case, is linked to a 10-membered ring with an associated epoxide function and is known as a germacranolide.

parthenolide (I)

The question arose as to whether parthenolide and/or other sesquiterpene lactones are responsible for any of the biological effects of feverfew that had been observed. The first experiments were aimed at identifying the components of feverfew responsible for its inhibitory effects on adrenaline-induced platelet serotonin release, and involved a rough fractionation of an extract of feverfew to look for such activity in the different fractions obtained (*33*). The results were quite conclusive in that all the active fractions, but none of those that did not inhibit serotonin release, contained compounds with an α–methylenebutyrolactone unit. Particular sesquiterpene lactones that were identified with this activity were parthenolide (I), canin (II), artecanin (III), secotanapartholide A (IV) and 3β-hydroxyparthenolide (V).

The ability of parthenolide to inhibit both platelet aggregation and serotonin release was confirmed in a direct comparison of feverfew extracts with the purified plant constituent itself (*34, 35*). Inhibition by parthenolide was at concentrations within the micromolar range. It was also found that both parthenolide and feverfew extract were particularly good inhibitors of the serotonin release induced when platelets were stimulated with phorbol esters, agents that activate protein kinase C. Later, it was found that parthenolide was the most potent of several sesquiterpene lactones tested as an inhibitor of serotonin release from platelets (*31*), and there was a close correlation between anti-secretory activity and the parthenolide content of plants grown from seeds derived from different geographical locations (*36*), as also between anti-secretory activity and the parthenolide content of several different commercial preparations of feverfew (*37*).

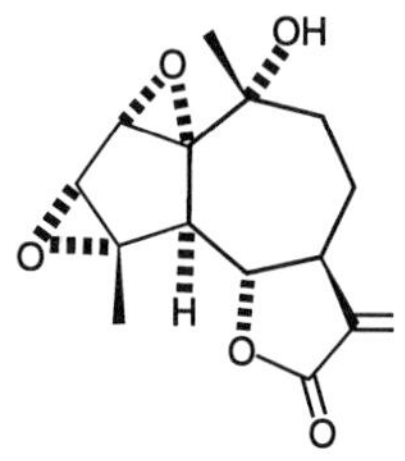

canin (II)

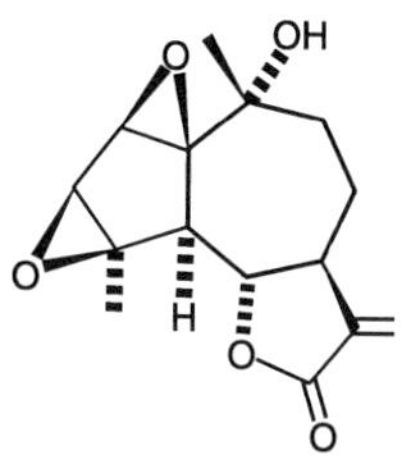

artecanin (III)

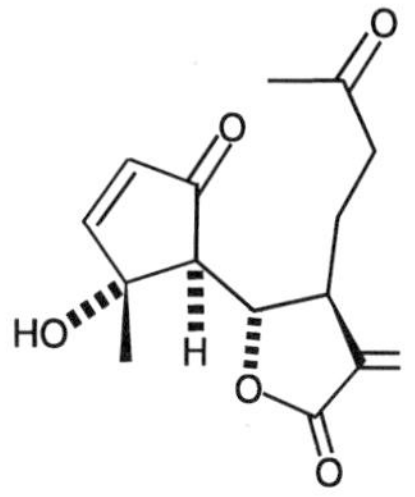

secotanapartholide A (IV)

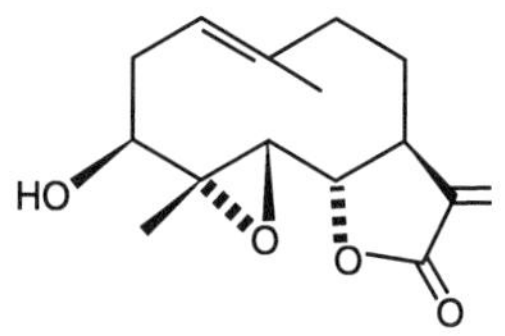

3β-hydroxyparthenolide (V)

Following on from the early demonstrations of effects of feverfew extracts on secretory activity in platelets and polymorphonuclear leukocytes, several other investigations were performed. Feverfew extracts were found to inhibit various white cell activities in addition to secretion. These include the phagocytic activity of polymorphonuclear leukocytes together with the associated chemiluminescence response, and also leukocyte aggregation (*38, 39*). Inhibition by feverfew of the release of histamine from mast cells was also observed (*40*). Monocyte adhesion was shown to be slightly reduced by a feverfew extract (*41*), but more prolonged incubation produced cytotoxic effects on mononuclear cells (*42*). Any or all of these effects could also be related to a beneficial effect of feverfew in arthritis and other inflammatory conditions.

Feverfew was found to inhibit platelet adhesion to collagen substrates (*43*) as well as reduce the extent of platelet aggregation (*24, 25, 27, 35*), and the platelet release reaction (*27, 35*). There was also an apparant protective effect on endothelial cells subjected to perfusion-induced injury (*44*). All this was taken to indicate that the herb may also have anti-thrombotic potential (*45*).

In view of the purported effectiveness of feverfew in migraine, there has been interest in its possible effects on the contractility of smooth muscle in blood vessels. Early studies conducted by Johnson et al indicated that feverfew extract produced a reduction in smooth muscle contractions induced by acetylcholine, serotonin, histamine, prostaglandin E_2 and bradykinin (see the patent application lodged by these authors in 1984) (*46*). More recently both feverfew extracts and parthenolide have been clearly shown to reduce the contractility of blood vessels to a number of agonists (*47-49*). The effect appears to be non-competitive and irreversible and possibly associated with cytotoxicity. Indirect evidence has been obtained for inhibition by parthenolide of serotonin release from rat stomach fundus and associated contractile responses (*50*).

The early observations were indicative of inhibitory effects of feverfew on arachidonate release via phospholipase A_2, without an effect on cyclo-oxygenase (*23-27*), and further evidence for an effect of feverfew on arachidonate release has also been obtained (*51, 52*). However, other work has suggested an additional direct effect on the enzymes cyclo-oxygenase and lipoxygenase (*53, 54*). These additional effects may be a consequence of using comparatively higher concentrations of parthenolide in the test-systems that were used. An extract of feverfew has been shown to inhibit voltage-dependent potassium channels in single smooth muscle cells (*55*). An effect on the cAMP content of aortic segments has also been observed (*44*). Heptinstall and colleagues have recently found that parthenolide modifies calcium mobilization in single cells labelled with Fluo-3 (*unpublished data*). Interestingly, and unexpectedly, its effect was to enhance calcium mobilization in response to platelet activating agents, rather than inhibit it. Since it has recently been demonstrated that one rare form of migraine, namely familial hemiplegic migraine, is caused by mutations in a gene for a calcium channel and there is speculation that a similar etiology may be involved in common types of migraine (*56*), it is tempting to suggest that parthenolide may help restore normal calcium mobilization in cells with defective calcium channels.

Very recently parthenolide was shown to inhibit the expression of inducible cyclo-oxygenase (COX-2) and pro-inflammatory cytokines (TNFα and IL-1) in macrophages. Parthenolide also inhibited protein tyrosine phosphorylation and this was correlated with its effect on expression on COX-2 and the cytokines (*57*). Even more recently parthenolide was shown to inhibit activation of NF-κB, a factor important in the transcription of genes for various inflammatory cytokines, including IL-6 (*58*). Thus inhibition of cytokine production may be another mechanism through which feverfew may produce its medicinal effects.

Parthenolide contains an α–methylenebutyrolactone group and so can react with nucleophiles by the well-known Michael addition process. The compound is particularly reactive towards sulphydryl groups with formation of addition complexes (*59*).

To see whether this type of reaction may be contributing to the biological effects of feverfew, the effects of pre-incubating an extract of the herb with cysteine or 2-mercaptopropionylglycine have been investigated. These are agents which themselves contain free sulphydryl groups and thus act to conserve free sulphydryl groups in the tissue under investigation. It was found that pre-incubation with either agents served to protect platelets from the inhibitory effects of feverfew on both aggregation and secretion (*60, 61*). In another approach direct measurements were made of the number of free sulphydryl groups in the acid-soluble fraction of platelets before and after incubation with pure parthenolide, and the results compared with the anti-secretory effects of parthenolide. The effects on the sulphydryl groups were shown to coincide with the effects on secretory activity. The expected consequences of sulphydryl group blockade on protein polymerisation and arachidonate metabolism were also observed. Thus further evidence was obtained that the effects of feverfew on platelets are via Michael addition between the free sulphydryl groups in platelets and the α-methylenebutyrolactone unit of the parthenolide molecule.

A reduction in the number of sulphydryl groups in leukocytes in parallel with inhibition of leukocyte function by feverfew has also been described (*38*). The effects of feverfew and of parthenolide on smooth muscle contractility are also neutralised by cysteine, again indication of SH-group involvement in its pharmacological effect (*49*). The effect of parthenolide on platelet calcium mobilization was also found to be dependent on the α-methylenebutyrolactone function since it is inhibited by cysteine (*Heptinstall et al, unpublished data*). Cysteine and other sulphydryl-containing agents also abolished the inhibitory efect of parthenolide on NF-κB activation, protein tyrosine phosphorylation, and COX-2 and cytokine expression (*57, 58*).

Overall, then, several biological mechanisms have been identified through which feverfew may produce medicinal benefit, and parthenolide appears to be the main constituent that is responsible for many of the biological effects that have been observed. The mechanism of action may involve parthenolide interacting with sulphydryl groups in cells and tissues.

What is the Evidence that Parthenolide might *not* be Involved?

Contrary to the earlier very positive studies in migraine prophylaxis, the most recent trial of feverfew did *not* confirm any therapeutic effect for the herb. This trial (*22*) differed from those conducted previously (*13, 14*) in that an ethanolic extract of the herb was used rather than whole leaf. However the reported levels of parthenolide in the capsules that were used (0.5 mg) compared very favourably with the levels present (0.54 mg) in capsules used in one of the previous studies (*14*). Thus, if parthenolide is indeed the active substance producing benefit in migraine, the amount present should have been sufficient for benefit to be seen. There appeared to be no problem with the stability of the parthenolide in the extract as levels remained constant throughout the course of the study.

In looking for reasons for the discrepancy, the authors comment that the most recent trial differed from previous trials in that all of the patients had never previously consumed feverfew, and suggest that pre-use of the herb may have confounded previous results. However, since in the trial performed by Murphy et al (*14*) significant benefit was achieved even in the subgroup of 42 patients who had not previously used feverfew as well as in the whole group, this is an unlikely explanation.

Another possibility suggested by the authors is that the extract may have lacked some essential ingredient other than parthenolide, required for feverfew's anti-migraine effect. They suggest that a compound such as chrysanthenyl acetate, present in much reduced amounts in the extract (0.017%) compared with dried whole plant material (0.25%), may be a suitable candidate, based on the reported ability of this essential oil to act as an inhibitor of prostaglandin synthesis (*54*). Other, as yet unidentified components of feverfew leaf, not extracted, or degraded in processing, may yet be proven to contribute to the plant's effectiveness in migraine; such compounds may act separately or concomitantly, perhaps in association with parthenolide and other sesquiterpene lactones.

Perhaps it is pertinent to note here that parthenolide is present in at least 32 plants in addition to feverfew (*62*). No other parthenolide-containing plant has gained the prominence of feverfew as a medicinal agent, although, some species of the Magnoliaceae family, which also contain parthenolide, have been reported as being used as febrifuges in Thailand (*63*). Parthenolide has also been identified as the active constituent of an extract of *Magnolia grandiflora* exhibiting tumour-inhibiting properties (*64*).

There are occasional reports that other non-sesquiterpene lactone components of feverfew may exert biological effects. One group has noted a pro-contractile activity in an extract prepared from feverfew powder that did not contain parthenolide or other butyrolactones (*65*), together with apparent inhibition of both thromboxane and leukotriene synthesis via both sesquiterpene lactone and non-sesquiterpene lactone components of feverfew (*66*). There is also a report of a biologically active lipophilic flavonol present in feverfew that may have anti-inflammatory properties (*67*).

A final point that needs to be made is that, despite a wide range of biological effects having been demonstrated for feverfew and/or parthenolide in *in vitro* test systems, it has proved difficult to detect the same effects *in vivo*. Effects that have been actively looked for in blood taken after administration of feverfew to patients and healthy volunteers are inhibition of platelet aggregation and of serotonin release (*68, 69*).

Part 4. What is the Status of Commercial Preparations of Feverfew?

Clearly, until the component(s) that are responsible for the beneficial effects of feverfew are known, it may not be possible to provide clear guidelines for manufacture of effective commercial preparations of the herb. However, what can be done is try to ensure that commercial preparations match as closely as possible those preparations that proved to be effective in clinical trials. The characteristics of those preparations were as follows:

1) The successful preparations contained pure feverfew leaf obtained from the feverfew chemotype that grows in the U.K.
2) The feverfew was identified as *Tanacetum parthenium* by a qualified botanist.
3) The preparations all contained parthenolide.
4) The mean amounts of pure leaf administered in two trials (*14, 16*) were 50 and 82 mg.
5) In one trial (*14*) the mean amount of parthenolide administered was determined to be 0.54 mg.

The earliest analyses (70) of commercial preparations of purported feverfew involved assessment of the ability of extracts of the preparations to inhibit serotonin release from stimulated blood platelets. The analyses were performed before parthenolide was recognised as being responsible for feverfew's anti-secretory activity. The results indicated that three "herbal" preparations contained lower amounts of material active in the assay system than might have been expected when compared with air-dried leaves taken from a fresh feverfew plant grown locally, and that two "homeopathic" preparations exhibited no anti-secretory activity. Also, there was no detectable activity in "Chinese chrysanthemum crystals" sold as being "like feverfew". Thus, this early study was sufficient to emphasize the variability in some of the commercial preparations available at that time. There is much misinformation transmitted in the popular literature and in the publicity of manufacturers of herbal products (71).

Three further studies of variability in anti-secretory activity and parthenolide in different plant products and commercial preparations have been performed (36, 37, 72). In the most exhaustive investigation (37), 22 commercial preparations were included in the analyses. The parthenolide content of the various products were determined using three separate physicochemical techniques and the values obtained compared with the platelet anti-secretory activity of extracts of the preparations. Similar values were obtained irrespective of the type of measurement that was performed. In most cases nothing was known about the authenticity of the product. Nor was anything known about the parts of the plant used in its manufacture. Of the 22 different capsules or tablets that were analysed, only 3 contained between 0.25 and 0.6 mg per unit, levels that are comparable with the trial capsules used by Murphy et al (14), 11 contained between 0.1 and 0.25 mg per unit, levels below the amounts of parthenolide present in those trial capsules, and in 8 of the preparations parthenolide was undetectable.

What is the Proper Basis for the Regulation and Control of Feverfew Products for Migraine?

Since therapeutic efficacy has only been clinically demonstrated for the parthenolide-containing leaf of feverfew grown in the U.K., the Canadian regulatory authorities stipulate that dried feverfew leaf products contain a minimum level of 0.2% parthenolide. This arbitrary level is about half the average level found in samples of feverfew leaf (0.42%) grown at the Chelsea Physic Garden (73) and used in the trial by Johnson et al (13), and about a third of that present in Nottingham grown feverfew (37). Despite widely held popular perception, the parthenolide criterion was established in an attempt to ensure use of the proper chemotype of *Tanacetum parthenium*. It is neither a reliable indicator of the botanical origin of feverfew products nor their expected activity.

At least three sesquiterpene lactone chemotypes of *Tanacetum parthenium* have been identified (74), which are likely to be a result of the biogenetic variability of sesquiterpene lactone synthesis (75). Mexican grown feverfew has been found to contain no parthenolide, the sesquiterpene lactone content being dominated by the isomeric eudesmanolides santamarin(e) (VI) and reynosin (VII), accompanied by lesser amounts of the guaianolides canin (II) and artecanin (III) (76).

santamarin (VI)   reynosin (VII)

Parthenolide was similarly not detected in Yugoslavian grown feverfew from which only eudesmanolides were reported: magnolialide (VIII), 8β-hydroxyreynosin (IX) and 1β-hydroxyarbusculin (X) accompanied by santamarin and reynosin (77). A trace amount of reynosin was reported (parthenolide:reynosin, 200:1) from Belgian grown feverfew (29).

magnolialide (VIII)   8β-hydroxyreynosin (IX)   1β-hydoxyarbusculin (X)

A recent confirmation that parthenolide is the main sesquiterpene lactone present in U.K. grown feverfew (37) was accompanied with a detailed structural elucidation of other, more minor, components including costunolide (XI, R=H), 3-hydroxycostunolide (XI, R=OH), epoxy-artemorin (XII), canin (II), artecanin (III), tanaparthin-α–peroxide (XIII), secotanapartholide A (IV) and secotanapartholide B (XIV).

There is a serious question concerning the presence of unusually structured compounds such as the trimeric chrysanthemonin, chrysanthemolide and partholide (78), which are not identified elsewhere as feverfew constituents and may be the result of degradation during protracted refluxing and subsequent processing. Likewise probably artefactual are the chlorine-containing sesquiterpene lactones from *Chrysanthemum parthenium* reported by Wagner et al (79) extracted from the same material with chlorinated solvent. The two isomeric chlorohydrins in question were not detected by Hylands, at the University of London, who provided the feverfew leaf and almost certainly resulted from opening of the epoxide ring of parthenolide by hydrogen chloride present in the chloroform used for extraction.

3-hydroxycostunolide (XI)   epoxyartemorin (XII)

tanaparthin-α-peroxide (XIII)   secotanapartholide B (XIV)

Regarding parthenolide stability, it has been found that a roughly 20% reduction in parthenolide level is caused under storage of dried feverfew leaf for a year at room temperature unprotected from light - a more than 50% reduction was observed after 2 years (*80*). Cool, dark storage is currently recommended for feverfew leaf and its products (*37, 68*).

Conclusions

It seems obvious that pending definite identification of the active principle(s) responsible for feverfew's migraine prophylactic activity, close attention should be paid to profiling the secondary metabolite composition of the different varieties of the plant which provide raw material for currently marketed products, particularly respecting essential oil chemotypes. The findings of the most recent trial of feverfew leaf extract containing apparently sufficient levels of parthenolide clearly suggest that the simple assurance of parthenolide content may not guarantee medicinal efficacy. Further clinical trials are needed in which an extract produced under mild conditions and carefully preserved is compared with dried leaf material. Interesting results might be obtained by comparing such an extract with leaves from the traditional U.K. grown feverfew, possibly another parthenolide-rich variety that is morphologically distinct from the U.K. feverfew, and a parthenolide-free variety such as is used in South America (*81*) for treating "female problems".

172

Note added in proof

Since going to press a further study has been published (*82*) describing the efficacy of feverfew leaf in migraine prophylaxis. A double-blind placebo controlled cross-over trial was conducted for a period of 4 months. The study involved 57 patients who had never received feverfew, and the plant was *Tanacetum parthenium* grown in Israel from seeds purchased in the Netherlands. Two groups of patients (Group A , n = 30 and Group B, n = 27) received feverfew (100 mg, equivalent to 0.2 mg parthenolide) for 2 months, Group A then continued to take feverfew for a further month followed by placebo, and Group B received placebo followed by feverfew. Migraine severity was determined using a numerical self-assessment pain scale and linked symptoms were recorded on a numerical analogue scale. Taking the results for all the patients together, at the end of the second month there was a dramatic reduction, compared with starting values, in pain intensity and in the severity of symptoms such as vomiting, nausea, sensitivity to noise and sensitivity to light. At the end of the third month, for Group B (on placebo) the pain and associated symptoms significantly worsened compared with Group A (on feverfew), and at the end of the fourth month Group A (now on placebo) significantly worsened compared with Group B (having returned to feverfew). The authors concluded: "the results of the trial provide convincing evidence that consuming feverfew leaf preparation prophylactically, can ease profoundly the pain intensity and the prevalence of the typical symptoms associated with migraine attacks". Certainly, taken together with the results of the previous trials of pure leaf material in migraine (*13,14*) the evidence that feverfew provides a means of reducing the severity of the condition does appear to be growing.

Literature Cited

1.   Tutin, T.G.; Heywood, V.H.; Burges, N.A.; Moore, D.M.; Valentine, D.H.; Walters, S.M.; Webb, D.A. *Flora Europaea*; Cambridge University Press: Cambridge, 1976; Vol. 4, p 171.

2.   Foster, S. *Feverfew*; Botanical Series No. 310; American Botanical Council: Austin, Texas, 1996.

3.   Nicholson, G. *Dictionary of Gardening*; Vol. 5; L. Upcott Gill: London, 1886-87.

4.   Hylands, D.M. *Migraine Matters* **1984**, *2*, 25-27.

5.   Culpeper, N. *Culpeper's Complete Herbal*; 1649, reproduced by W. Foulsham & Co. Ltd.: London.

6.   Johnson, E.S. Mims Magazine **1983**, *May 15*, 32.

7.   Berry, M.I. *Pharmaceutical J.* **1984**, *232*, 611-614.

8.   Clair, C. *Of herbs and spices*; Abelard-Shuman Ltd.: London, 1961; p 165.

9.   Gerard, J. *The herball, generall historie of plantes*; 2nd Ed.; Adams Islip: London, 1597.

10.   Johnson, S. *Feverfew. A tradition herbal remedy for migraine and arthritis*; Sheldon Press: London, 1984.

11.   Hancock, K. *Feverfew. Your headache may be over*; Keats Publishing: New Canaan, CT, 1986.

12. Mitchell, J.C.; Geissmann, T.A.; Dupuis, G.; Towers, G.H.N. *J. Invest. Dermatol.* **1971**, *56*, 98-101.

13. Johnson, E.S.; Kadam, N.P.; Hylands, D.M.; Hylands, P.J. *Br. Med. J.* **1985**, *291*, 569-573.

14. Murphy, J.J.; Heptinstall, S.; Mitchell, J.R.A. *Lancet* **1988**, *ii*, 189-192.

15. International Headache Society Committee on Clinical Trials in Migraine. Guidelines for controlled trials of drugs in migraine. 1st. Ed. *Cephalalgia* **1991**, *11*, 1-12.

16. Pattrick, M.; Heptinstall, S.; Doherty, M. *Annals Rheum. Dis.* **1989**, *48*, 547-549.

17. Johnson, E.S.; Kadam, N.P.; Anderson, D.; Jenkinson, P.C.; Dewdney, R.S.; Blower, S.D. *Human Toxicol.* **1987**, *6*, 533-534.

18. Anderson, D.; Jenkinson, P.C.; Dewdney, R.S.; Blower, S.D.; Johnson, E.S.; Kadam, N.P. *Human Toxicol.* **1988**, *7*, 145-152.

19. *Martindale's Pharmacopeoia*; Reynolds, J.E.F., 30th Ed.; Royal Pharmaceutical Society: London, 1996.

20. *British Herbal Pharmacopeoia*; 4th Ed.; Scientific Committee of the Herbal Medicine Association: 1996.

21. *Feverfew: for long-term prevention of migraine and other recurrent headaches*; British Migraine Association: Byfleet, U.K.

22. de Weerdt, C.J.; Bootsma, H.P.R.; Hendriks, H. *Phytomed.* **1996**, *3*, 225-230.

23. Collier, H.O.J.; Butt, N.M.; McDonald-Gibson, W.J.; Saeed, S.A. *Lancet* **1980**, *ii*, 922-923.

24. Makheja, A.M.; Bailey, J.M. *Lancet* **1981**, *ii*, 1054.

25. Makheja, A.M.; Bailey, J.M. *Prostaglandins Leukot. Med.* **1982**, *8*, 653-660.

26. Thakkar, J.K.; Sperelaki, N.; Pang, D.; Franson, R.C. *Biochim. Biophys. Acta* **1983**, *750*, 134-140.

27. Heptinstall, S.; White, A.; Williamson, L.; Mitchell, J.R.A. *Lancet* **1985**, *i*, 1071-1074.

28. Hanington, E.; Jones, R.J.; Amess, J.A.L.; Wachowicz, B. *Lancet* **1981**, *ii*, 720-723.

29. Bohlmann, F.; Zdero, C. *Phytochem.* **1982**, *21*, 2543-2549.

30. Blakeman, J.P.; Atkinson, P. *Physiol. Plant Pathol.* **1979**, *15*, 183-192.

31. Hewlett, M.J.; Begley, M.J.; Groenewegen, W.A.; Heptinstall, S.; Knight, D.W.; May, J.; Salan, U.; Toplis, D. *J. Chem. Soc., Perkin Trans.* **1996**, *1*, 1979-1986.

32. Brown, A.M.G.; Lowe, K.C.; Davey, M.R.; Power, J.B.; Knight, D.W.; Heptinstall, S. *Phytochem. Anal.* **1996**, *7*, 86-91.

33. Groenewegen, W.A.; Knight, D.W.; Heptinstall, S. *J. Pharm. Pharmacol.* **1986**, *38*, 709-712.

34. Heptinstall, S.; Groenewegen, W.A.; Knight, D.W.; Spangenberg, P.; Lösche W. In *Current problems in neurology: 4. Advances in Headache Research, Proceedings of the 6th. International Migraine Symposium 1987*; Clifford Rose, F.; John Libbey and Co. Ltd.: London, pp 129-134.

35. Groenewegen, W.A.; Heptinstall, S. *J. Pharm. Pharmacol.* **1990**, *42*, 553-557.

174

36. Marles, R.J.; Kaminski, J.; ,Arnason, J.T.; Pazos Sanou, L.; Heptinstall, S.; Fischer, N.H.; Crompton, C.W.; Kindack, D.G.; Awang, D.V.C. *J. Nat. Prod.* **1992**, *55*, 1044-1056.
37. Heptinstall, S.; Awang, D.V.C.; Dawson, B.A.; Kindack, D.; Knight, D.W.; May, J. *J. Pharm. Pharmacol.* **1992**, *44*, 391-395.
38. Lösche, W.; Michel, E.; Heptinstall, S.; Krause, S.; Groenewegen, W.A.; Pescarmona, G.P.; Thielmann, K. *Planta. Med.* **1988**, *54*, 381-384.
39. Williamson, L.M.; Harvey, D.M.; Sheppard, K.J.; Fletcher, J. *Inflammation* **1988**, *12*, 11-16.
40. Hayes, N.A.; Foreman, J.C. *J. Pharm. Pharmacol.* **1987**, *39*, 466-470.
41. Krause, S.; Arese, P.; Heptinstall, S.; Lösche, W. *Arzneim.-Forsch./Drug Res.* **1990**, *40*, 689-692.
42. O'Neill, L.A.J.; Barrett, M.L.; Lewis, G.P. *Br. J. Clin. Pharmacol.* **1987**, *23*, 81-83.
43. Lösche, W.; Mazurov, A.V.; Heptinstall, S.; Groenewegen, W.A.; Repin, V.S.; Till, U. *Thromb. Res.* **1987**, *48*, 511-518.
44. Voyno Yasenetskaya, T.A.; Lösche, W.; Groenewegen, W.A.; Heptinstall, S.; Repin, V.S.; Till, U. *J. Pharm. Pharmacol.* **1988**, *40*, 501-502.
45. Lösche, W.; Mazurov, A.V.; Voyno-Yasenetskaya, T.A.; Groenewegen, W.A.; Heptinstall, S.; Repin, V.S. *Folia Haematologia* **1988**, *115*, 181-184.
46. Johnson, E.S.; Hylands, P.J.; Hylands, D.M. GB Patent 1982; 82/14572.
47. Barsby, R.W.J.; Salan, U.; Knight, D.W.; Hoult, J.R.S. *Lancet* **1991**, *338*, 1015.
48. Barsby, R.W.J.; Salan, U.; Knight, D.W.; Hoult, J.R.S. *J. Pharm. Pharmacol.* **1992**; *44*, 737-740.
49. Hay, A.J.B.; Hamburger, M.; Hostettmann, K.; Hoult, J.R.S. *Brit. J. Pharmacol.* **1994**, *112*, 9-12.
50. Bejar, E. *J. Ethnopharmacol.* **1996**, *50*, 1-12.
51. Lösche, W.; Groenewegen, W.A.; Krause, S.; Spangenberg, P.; Heptinstall, S. *Biomed. Biochim. Acta* **1988**, *47*, S241-S243.
52. Keery, R.J.; Lumley, P. *Br. J. Pharmacol.* **1986**, *89*, 834P.
53. Capasso, F. *J. Pharm. Pharmacol.* **1986**, *38*, 71-72.
54. Pugh, W.J.; Sambo, K. *J. Pharm. Pharmacol.* **1988**, *40*, 743-745.
55. Barsby, R.W.J.; Knight, D.W.; McFadzean, I. *J. Pharm. Parmacol.* **1993**, *45*, 641-645.
56. Ophoff, R.A.; Terwindt, G.M.; Vergouwe, M.N.; Van Eijk, R.; Oefner, P.J.; Hoffman, S.M.G.; Lamerdin, J.E.; Mohrenweiser, H.W.; Bulman, D,E.; Ferrari, M.; Haan, J.; Lindhout, D.; Van Ommen, G.J.B.; Hofker, M.H.; Ferrari, M.D.; Frants, R.R. *Cell* **1996**, *87*, 543-552.
57. Bork, P.M.; Lienhard-Schmitz, M.L.; Kuhnt, M.; Escher, C.; Heinrich, M.; *FEBS Lett.* **1997**, *402*, 85-90.
58. Hwang, D.; Fischer, N.H.; Jang, B.C.; Tak, H.; Kim, J.K.; Lee, W. *Biochem. Biophys. Res. Comm.* **1996**, *226*, 810-818.
59. Kupchan, S.M.; Fessler, D.C.; Eakin, M.A.; Giacobbe, T.J. *Science* **1970**, *168*, 376-377.
60. Heptinstall, S.; Groenewegen, W,A.; Spangenberg, P.; Lösche, W. *Folia Haematologia* **1988**, *115*, 447-449.

61. Heptinstall, S.; Groenewegen, W.A.; Spangenberg, P.; Lösche, W. *J. Pharm. Pharmacol.* **1987**, *39*, 459-465.
62. Farnsworth, N. *Napralert*; University of Illinois at Chicago: 1991.
63. Ruangrungsi, N; Rivepiboon, A.; Lange, G.L.; Lee, M.; Decicco, C.P.; Picha, P.; Preechanukool, K. *J. Nat. Prod.* **1987**, *50*, 891-896.
64. Wiedhopf. R.M.; Young, M.; Bianchi, E.; Cole, J.R. *J. Pharm. Sci.* **1973**, *62*, 345.
65. Barsby, R.W.J.; Salan, U.; Knight, D.W.; Hoult, J.R.S. *Planta. Med.* **1993**, *59*, 20-25.
66. Sumner, H.; Salan, U.; Knight, D.W.; Hoult, J.R.S. *Biochem. Pharmacol.* **1992**, *43*, 2313-2320.
67. Williams, C,A.; Hoult, J.R.S.; Harborne, J.B.; Grennham, J.; Eagles, J. *Phytochem.* **1995,** *38*, 267-270.
68. Groenewegen,W.A. *PhD Thesis*; University of Nottingham: 1988.
69. Biggs, M.J.; Johnson, E.S.; Persaud, N.P.; Ratcliffe, D.M. *Lancet* **1982**, *ii*, 776.
70. Groenewegen, W.A.; Heptinstall, S. *Lancet* **1986**, *i*, 44-45.
71. Awang, D.V.C. *HerbalGram* **1993**, *29*, 34-36; 66.
72. Awang, D.V.C.; Dawson, B.A.; Kindack, D.G.; Crompton, C.W.; Heptinstall, S. *J. Nat. Prod.* **1991**, *54*, 1516-1521.
73. Jessup, D.M. *PhD Thesis*; University of London: 1982.
74. Awang, D.V.C. *Proceedings of the 57th Annual Congress of the French Canadian Association for the Advancement of Science, 15-19 May, 1989*; University of Quebec at Montreal: pp 1-24.
75. Seaman, F.C.; Mabry, T.J. *Biochem. Systematics Eco.* **1979**, *7*, 105-114.
76. Romo, J.; Romo de Vivar, A; Trevino, R.; Joseph-Nathan, P.; Diaz, E. *Phytochem.* **1970**, *9*, 1615.
77. Stefanovic, M; Ristic, N.; Djermanovic, M.; Mladnovic, S. *J. Serb. Chem. Soc.* **1985**, *50*, 435.
78. Hylands, D.M.; Hylands, P. *J. Phytochem. Soc. Eur. Symp.* **1986**, 17.
79. Wagner, H.; Fessler, B.; Lotter, H.; Wray, V. *Planta Med.* **1988**, *54*, 171-172.
80. Smith, B.R.M.; Burford, M.D. *J.Chromatog.* **1992**, *627*, 255-261.
81. Cáceres, A. Personal communication.
82. Palevitch, D.; Earon, G.; Carasso, R. *Phytother. Res.* **1997**, 11, 508-511.

Chapter 14

Garlic: A Review of Its Medicinal Effects and Indicated Active Compounds

Larry D. Lawson

Murdock Madaus Schwabe, Inc., 10 Mountain Springs Parkway,
Springville, UT 84663

Numerous clinical trials with garlic cloves and standardized garlic powder tablets leave little doubt that modest amounts of garlic have significant cardiovascular effects by reducing serum cholesterol, blood pressure, and platelet aggregation. Epidemiological and animal studies strongly indicate significant anticancer effects, particularly for the intestinal tract. Furthermore, its intestinal and topical antimicrobial activities have been its longest recognized effects. Identification of the compounds essential to the activity of garlic, mostly ascribed to its high content of sulfur compounds, has only been partially resolved. So far, the thiosulfinates, of which allicin is 70-80%, are the only compounds with reasonably proven activity at levels representing normal amounts of garlic consumption. They are clearly responsible for the antimicrobial effects. Several evidences also indicate that they are essential to most of the hypolipidemic, antithrombotic, antioxidant, and hypoglycemic effects of garlic, and for some of its anticancer effects. However, because the thiosulfinates are rapidly metabolized and since their active metabolites have not yet been identified, little is known about their mechanism of action. The compounds responsible for the hypotensive effects and much of the anticancer and immune effects of garlic remain unknown. Until they are known, it is best to consume garlic in whole form, fresh or dried.

Garlic (*Allium sativum* L.) has been used as a medicine for more millennia and by more cultures than perhaps any other plant. Even today, garlic cloves are commonly used as a medication by much of the world, especially in eastern Europe and Asia, while garlic pill supplements are popular in western Europe and growing in popularity in the U.S. In Germany, where most of the clinical research on garlic has taken place and where garlic tablets are the leading non-prescription phytomedicine, 8% of the population regularly consumes garlic supplements, ranking in sales with leading prescription drugs *(1)*. In the U.S., where the total garlic consumption in

1994 was 438 million pounds (about 2.5 grams per person per day) *(2)*, garlic ranks in second place behind Echinacea among the best selling herbal supplements *(3)*.

One of the major reasons for the popular medicinal use of garlic is the large amount of scientific research conducted in the 20th century that has confirmed much of the traditional uses of garlic. As of the end of 1996, 1158 pharmacological studies and 650 chemical studies have been published on garlic and garlic products, making it one of the most researched medicinal plants. The pharmacological studies (1202 as of mid-1997; see Table I) have included 208 human studies and have focused mainly on cardiovascular (344), anticancer (221), antimicrobial (252), and antioxidant (60) effects. In addition, 105 toxicology studies and 42 studies on the metabolism of garlic compounds have been published, for a total of 1990 studies on garlic, not including botanical and agricultural studies. Over 40 nations have contributed to this vast literature, with Germany, India, and the U.S. accounting for 63% of the total.

Table I. Pharmacological Studies on Garlic (1900 - 1996)[a]

Effect	Total studies	Human studies
Cardiovascular	344	104 (2)[b]
Blood lipids	179	62
Blood pressure	78	18
Blood fibrinolysis, coagulation, flow	51	18
Platelet aggregation	76	6
Atherosclerosis (plaques)	23	2
Antimicrobial	252	35
Cancer	221	12 (10)[b]
Antioxidant	60	4
Hypoglycemic	28	3
Immune stimulation	15	3
Antiinflammatory	11	1
Gastrointestinal disturbances	43	27
Respiratory	11	5
Heavy metal antidote	28	1
18 Other effects	145	12
Total pharmacological studies	1158	208

[a] Includes only actual studies: human (in vivo), animal, and in vitro studies. Not included are reviews, commentaries, patents, pharmacopoeias. In addition, 44 studies were published in the first half of 1997, 48% of which were on cancer effects.
[b] Epidemiological studies.

The most extensive review of the scientific and historical literature on garlic was published in 1996 by Koch and Lawson *(4)*, a work which includes 2250 garlic references and details the evidence for the identity of the main active compounds for each pharmacological effect. A summary of the book has also been published *(5)*. Other relatively recent reviews includes those of Block on the chemistry of garlic's sulfur and selenium compounds *(6-8)*, Lawson *(9)* on analysis and cardiovascular effects, Sendl *(10)* on analysis of garlic and wild garlic, Reuter *(11)* and Srivastava

(12) on pharmacological effects, Agarwal *(13)* on cardiovascular effects, and short broad reviews by Bradley *(14)*, Leung & Foster *(15)*, and Weiss *(16)*. Older reviews have also been listed by Lawson *(9)*. There is also a new book for general readership by Fulder *(17)* that is very commendable.

To be able to consistently evaluate the pharmacological effects of garlic and to be able to identify the active compound(s) for each effect, a thorough understanding of its composition, and the chemical changes that occur during the production of the commercial products that are often used in the studies, is essential. The purpose of this review is to critically summarize the evidence for the various pharmacological effects of garlic and to present the evidence for the active compound(s) for each effect at doses representing a typical level of garlic consumption. The main emphasis of the review will be upon whole garlic (fresh or dried) and will only touch modestly on commercial oil or extract products.

Medicinal History

The medicinal use of garlic, which began about 5000 years ago (reviewed in *2, 18, 19*), was first recorded by both the Sumerians of Mesopotamia (region of the Tigris and Euphrates rivers) as well as by the people of ancient India. In India it became a part of the popular Ayruvedic medicine and first became recorded in an Ayruvedic medical textbook in 500 AD from much older texts. Garlic was also of great importance in Egypt, where it has been found preserved in Pharaohs' tombs and where its extensive use by pyramid builders was inscribed on the Great Pyramid of Cheops. The Egyptian *Ebers Codex* of 1550 BC mentions the use of garlic in 22 medical formulas. As recorded in the Bible (Numbers 11:5), the Israelites (Jews) mourned for it when they left Egypt about 1400 BC. The popularity of garlic among the Jews has remained even until today, having survived their scattering among other nations, and has, in fact, influenced other nations to use it. China has long used garlic for medicinal purposes, being first recorded in a Chinese medical text about 500 AD. The Greeks used garlic to strengthen Olympic athletes and to treat battle wounds. The Greek physician Hippocrates (460-370 BC), who is considered the "father of medicine," recommended garlic for infections, pneumonia, cancer, digestive problems, increased urine secretion, and improved menstrual flow. Dioscorides, a Greek who lived in the first century AD, and is called the "father of pharmacy," recommended crushed garlic for snake bites, rabid dog bites, coughs, clearing arteries, infections, and leprosy. Some of the most important Roman medical writers (Pliny the Elder, 23-79 AD; Celsus, 25 BC- 50 AD; Galen, 129-199 AD) made similar recommendations. Galen, who was the personal physician of emperor Marcus Aurelius and whose writings greatly influenced Western and Arabic (Unani Tibb system) medicine for a thousand years, called garlic the "theriac (cure-all) of the peasant."

Around 1150, St. Hildegard of Bingen, a German nun who wrote two medical books, strongly encouraged eating raw garlic to improve health and cure the sick. The London College of Physicians recommended garlic to treat poisons, bites, edema, ulcers, and toothaches, as well as for the great plague of London in 1665. Sydenham (1624-1689), a leading English physician, used garlic to cure smallpox. Cholera epidemics of the 1850s in France and Bulgaria were cured with garlic. In

1858, Louis Pasteur showed that garlic and onions could kill the cause of infectious diseases - germs. Albert Schweizer in the early and mid-1900s used garlic in Africa to cure cholera and typhoid fever. In World War I, garlic was widely used in Europe, especially England, to treat battle wounds and dysentery. In World War II, Russia, where garlic use has long been popular, used it again for soldiers' wounds when they ran out of penicillin, which resulted in the nickname of "Russian penicillin."

General Composition of Garlic Cloves

The known composition of the more abundant compounds of garlic cloves are listed in Table II. Although garlic is commonly eaten for both its flavor and health benefits, its content of vitamins and minerals, while average compared to other vegetables on a per weight basis, are much too low (less than 2% of the daily need) at levels normally consumed (2-4 g, or a typical clove) to account for its known health effects. Some unique features of garlic are its low moisture content (62-68% compared to 80-90% for most fruits and vegetables); its high content of fructans, fructose polymers of 10-60 units *(20, 21)* that constitute about 65% of its dry weight; its high content of common free amino acids, which is similar to its protein content and is strongly dominated by arginine; and its very low content of lipids and other oil-soluble compounds. However, its most unique feature is that of its high content of organosulfur compounds, 99.5% of which contain the sulfur amino acid cysteine, even though cysteine itself is absent. The sulfur content of garlic (about 3 mg/g) is four times greater than that of other high sulfur-containing vegetables and fruits, such as onion, broccoli, cauliflower and apricots *(22)*.

Table II. Composition of Garlic Cloves (mg/g fresh weight)

Water	620-680	Adenosine (0 before crush)	0.1 (8 hr)
Water solubles	310-370	Saponins	0.4-1.1
Carbohydrates	260-300	Vitamins	0.15
Fructans	220-250	Ascorbic acid	0.14*
Fiber	15	Thiamine	0.002*
Protein	15-21*	Riboflavin	0.0008*
Amino acids (free, common)	10-15	Minerals	7
Arginine	5-8	Potassium	4.4*
Organosulfur compounds	11-35	Phosphorus	1.8*
Cysteine sulfoxides	6-19	Calcium	0.24*
γ-Glutamylcysteines	5-16	Magnesium	0.18*
S-alkenyl cysteines	0.01-0.03	Sodium	0.11*
Scordinins	0.03	Iron	0.02*
γ-Glutamylphenylalanine	0.4-1.1	Chromium	0.0005*
Lipids	1-2	Selenium	0.0002*
β-Sitosterol	0.015	Germanium	0.00004*
Phenolic acids	0.04	Sulfur	2.3-3.7
Phytic acid	0.8	Nitrogen	6-13

Data from Lawson *(22)*, which also includes additional less abundant compounds.
* Less than 2% of the U.S. Recommended Dietary Allowance in a 3-4 gram clove.

Garlic is also one of the highest selenium-containing foods on a per gram basis *(23)*. Even though the amount of selenium in a single clove is very small, the selenium content of garlic (and likewise the anticancer effects of garlic) can be increased dramatically (up to 2500-fold), when grown in a selenium-enriched soil *(24, 25)*, because selenium, being in the same elemental family as sulfur, will replace sulfur in garlic's cysteine compounds. The most abundant selenium compound in normal garlic is selenocysteine (cys-SeH); whereas in selenium-enriched garlic it is *Se*-methyl selenocysteine *(8, 26, 27)*.

The Sulfur Compounds of Garlic Cloves

The large majority of the analytical and pharmacological research on garlic has focused on its sulfur compounds. This has been the case not only because of their uniquely high abundance in garlic, but also because they are the only known compounds in garlic that have pharmacological activity at doses representing typical levels of garlic consumption, and probably because of the long-recognized pharmacological activity of sulfur-containing drugs such as penicillin, probucol, thiazide, and captopril.

Table III. The Principal Organosulfur Compounds in Whole and Crushed Garlic

Compound	Whole	Crushed
	(mg/g fresh weight)	
S-(+)-Alkyl-L-cysteine sulfoxides		
Allylcysteine sulfoxide (alliin)	6-14	nd
Methylcysteine sulfoxide (methiin)	0.5-2	nd
trans-1-Propenylcysteine sulfoxide (isoalliin)	0.1-1.2	nd
Cycloalliin	0.5-1.5	0.5-1.5
γ-L-Glutamyl-S-alkyl-L-cysteines		
γ-Glutamyl-S-*trans*-1-propenylcysteine	3-9	3-9
γ-Glutamyl-S-allylcysteine	2-6	2-6
γ-Glutamyl-S-methylcysteine	0.1-0.4	0.1-0.4
Alkyl alkanethiosulfinates		
Allyl 2-propenethiosulfinate (allicin)	nd	2.5-4.5
Allyl methyl thiosulfinates (2 isomers)	nd	0.3-1.5
Allyl *trans*-1-propenyl thiosulfinates (2 isomers)	nd	0.05-1.0
Methyl *trans*-1-propenyl thiosulfinates (2 isomers)	nd	0.02-0.2
Methyl methanethiosulfinate	nd	0.05-0.1

Nearly all (95%) of the sulfur in intact garlic cloves is found in two classes of compounds in similar abundance—the S-alkylcysteine sulfoxides and the γ-glutamyl-S-alkylcysteines (see Table III and Figure 1a) *(9, 22)*. Both types of compounds are substituted at the sulfur atom with either allyl (2-propenyl), isoallyl (*trans*-1-propenyl), or methyl groups. The most abundant sulfur compound in garlic is alliin

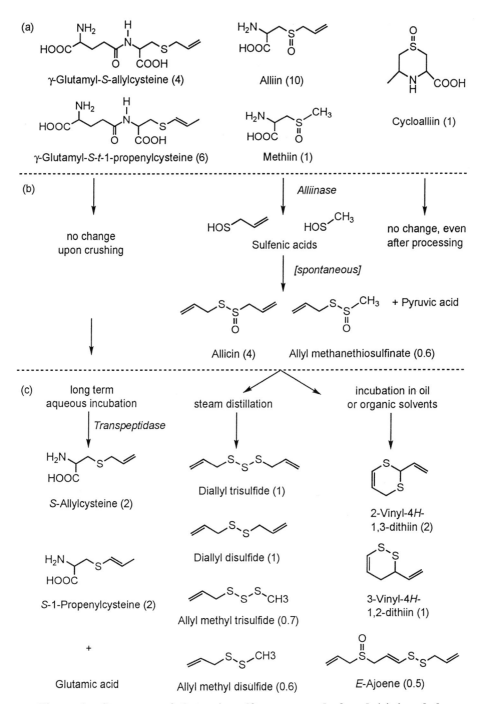

Figure 1. Structures of the main sulfur compounds found (a) in whole garlic cloves and (b) after crushing the cloves and (c) after processing to oils or by aging. Typical quantities ( ) are given as mg/g cloves.

(*S*-allylcysteine sulfoxide), which is typically present at 10 mg/g fresh wt. or 30 mg/g dry wt. (since garlic consistently contains about 65% water, dry weight values are usually three times the fresh weight values) and accounts for about 80% of the total cysteine sulfoxides. It is absent in onions, is less abundant in several other *Allium* species [*A. ursinum* (wild garlic), *A. ampeloprasum* (elephant garlic), *A. tuberosum* (garlic chives/Chinese chives)], and only rarely occurs in small amounts in other plants. The biosynthesis of alliin appears to involve serine and an unknown source of the *S*-allyl group to form the intermediate, *S*-allylcysteine, which is rapidly oxidized to alliin and found only in trace amounts *(22)*. However, the γ-glutamylcysteines act as reserve sources of additional alliin and isoalliin during wintering or sprouting due to γ-glutamyl transpeptidase hydrolysis and subsequent oxidation *(28)*. Although the cysteine sulfoxides are very important parent compounds to many of the pharmacological effects of garlic, the γ-glutamylcysteines have as yet no well-established effect, partly because they have rarely been used in pharmacological studies.

Natural Variation of Garlic's Sulfur Compounds. All plant species vary to some degree in their content, due to soil location and composition, climate, variety differences, harvest date, and post-harvest handling. The extent of these differences, especially for active or suspected active compounds, is important to monitor since they will affect the pharmacological effects. About 85% of the alliin and other cysteine sulfoxides of a garlic plant are found in the bulb, with about 12% in the leaves and 2% in the roots; however, the γ-glutamylcysteines are found only in the bulbs *(22)*. The amount of alliin and γ-glutamylcysteines present in the bulbs increases several-fold in the four weeks prior to harvest time *(29, 30)*. Furthermore, we have found that alliin increases about 25% during the typical curing process (whole plants dried in the shade for at least two weeks) and that extending the normal harvest date ("normal" was judged by appearance factors by two long time commercial growers who independently chose the same date) by 2 weeks—until the plants are almost completely brown—increases the content of these compounds an additional 20% on a dry weight basis (L. Lawson and G. Reynolds, unpublished). Therefore, a careful choice of the harvest date can be very important. In a study with 16 strains of softneck garlic (*A. sativum* var. *sativum*), the type grown in California and most commonly found in grocery stores, and 53 strains of hardneck or topset garlic (*A. sativum* var. *ophioscorodon*) grown on the same 1-acre farm outside Troy, New York by G. Reynolds, there was a 1.8-fold and 2.7-fold variation, respectively, in alliin content and a 1.5-fold and 4.2-fold variation in content of γ-glutamylcysteines *(22)*. Another study of six purchases at grocery stores over four months showed a 2.0-fold variation in alliin *(31)*.

Enzymatic Formation of the Thiosulfinates upon Crushing Garlic. When garlic cloves are cut, crushed, or chewed (or when the powder of dried cloves becomes wet in a non-acid environment), the cysteine sulfoxides, which are odorless and insoluble in organic solvents, are very rapidly converted to a new class of compounds, the thiosulfinates (see Table III and Figure 1b). The thiosulfinates contain two sulfur atoms (being derived from two cysteine sulfoxide molecules with the loss of pyruvate and ammonia), are more soluble in organic solvents than in water, and are somewhat volatile, which property makes them responsible for the odor of freshly chopped garlic. The formation of thiosulfinates takes place when the cysteine sulfoxides, which are located only in the clove mesophyll storage cells, come in contact with the enzyme alliinase or alliin lyase (EC 4.4.1.4), which is located only in the vascular bundle sheath cells *(32)*. Alliinase is a glycoprotein of MW 55,000 and requires pyridoxal phosphate. It is one of the two most abundant proteins in garlic *(33, 34)* and is active at pH 4.5-8, but is immediately and irreversibly inhibited at acidic pH values below 3.5 *(35-37)*. It is also effectively inhibited by 10 mM amino-oxyacetate and by cooking, but is poorly inhibited by alcohols *(22)*. There appears to be at least two forms of alliinase in garlic, one that is alliin-specific and one that is methiin-specific, the latter being considerably easier to inhibit *(28)*.

With the exception of ring-structured cycloalliin (formed by cyclization of isoalliin), all of the cysteine sulfoxides are lysed by alliinase to form the very transitory sulfenic acids, which self-condense to form the thiosulfinates. Due to the very large abundance of alliinase (10 mg/g fresh *(38)*), the rate of formation of the thiosulfinates is extremely rapid, being complete upon crushing in under 10 seconds for alliin and isoalliin and in about 60 seconds for methiin *(22, 36)*. Because the thiosulfinates result from condensation of three types of sulfenic acids, a total of nine thiosulfinates can be formed; however, only eight are found (Table III), since 1-propenyl 1-propenethiosulfinate is too unstable to exist and has been shown to rapidly form zwiebelanes in onions, where 1-propenyl cysteine sulfoxide (isoalliin) is much more abundant *(39)*. Due to the abundance of alliin, the main thiosulfinate formed upon crushing garlic is allicin, which varies in abundance from 60 to 90% of the total thiosulfinates, with 75% being typical. Most of the methiin is converted to two allyl methyl thiosulfinate regioisomers, allyl methanethiosulfinate and methyl 2-propenethiosulfinate, the former being twice as abundant as the latter *(31, 40)*.

Stability of the Thiosulfinates. The thiosulfinates are self-reactive compounds that can be quite unstable, depending upon the environment (solvent), temperature, and concentration. Since they undergo self-reaction by both monomolecular and bimolecular mechanisms *(6, 41)*, both dilution and the presence of solvents that hydrogen bond with the oxygen atom (water and, to a lesser extent, alcohols) greatly improve their stability. For example, in the absence of solvent or in the presence of low-polarity solvents (hexane, diethyl ether), the half-life of allicin at room temperature is 2-16 hours; however, in crushed garlic (or in garlic juice) it is 2.4 days, increasing to 22 days upon 10-fold dilution with water (60 days upon dilution with 1 mM citric acid; 4 days at 37°C in water), and increasing another 20-fold at 4°C *(20)*. Although allicin is too unstable to be present in any commercial product, it is very sufficiently stable for common home uses (e.g., chopped cloves in a salad or sandwich or stored in a refrigerator).

Composition of Garlic Odor. The odor of both chopped garlic and the breath after eating garlic has always been a salient feature of garlic, both positively and negatively. The odor of fresh cut garlic is mainly due to allicin, although diallyl disulfide becomes the dominant compound after 30 minutes, while the odor of cooked garlic is due mostly to diallyl and allyl methyl trisulfides and lesser amounts of their disulfides *(22, 42)*. When fresh garlic is eaten, the initial breath odor is due mainly to allyl mercaptan; however, this represents odor from the throat and disappears in about one hour *(43, 44)*. The odor that comes from the lungs *(45)* rises slowly and lasts over 24 hours. It consists mainly of allyl methyl sulfide (87% of the sulfur compounds at nine hours) and dimethyl sulfide (11%) and is due to metabolism of the thiosulfinates *(43,46, 47)*. Dimethyl selenide (methyl-Se-methyl) is also found in the breath at levels that may add to the odor *(43)*. Although the odor of garlic is carried into human breast milk, babies seem to prefer the taste since they nurse longer than when their mothers don't eat garlic *(48, 49)*.

The Sulfur Compounds of Commercial Garlic Products

In addition to the fresh or cooked cloves, garlic is frequently consumed as a spice (garlic powder and garlic salt), as pickled cloves, and as supplements in pill form (garlic powder tablets and capsules; steam-distilled oil capsules; oil-macerate capsules; aged extract tablets, capsules, and liquid). Although the oil products and the aged extracts have all been thought to be extracts, they really should be called transformation products, since none of the main sulfur compounds of these products are significantly found in whole or crushed cloves. These transformations will now be discussed.

Transformation of the Thiosulfinates upon Commercial Processing. The compounds into which the thiosulfinates are transformed depend upon the medium and temperature. In the presence of water, diallyl trisulfide, diallyl disulfide, and allyl methyl trisulfide are the principal products (Figure 1c). Upon steam-distillation of crushed cloves, a commercial oil is produced, in which as many as 30 different sulfides have been found, comprising diallyl (57%), allyl methyl (37%), and dimethyl (6%) mono- to hexasulfides, along with trace amounts of the hepta- and octasulfides and small amounts of the allyl 1-propenyl and methyl 1-propenyl di-, tri-, and tetrasulfides *(50)*. Upon incubation at room temperature in organic solvents, such as hexane, ether, or triglyceride oils (oil-macerated products), two additional types of compounds are formed: the ring-structured vinyldithiins, which are the main compounds formed (70-80%), and lesser amounts (12-16%) of ajoene (*E*,*Z*-4,5,9-trithiadodeca-1,6,11-triene 9-oxide) *(50-52)*. Some allyl sulfides (diallyl trisulfide, allyl methyl trisulfide and diallyl disulfide) are also formed (4-18%). Except for ajoene, none of the thiosulfinate transformation products retain the oxygen atom. This results in greater volatility (odor) and less solubility in water (e.g., allicin solubility in water is 1%, while for diallyl disulfide it is 0.005% and for diallyl trisulfide it is 0.0006% *(22, 53)*.

Processing Changes During Long Term Incubation (Aging). Among the various types of garlic products, both aged extracts and pickled cloves involve long term incubation. The aged extracts employ chopped cloves that are incubated in dilute (20%) ethanol for 18-20 months prior to drying of the extract *(54, 55)*, while pickled cloves are incubated and permanently packaged in vinegar (5% acetic acid), with the incubation time being dependent on when the cloves are eaten *(22)*. The compositional changes that take place in these products over a two-year period, especially in the aged extracts, are considerable, as shown in Table IV. The only sulfur compound that is not affected by time is cycloalliin.

For chopped cloves aged in dilute ethanol, the main changes are (1) the initial loss of alliin to thiosulfinate formation, (2) the complete loss of thiosulfinates after 3 months (converted to volatile allyl sulfides which evaporate almost completely), and (3) complete hydrolysis of the γ-glutamylcysteines to the theoretical amounts of S-allylcysteine, S-1-propenylcysteine (the main sulfur compounds present after 3 months) and glutamic acid (see Figure 1c). There are also substantial increases in cystine (due to protein hydrolysis) and S-allylmercaptocysteine (probably due to the reaction of allicin with protein-derived cysteine). The content of S-allylcysteine remains constant after 3 months, but S-1-propenylcysteine steadily decreases.

Table IV. Compositional changes during aging of a 20% ethanol extract of chopped garlic or of garlic powder in acetic acid (pickling).[a]

Compound	Incubation time (months)				
	0	1	3	12	24
Chopped cloves in 20% ethanol	(mg/g dry extract)				
Alliin	5.0	3.2	2.8	2.9	2.7
Allicin	8.3	4.1	0.4	0	0
Allyl methyl thiosulfinates	2.1	1.3	0.4	0	0
Allyl sulfides	0.19	0.14	0.12	0.09	0.08
Cycloalliin	3.5	3.6	3.5	4.0	3.6
γ-Glutamyl-S-allylcysteine	12.7	5.8	1.1	0	0
S-Allylcysteine	0.2	5.9	7.2	7.1	7.2
γ-Glutamyl-S-1-propenylcysteine	15.9	3.4	0.5	0	0
S-1-Propenylcysteine	0.5	6.7	8.1	6.5	4.4
S-Allylmercaptocysteine	0.01	0.6	1.2	1.7	1.9
Cystine	0.07	0.5	0.8	0.9	1.2
Total main sulfur compounds	48	34	26	23	21
Glutamic acid	1.1	9.7	14.2	15.8	16.2
Arginine	25	27	28	30	33
Garlic powder in 5% acetic acid					
Alliin	15.3	13.3	12.2	10.4	8.1
Allicin	0	0	0	0	0
γ-Glutamyl-S-allylcysteine	11.9	11.1	10.4	8.2	5.9
γ-Glutamyl-S-1-propenylcysteine	8.1	6.4	3.6	0.7	0.2
S-Allylcysteine	0.3	0.5	0.6	1.2	1.7
S-1-Propenylcysteine	0.1	0.2	0.2	0.2	0.2

[a] Whole cloves were chopped into small pieces (2x2x1 mm) and placed into a 20% ethanol solution (12 mL/g) in a closed container and stored at room temperature, with samples being removed at the indicated times. Very similar results were also found at 3 mL/g and when using water only. Commercial garlic powder was pickled by incubation in 5% acetic acid at 36 mL/g. The zero time values were measured after 24 h (22, 56).

When garlic powder is incubated in 5% acetic acid (simulates whole cloves in vinegar; similar results were found in commercial products), no thiosulfinates are formed because the acidic pH (3.5) inhibits alliinase. The acidic environment also greatly slows down the hydrolysis of the γ-glutamylcysteines, but it adversely affects the S-alkenylcysteines, particularly the S-1-propenyl compounds, since only 2% and 42% of the expected S-1-propenylcysteine and S-allylcysteine, respectively, were found at 24 months.

Composition of Commercial Garlic Products. Table V compares the composition of several brands of the various types of commercial garlic products and supplements with garlic cloves. As can be seen for the range values given, there is a much larger variation in the composition among brands of whole powder tablets (3-fold for standardized products, 35-fold for non-standardized) and oils (50-fold) than among cloves (2- to 3-fold), reflecting differences in manufacturing practices and indicating the need for product labels to declare a specific content/yield value of a compound that represents that product. For the brands that do make specific compound value claims, the values have been found to almost always be fairly accurate.

Garlic Powder Products. Garlic powder is the product most identical with fresh cloves (Table V) since it has only been dehydrated at low oven temperatures (50-60 °C)—at these temperatures there is only a 5-15% loss of allicin-yield *(22)*— and then pulverized; however, the amount of alliin present can vary considerably depending on the care used in slicing and handling of the cloves. The slices can be dried faster if the slices are thinner, but more slicing increases the alliin loss and consequently the odor (allyl sulfides) of the powder. Hence, spice powders have an alliin content (or allicin yield) that is typically about 50% less than that of the powders used to make quality alliin/allicin-standardized tablets *(22)*. Of 29 recently purchased U.S. brands of garlic powder supplements, 11 make specific claims of allicin-yield. The allicin-yield of garlic powder products is fairly stable, with an average 5-year loss of 36% *(22)*.

A very important aspect of the effective quality of garlic powder products is their ability to form allicin after consumption, given the fact that alliinase is rapidly and irreversibly inhibited at the acidic levels typically found in the stomach (pH 1.5-3) *(36, 37)*. When garlic cloves are consumed, allicin formation is not a problem, since it is stable to acid and is formed within 6 seconds of chewing, well before reaching the stomach. Therefore, it is essential that garlic powder products be protected with an acid-resistant coating, either by an enteric-coating or other means. An acid-resistant coating also decreases the breath odor by delaying all allicin formation and release of any allyl sulfides until the tablet is past the stomach. About two-thirds of the U.S. powder supplements claim some type of acid-resistance.

The effectiveness of garlic products in forming allicin in the body is best estimated as the *effective allicin yield*, which is defined in general terms as the amount of allicin formed under simulated gastrointestinal conditions, or in more specific terms as: the amount of allicin formed (or alliin lost) after the products have been agitated at 37 °C in simulated gastric fluid (pH 1.5) for one hour followed by addition of simulated intestinal fluid (pH 7.5) and continued agitation for two hours, according to a modification of U.S.P. protocol #701 for evaluating tablet disintegration (see Table 3.19 in *(22)*). Of 20 U.S. products tested, eight brands were found to have a *percent effective allicin yield* (yield under simulated gastrointestinal conditions compared to yield when products are dissolved in water) of less than 1%, while only six had effective yields of greater than 70% *(22)*. The effective allicin yield should be the standard of quality for all garlic powder supplements.

Table V. Principal Sulfur-Containing Constituents of Commercial Cloves, Pickled Cloves, Garlic Powders, and Garlic Supplements[a]

Product	Constituent and amount (mg/g product)	Alliin-derived compounds[b]
Garlic cloves	alliin (**10**: 6-14), γ-glutamylcysteines[c] (**10**: 5-15) allicin yield (**3.5**: 2.5-5.1)	**4.8** (3.5-8)
Pickled cloves	alliin (**3.4**: 2.0-4.2) γ-glutamylcysteines (**3.3**:3-4)	**0.0**
Powders[d] for spices	alliin (**11.5**: 10-17), γ-glutamylcysteines (**26**:12-35) allicin yield (**4.5**: 3-7)	**7.3** (4-11)
Powders[d] for tablets	alliin (**21**:7-29), γ-glutamylcysteines (**29**: 14-40) allicin yield (**8.5**: 3-11)	**12.5** (4-17)
Powder tablets		
(allicin-standardized)	alliin (**13**: 7-24), γ-glutamylcysteines (**22**: 7-32) allicin yield (**4.2**: 1.3-8.9)	**5.6** (1.4-12)
(non-standardized)	alliin (**7**: 0.4-14), γ-glutamylcysteines (**12**: 2-31) allicin yield (**1.9**: 0.1-5.7)	**2.6** (0.1-8)
Steam-distilled oil capsules	diallyl disulfide (**1.0**: 0.05-2.8) diallyl trisulfide (**0.7**: 0.04 -2.0) allyl methyl trisulfide (**0.6**: 0.03-1.7)	**3.8** (0.2-11)
Oil-macerate capsules	vinyldithiins (**1.1**: 0.1-4.7), ajoene (**0.2**: 0.02-1.1) diallyl trisulfide (**0.11**: 0.02-0.45)	**1.5** (0.4-6.0)
Aged extract tablets/capsules	alliin (**0.3**: 0.2-0.4) γ-glutamylcysteines (**0.34**: 0.2-0.5) γ-glutamyl-*S*-allylcysteine (**0.25**: 0.1-0.4) *S*-allylcysteine (**0.6**: 0.5-0.7) *S*-allylmercaptocysteine (**0.15**: 0.1-0.2)	**0.15** (0.1-0.2)

[a] Values are given as the **mean** and ranges for several (7-15) brands, or lots of a single brand (aged extract), of each type of product. Adapted from Lawson, 1996 *(22)*.
[b] Total alliin-derived compounds includes allicin and other allyl thiosulfinates (after addition of water), allyl sulfides, vinyldithiins, ajoene, and *S*-allylmercaptocysteine.
[c] The γ-glutamylcysteines values for all products are the sum of the *S*-allyl and *S-trans*-1-propenyl compounds.
[d] To compare powders to cloves, divide powders by three since cloves typically contain 65% water.

Garlic Oil Products. Unlike most plant oils, commercial garlic oils are not actually present in garlic cloves. They are the result of converting the water-soluble thiosulfinates of crushed cloves to oil-soluble sulfides by the use of steam (steam-distilled oil, often erroneously called garlic essential oil) or by incubation in a common plant oil (oil-macerate), such as soybean oil. The steam-distilled oils have been prepared for the past 150 years and are very common, while the oil-macerates, long popular in Europe, have only been in the U.S. (only three brands known) for a few years *(22)*. The composition of these oils has been described previously (see *Transformation of the Thiosulfinates upon Processing*), and the ranges among brands are shown in Table V. All commercial garlic oils have been highly diluted with common plant oils to achieve a content of thiosulfinate transformation products (sulfides) that represents the amount of thiosulfinates in crushed cloves from which they were formed, a standard which has been met by most steam-distilled oils, but only by a few oil-macerates (see Table 3.21 of *(22)*). Currently, no steam-distilled oil product claims a specific amount of sulfides, but 8 out of 22 brands claim a specific amount of "pure garlic oil." Of 11 brands of oil-macerates, only one makes a claim for specific compounds (0.9 mg vinyldithiins and 0.15 mg ajoene per 280 mg capsule). The steam-distilled oils products are stable for at least five years, while the oil-macerates are less stable, due to instability of ajoene after about 18 months (see Table 3.24 of *(22)*).

The only garlic-derived compounds in garlic oils are the thiosulfinate-derived (95% alliin-derived) compounds, compounds which have preserved the important dithioallyl (allyl-SS-) moiety of allicin and the other allyl thiosulfinates. Hence, quality garlic oils represent much of the pharmacological activity of crushed garlic and are useful in providing evidence for the pharmacological activity of allicin.

Aged Extracts and Pickled Cloves. Aged extracts represent a successful attempt to remove the odor-forming ability (alliin, alliin-derived compounds, and active alliinase) of garlic. However, most of the sulfur compounds of garlic are lost in the process (Table V). Their method of preparation and the chemical changes that take place during the aging period were previously described (see *Processing Changes During Long Term Incubation*). They are sold in both dry form and as a liquid containing 10% ethanol. Even though major changes occur with most of the sulfur compounds during the aging period (Table IV), the amount of total S-allylcysteine (γ-glutamyl-S-allylcysteine plus S-allylcysteine) should be nearly the same, on a mole basis, as was present originally in the cloves *(22)*; however, based on a typical total S-allylcysteine content in cloves (41 μmole/g dry wt.), the total S-allylcysteine content in commercial aged extracts (7.8 μmole/g dry wt. after correcting for 40% excipients) is only 19% of what would be expected, indicating considerable manufacturing losses or the use of cloves of unusually low content of γ-glutamylcysteines. The commercial aged products claim to be standardized on S-allylcysteine, but no specific or even minimum amount has ever been declared.

Pickling whole cloves also inactivates alliinase and prevents thiosulfinate and odor formation with only a comparatively modest loss (Tables IV and V) of the original sulfur compounds. Hence, commercial pickled cloves have a total S-allylcysteine content of about 22 μmole/g dry wt. Most of the loss is due to diffusion into the pickling solution.

A much shorter and simpler method of preparing a non-odor producing garlic product in which there is little loss of sulfur compounds can be achieved by cooking whole cloves (steam, microwave, or boil) to inactivate alliinase prior to consumption or prior to drying and pulverization. It should be kept in mind, however, that alliinase-inactivated garlic products, while producing little odor, do not possess some of the medicinal benefits of fresh garlic, due to the lack of thiosulfinate yield.

Analysis of Garlic and Standardization of Garlic Products. Allicin is the most important compound upon which to judge the quality of garlic cloves and garlic powders, since thiosulfinates are the only identified compounds from garlic that are reasonably well proven to be essential to the pharmacological activities of garlic at levels representing normal garlic consumption (2-4 g/day). Even though allicin may not be responsible for some of the effects of garlic (see *Summary of the Evidence for the Active Compounds*), it still serves as an important marker compound even for these effects—products that are high in allicin yield are also high in other garlic compounds—until the respective active compounds for these effects are identified.

The allicin yield of garlic cloves (homogenized with 10 mL water/g) and powder (30-60 mL water/g) can only be reliably analyzed by HPLC, preferably by elution with methanol/water (1:1) *(31, 40, 57)*. The allicin standard can be synthesized from pure diallyl disulfide (page 56 in ref. *22*) or isolated from cloves or powder by TLC *(31)* and then quantitated by the cysteine-depletion method *(58)*. Since 1 mg allicin is formed from 2.185 mg of alliin, allicin can also be quantitated by adding a known amount of alliin to a filtered garlic homogenate and measuring the amount of increase in the allicin peak area *(59, 60)*. Total thiosulfinates can be determined by the cysteine-depletion method. The alliin content of cloves or products can also be a valuable measure of quality, as long as the activity of alliinase is verified (by rapid depletion of alliin or rapid increase of allicin). The alliin content can be measured by HPLC analysis *(61)* after microwaving cloves for 30 seconds or homogenizing cloves or powder in 10 mM amino-oxyacetate to inhibit alliinase (alcohols have also been used, but do not give complete inhibition). Alliin standard has recently become commercially available (LKT Labs, St. Paul, MN; Indofine, Somerville, NJ; and Extrasynthese, Lyon Nord, France). γ-Glutamylcysteines and S-alkenylcysteines are also analyzed by reversed-phase HPLC *(28)*. The γ-glutamyl-cysteines are not commercially available, but they can be isolated from garlic cloves (*S*-allyl) or onion seeds (*S*-1-propenyl) *(28)*. They can be quantitated by glutamate release upon treatment with γ-glutamyl transpeptidase, and γ-glutamyl-*S*-allylcysteine can also be assayed by *S*-allylcysteine release with the transpeptidase *(28)*. *S*-Allylcysteine (deoxyalliin) can be synthesized *(28)* or purchased (LKT Labs). *S*-1-Propenyl-cysteine is not commercially available.

The sulfides content of steam-distilled garlic oil can be standardized upon diallyl disulfide plus diallyl trisulfide (about 45% of the total sulfides), both of which have recently become available in pure form (>98%, LKT Labs), as determined by HPLC or GC *(50)*. Other sulfides in the oil can be estimated using the relative extinction coefficients *(50)*. The main compounds of the oil-macerate products (vinyldithiins and ajoene) can be analyzed by HPLC *(50, 52)*, but standards are not yet commercially available and are somewhat difficult to prepare in pure form *(50, 52)*. Crude standardization of the oil-products can be achieved by total sulfur content.

Absorption and Metabolic Fate of the Sulfur Compounds

There are only a few reports on the absorption, metabolism, and excretion of garlic's sulfur compounds, insufficient to determine how these compounds might function in the body. For example, until it is known what metabolic form of allicin actually reaches target cells, in vitro studies to determine its mechanism of action have limited use. Furthermore, no marker compound for any of garlic's sulfur compounds have yet been found in human blood.

Allicin is well-absorbcd, as indicated by a persistent garlic odor on the breath, skin, and amniotic fluid (62) of persons after consumption of fresh cloves. An animal study with ^{35}S-labeled allicin showed at least 79% absorption within 30-60 minutes after oral intake and 65% urinary excretion of the allicin metabolites within 72 hours (63). Furthermore, the antithrombotic activity of fresh garlic (due to allicin, see *Antithrombotic Effects*) is the same in rats whether given orally or intraperitoneally (64). Additionally, as will be presented later, substantial absorption of allicin has been indicated in people, since oral consumption of pure allicin has been shown to significantly increase overall body catabolism of triglycerides.

The metabolic fate of allicin in the body is not well understood; however, our current understanding of what probably takes place up to a certain point is summarized in Figure 2. Neither allicin nor its common transformation products—diallyl sulfides, vinyldithiins, ajoene—can be found in the blood or urine, nor can their odor be detected in the stool, after consuming large amounts of garlic (up to 25 g) (53) or of pure allicin (60 mg) (unpublished), indicating that it is rapidly metabolized to new compounds; however, body metabolites of allicin have not yet been identified. The exception to this statement is the presence of allyl methyl sulfide and much lesser amounts of diallyl disulfide in the breath after garlic consumption (43, 46). It has been assumed that these sulfides originate from the thiosulfinates, rather than from other *S*-allyl compounds in garlic. We have recently confirmed this assumption by showing that consumption of pure allicin also results in substantial amounts of allyl methyl sulfide in the breath, accounting for about 10% of the allicin consumed, and demonstrating that the methyl group comes from the body, probably as the reaction product between allyl mercaptan and S-adenosylmethionine (see Figure 2) (unpublished). We also found that consumption of microwave-cooked garlic did not result in allyl methyl sulfide in the breath, showing that the allyl group can only come from allyl thiosulfinates and not from alliin or γ-glutamyl-S-allylcysteine.

In isolated fresh whole human blood, allicin is very rapidly metabolized by the blood cells to allyl mercaptan (allyl-SH) (Table VI) (9, 65). This also appears to occur in the epithelial cells of the throat, since consumption of crushed garlic results in the immediate (15 sec), but short-lived, presence of allyl-SH in the breath (43, 44). However, allyl-SH is probably not the final effective metabolite of allicin, since this highly odorous compound has not been found in measurable quantities in the blood, stool, or urine after oral consumption of 150 mg of allyl mercaptan—nor was it detected, except in the breath, by the most sensitive of all detectors, the sense of smell (limit of detection: 0.1 μg/mL urine) (Lawson & Wang, unpublished). Good possibilities for the identity of the active metabolite of allicin after allyl mercaptan formation would be that of allylsulfinic acid (2-propenesulfinic acid) or allylsulfonic acid (2-propenesulfonic acid). These compounds are proposed because a similar compound, cysteine, which also has a thiol (SH) group, is metabolized in the body to β-sulfinylpyruvate, which is a sulfinic acid, and to lesser amounts of taurine, which is a sulfonic acid. Furthermore, rat studies with ^{35}S-labeled allicin have shown that most of the ^{35}S-labeled metabolites are highly polar, indicating that they have been oxidized *(63)*, and diallyl sulfide has been shown to be completely metabolized in rats to the oxygenated products, diallyl sulfoxide (DASO) and diallyl sulfone (DASO$_2$) *(66)*.

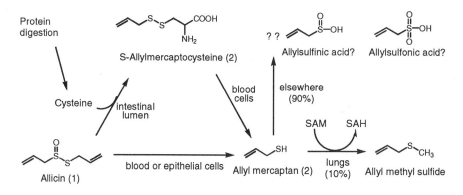

Figure 2. Current understanding of the metabolic fate of allicin in humans.
Abbreviations: SAM, *S*-adenosylmethionine; SAH, *S*-adenosylhomocysteine.

Many of the transformation products of allicin that are present in commercial garlic oils are also metabolized initially to allyl mercaptan (Table VI). This is an important observation because it indicates that pharmacological studies with the garlic oils, which contain almost exclusively allicin-derived compounds, have direct implications on the effects of allicin itself.

Alliin from cooked garlic is rapidly absorbed and excreted and can be partially metabolized to diallyl disulfide in animal liver *(63, 67, 68)*, but it is probably not significantly metabolized to diallyl disulfide in humans, since, as was discussed above, consumption of alliin-abundant cooked garlic does not result in the presence of allyl sulfides in the breath. Since the abundant γ-glutamyl-*S*-alkenylcysteines are structurally similar to glutathione, they are probably absorbed intact and then hydrolyzed in the kidney by γ-glutamyl transpeptidase to *S*-allylcysteine and *S*-1-propenylcysteine. Metabolites of these compounds have been found in human urine after garlic consumption and include N-acetyl-*S*-allylcysteine, N-acetyl-*S*-(2-carboxypropyl)-cysteine, N-acetylcysteine, and hexahydrohippuric acid *(69, 70)*. The amount of N-acetyl-*S*-allylcysteine found in a 24 h urine collection accounts for about 25 mole% (own calculation) of the γ-glutamyl-*S*-allylcysteine consumed *(70)*.

Table VI. Fate of Allicin and its Derived Compounds in Whole Blood.[a]

Compound (0.5 mM)	Half-life in blood (minutes)	Reaction product (moles/mole compound)
Allicin	<1	allyl-SH (1.6)
Ajoene	1	allyl-SH (0.8)
S-Allylmercaptocysteine	3	allyl-SH (0.9)
Diallyl trisulfide	4	allyl-SH (0.8)
Diallyl disulfide	60	allyl-SH (0.8)
Diallyl sulfide	NR	
1,2-Vinyldithiin	15	unknown
1,3-Vinyldithiin	NR	
Allyl-SH	NR	
Alliin	NR	

[a] In fresh whole human blood kept at 37°C (9, 65). NR (<10% decrease in 2 hr).

An excellent study on the metabolic fate of large doses of S-allylcysteine in three different animals (rat, mouse, and dog) showed that it is rapidly absorbed and is more abundant initially in several tissues, especially kidney, than it is in the blood (71). Its half-life in blood plasma (0.8 to 10.3 hours) and distribution among urinary metabolites varied greatly among the types of animals. The study showed that the bioavailability of S-allylcysteine decreased linearly with decreased dose, from 98% at 50 mg/kg body weight to 77% at 25 mg/kg to 64% at 12 mg/kg. Since the highest amount of S-allylcysteine that would be ingested from consuming five grams of garlic or any garlic product would not be more than 0.05 mg/kg, the actual bioavailability of S-allylcysteine may be very small, indicating a possible advantage of S-allylcysteine being present in garlic mainly in the γ-glutamyl form.

Pharmacological Effects and Indicated Active Compounds

The number of publications on the pharmacological effects of garlic cloves and garlic products is truly impressive and includes 1202 in vivo and in vitro studies, of which 208 are human studies, as mentioned in the Introduction and Table I. It is not possible in this short review to discuss each effect in detail; however, a very thorough review of nearly all these studies has recently been published by Reuter et al. (72). After presenting the evidence for each pharmacological effect of garlic, the evidence for the main active compound for each effect is also discussed.

In discussing possible active compounds, it is important to note that the expression "active compound" does not necessarily mean that the compound under consideration is actually causing the effect at the site of action, but rather that the initial presence of the compound is necessary for the effect. This distinction is needed because the metabolic fate of garlic's compounds, as with many herbs, is unknown and since metabolites of the compounds may well be causing the effects. Therefore, the terms "essential to" or "responsible for" or "initially responsible for" are often, but not always, used to more accurately describe an "active compound."

Lipid-Lowering Effect. The most thoroughly studied effect of garlic is that of its ability to favorably influence elevated serum lipids at easily achievable doses. In addition to the 56 animal studies that have been conducted with garlic clove homogenates, allicin, and garlic-derived oils since 1933, 62 human studies have been reported on the lipid-lowering effects of fresh garlic and garlic products, with most of the product studies having been conducted with alliin/allicin-standardized garlic powder tablets (reviewed by Reuter et al. *(72)*). The use of standardized tablets has made it possible to conduct dose-consistent and placebo-controlled studies. Seven open trials with 3-10 g of fresh garlic daily for 3-8 weeks gave an average 16% decrease in cholesterol and 30% decrease in triglycerides (Table VII). Twenty-three trials (13 placebo-controlled and 10 open) from 1986-1994 involving 4600 persons for 6-12 weeks using standardized (3.6-5.4 mg allicin yield), acid-protected garlic powder tablets at a dose (0.6-0.9 g powder) equal to 1.8-2.7 g of fresh garlic, resulted in an average decrease in serum cholesterol of 10.3% (controlled) and 11.3% (open) and in serum triglycerides of 8.5% and 15%, respectively *(72)*.

Table VII. Effects of Garlic and Garlic Products on Elevated Serum Lipids

Study type (# of human studies)	Daily dose	Garlic equivalent	Cholesterol	Triglycerides
Fresh cloves (7)	3-10 g	3-10 g	↓ 16%	↓ 30%
Powders				
•Allicin-standardized tablets[a]				
- placebo-controlled (18)	0.6-0.9 g	2-3 g[b]	↓ 9%	↓ 9%
- non-controlled (10)	0.6-0.9 g	2-3 g[b]	↓ 11%	↓ 11%
•Non-standardized tablets (4)	0.6-1.4 g	2-4 g[b]	↓ 4%	↓ 7%
Oils				
• steam-distilled (2)	10-18 mg	2.5-4.5 g[c]	↓ 10%	↓ 12%
• macerated (4)	2-15 mg	0.5-4 g[c]	↓ 14%	↓ 19%
Aged extracts (2)[d]	---	---	---	---

[a] "Allicin-standardized" means there is a label claim for a specific allicin yield.
[b] Based on the moisture content of whole cloves (65%).
[c] Based on the amount of thiosulfinates needed to produce the oils.
[d] See item 2 under *Evidence that Allicin is Responsible ... .*

Since 1994, five additional placebo-controlled studies have been conducted with alliin/allicin-standardized tablets *(73-78)*, giving an overall cholesterol-lowering effect of 9.1% for the 18 controlled studies, which involved a total of 515 treatment and 557 placebo persons. The effects are greater in individuals with higher initial lipid levels. Of the six controlled and eight open studies that examined lipoproteins, good decreases in LDL-cholesterol were found (8.3% and 14.3%), but only marginal increases in HDL-cholesterol were found (1.3% and 5.3%). A recent diet-monitored placebo-controlled study showed that simultaneous consumption of a standardized garlic powder tablet (0.9 g daily) would reverse the LDL-cholesterol raising effects of fish oil from an increase of 8.5% to a decrease of 9.5% *(75)*. Two of the recent controlled trials in which the persons were advised to simultaneously reduce fat intake to 30% ("step-1-diet") failed to find an effect on serum lipids *(73, 78)*. Several of the studies have been combined into meta-analyses, which concluded that consumption of the equivalent of one-half to one clove per day results in a 9-12% decrease in serum cholesterol *(78-80)*.

An important clinical trial with the same alliin\allicin-standardized garlic powder tablets compared the efficacy of 0.9 g of garlic powder with 0.6 g of bezafibrate, the most frequently prescribed lipid-lowering drug in Germany, on 94 hyperlipidemic patients who were also advised to adhere to the"step-1 diet" *(81)*. The study revealed very substantial and significant (p<0.001) effects in 4-12 weeks and showed that garlic powder was equally as effective as bezafibrate in lowering total cholesterol (initially 284 mg/dL; 10 and 12% decreases in 4 weeks; 25 and 27% decreases in 12 weeks) and LDL-cholesterol (initially 198 mg/dL; 11 and 15% decreases in 4 weeks; 32 and 33% decreases in 12 weeks) and only slightly less effective in raising HDL-cholesterol (initially 35 mg/dL; 11 and 14% increases in 4 weeks; 51 and 58% increases in 12 weeks). The fact that the patients also showed a bias against the effectiveness of garlic compared to the more popular bezafibrate makes the results for garlic even more significant. [Note: The fibrate and statin serum lipid-lowering drugs have been implicated to increase cancer rates at normally prescribed doses *(82)*—however, a major five-year study with simvastatin showed a significant decrease in total death *(83)*—while modest amounts of garlic appear to have substantial anticancer effects (see *Anticancer effects*).]

An interesting new evidence that garlic decreases blood lipids, particularly triglycerides, comes from the finding that consumption of a single dose of raw garlic causes a substantial increase in the level of acetone in the breath, an effect which reaches its peak at about 24 hours *(46)*. Hence, one mechanism by which garlic decreases body lipids is by increasing triglyceride catabolism.

Evidence that Allicin is Responsible for Most of the Serum Lipid-Lowering Effects of Garlic Cloves. There is considerable evidence that allicin is essential for most, but not all, of the lipid-lowering effects of garlic, particularly at low daily doses (2-3 g, representing allicin at 0.05-0.15 mg/kg body weight) of garlic consumption. The evidence has been slow in developing, mainly because the thiosulfinates, due to their somewhat unstable nature and strong taste, have not yet been employed in pure form in human studies, with one exception given below. The evidence is as follows:

1. *Human breath-acetone studies demonstrate increased triglyceride catabolism by garlic and pure allicin, but not by alliinase-inhibited or alliin-free garlic.* The most direct and quickest means to observe the effect of garlic on lipid metabolism is by measuring the 48-hour rise in human breath levels of acetone, which rise is due to increased catabolism of fatty acid-containing lipids, mainly triglycerides *(46)*. In the recent studies shown in Figure 3 (Lawson, L.D. & Wang, Z.J., unpublished), we have demonstrated that consumption of a seven-gram clove of crushed garlic (yielding 38 mg of allicin or 59 mg total thiosulfinates), as well as of the freeze-dried powder from the same cloves, causes substantial increases in breath

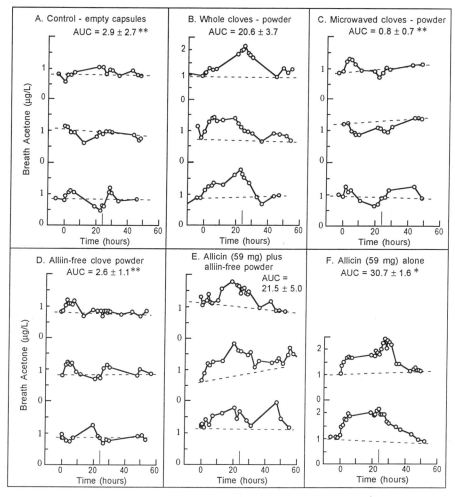

Figure 3. Evidence that allicin (thiosulfinates) plays a major role in the effect of garlic on decreasing body lipids, as indicated by human breath acetone levels. Gelatin capsules (14; 2 every 10 min) were consumed with yogurt over 1 hr starting at time zero and breath acetone measured by GLC. The capsules were either (**A**) empty controls, or they contained (**B**) 2.7 g powder of freeze-dried whole cloves yielding 38 mg of allicin or 59 mg of total thiosulfinates (made from 7 g of fresh cloves; ave. clove wt. 7.6 g), or (**C**) 2.7 g powder of the same cloves that were microwaved (to inactivate alliinase) prior to freeze-drying, or (**D**) 2.7 g of alliin/allicin-free powder prepared by repeated wetting and freeze-drying of powder **B**, or (**E**) 2.7 g of alliin-free powder to which 59 mg of pure (99%) allicin was added (equal to the total thiosulfinates of powder **B**), or (**F**) 59 mg of freshly prepared pure allicin (noticeable effects were also found with 7 mg allicin, not shown). The powders were mixed with 6.5 mL of water before adding to the capsules. The results of 2-3 experiments per treatment are shown. Dashed lines indicate baseline. AUC is the area under the curve to baseline (mean ± s.d.) for 0-48 h. Significant difference from **B** is indicated by * (P<0.02) or ** (P<0.001). Maximal effects typically occur near 24 hr (vertical lines).

acetone, but that removal of alliin or inhibition of alliinase by microwave cooking eliminates all of the activity. When pure synthetic allicin alone was used or when the allicin was added to the alliin-free garlic at the same original thiosulfinate level, all of the activity was restored. Furthermore, when alliinase-protected enteric-coated garlic powder tablets were also consumed at the same thiosulfinate release level (59 mg) the same amount of breath acetone release was found (not shown), demonstrating the effectiveness of such preparations.

2. *Clinical trials with garlic products not yielding allicin.* The only clinical trials with spray-dried garlic powder (0.7 g/day) failed to show an effect, even when initial cholesterol values were very high (280 mg/dL); however, spray-drying causes nearly complete loss of alliin and allicin-yield *(84, 85)*.

In a recent crossover clinical trial with a dry aged extract which also contains very little alliin and no active alliinase, a dose (7.2 g/day for 5 months) 10 times larger and for a considerably longer time than that needed in the trials with allicin-yielding garlic powder tablets produced only a modest 6% decrease in serum cholesterol and no change serum triglycerides *(86)*. A smaller amount (4 mL daily) of a liquid form of this product, which contained 10% ethanol *(9)*, has been shown to significantly lower serum cholesterol (11%) and triglycerides (15%) by 5-6 months; however, the product caused significant increases in serum lipids in the first two months, an effect not seen with cloves or powders, and the placebo did not contain ethanol *(87)*. It is not possible to compare these two clinical trials with aged extracts to each other or to garlic cloves, since neither were defined by content of any characteristic compounds or by the amount of garlic represented *(88)*.

3. *Animal studies with allicin and with alliinase-inhibited or allicin-depleted preparations.* A recent study with cholesterol-fed rabbits showed that feeding pure allicin at 3 mg/kg for 10 weeks resulted in very significant decreases in serum cholesterol and LDL-cholesterol and increase in HDL-cholesterol, although there was little change in triglyceride *(89)*. A study with normolipidemic rats showed significant decreases in serum cholesterol and triglyceride when fed allicin at 100 mg/kg for 8 weeks *(90)*. Furthermore, fresh garlic has been shown to clear the lipid-turbidity of rabbits fed a high fat diet, but boiled garlic (alliinase inactivated) failed to do so *(91)*. Fresh garlic juice, but not old garlic juice (allicin evaporated) decreased atherosclerotic lesions in cats *(92)*. Lastly, removal of the alliin and thiosulfinates from freeze-dried garlic by distillation completely eliminated the substantial cholesterol-lowering effects of the dried garlic in cholesterol-fed rats, although the oil from the distillation was effective *(93)*.

4. *Clinical and animal trials with allicin-derived garlic oils at doses equivalent to the allicin released from garlic cloves and powder.* As mentioned previously, all garlic oils are derived from the thiosulfinates, mainly allicin. These allicin-derived oils have been shown to be as effective as garlic cloves and garlic powder at similar allicin equivalents (Table VII). Clinical trials with fairly low doses of allicin-derived garlic oils (0.05-0.25 mg/kg, approximately equivalent to the allicin from 1-4 g of raw garlic) have demonstrated a 10-14% reduction in serum cholesterol *(94-99)*, a 12-36% reduction in serum triglycerides, and a 35% decrease in heart attacks *(100)*. Furthermore, a single-dose study found that the ether-extracted oil of homogenized garlic lowered serum cholesterol levels equally as well as the homogenate *(101)*.

Animal studies with garlic oils have also shown substantial cholesterol lowering with steam-distilled oil at 0.2 mg/kg *(102, 103)* and ether-extract oils at 0.5 mg/kg *(104, 105)*. Since both allicin and the main compounds in allicin-derived oils, particularly in the steam-distilled oil, are metabolized in blood to the same compound (see *Absorption and Metabolic Fate*), and since these oils contain no other garlic-derived compounds, these studies provide important evidence for the lipid-lowering activity of allicin. Furthermore, the finding of similar effects in both animals and humans at similar low dose levels provides additional evidence that the effects of garlic on serum cholesterol-lowering found in the clinical trials are real.

5. *In vitro and in vivo inhibition of cholesterol biosynthesis.* The work of Gebhardt *(106-109)* with rat liver cells has shown that allicin, diallyl disulfide, and ajoene are potent inhibitors of cholesterol biosynthesis at low concentrations (5-17 μM), while alliin is inactive. Since metabolites of these compounds, rather than the compounds themselves, actually reach the liver, it is uncertain if they cause direct inhibition of cholesterol synthesis in vivo. Gebhardt has also shown that allicin can decrease cellular biosynthesis of cholesterol by acting on HMGCoA-reductase levels without inhibiting the enzyme directly *(106)*. He has also shown that allicin increases the production of adenosine-generated cyclic-AMP and has concluded that allicin may decrease HMGCoA-reductase by modifying signal transduction pathways *(110)*. Such an effect could account for activity at low doses and for better effects on hyperlipidemics. The activity of liver HMGCoA-reductase has also been shown to be significantly decreased in rats fed a diet of 2.5% garlic powder *(111)*.

6. *No other identified compound* in whole or crushed garlic has indications of significant activity at doses present in normal levels of garlic consumption.

Antithrombotic Effects. The ability of crushed garlic cloves to prevent blood platelets from aggregating (thus reducing the risk of blocked arteries) has been well-established in numerous in vitro and in vivo studies (reviewed in Reuter et al.*(72)*). Eight animal and six human studies have shown in vivo that garlic *(112)*, allicin-standardized garlic powder *(113)* and allicin-derived garlic oils *(96, 115, 116)* at modest doses (equivalent to 2-3 g garlic/person) can prevent platelet aggregation and thrombus formation.

Direct in vivo studies with compounds present in whole or crushed cloves have not been conducted; however, both human and animal studies indicate allicin to be responsible for most of the effect in vivo, since garlic cloves, powders and garlic oils all have similar effects at similar levels of allicin yield or content of allicin-derived compounds *(72)*. Furthermore, serum levels of prothrombotic thromboxane in rats was shown to be significantly decreased by a low oral dose of fresh garlic (equal to 3 g for a person), but not by alliinase-inhibited (boiled) garlic *(64)*. In vitro studies with whole blood clearly show the thiosulfinates (allicin being somewhat more active than the other thiosulfinates) to be responsible for nearly all the activity, with little activity for alliin, and none for *S*-allylcysteine or adenosine, although adenosine is active in the absence of red blood cells. Allicin-derived garlic oils are also effective in vitro, with ajoene, diallyl trisulfide, and 1,2-vinyldithiin having similar activity to allicin *(53)*.

Effects on Fibrinolysis, Blood Coagulation and Blood Flow. In close connection with the platelet antiaggregatory effects of garlic are its effects on increasing fibrinolysis (fibrin breakdown), blood clotting time, and peripheral blood flow rate. The effect on fibrinolysis is one of the more dramatic effects of garlic consumption in people, with noticeable effects occurring in a few hours. To date, 11 clinical trials (3 controlled) have been reported on the fibrinolytic effects of garlic powders, cloves, and allicin-derived garlic oils, revealing an average 70% increase in fibrinolysis for a dose equivalent representing 2-6 g of garlic *(72)*. Improved flow of erythrocytes through the capillaries and decreased plasma viscosity by standardized garlic powder tablets (800 mg dose) has been clearly demonstrated by a novel non-invasive sub-cutaneous (under the fingernails) technique *(117, 118)*. The effects on increasing blood clotting time have been sufficient for surgeons to ask patients to not eat garlic prior to surgery *(119, 120)*. The active compounds have not been directly proven, nor has cooked (alliinase-inhibited) garlic been tested at a moderate dose, but the effect appears to be mostly due to allicin since the cloves, powder, and allicin-derived garlic oils are equally effective in clinical trials at similar levels of allicin equivalence. However, other active compounds are present, since, at high doses (35 g daily), both raw garlic and fried garlic (alliinase-inhibited) enhanced fibrinolysis almost equally *(121)*. Cycloalliin (Figure 1 and Table III) may account for the activity at higher doses, as it has been shown in two human studies to substantially increase fibrinolysis at 125-250 mg/person, the amount present in about 100 g of garlic *(122, 123)*.

Blood Pressure. Garlic can exert mild hypotensive effects, as shown in 14 human studies (10 placebo-controlled) with allicin-standardized garlic powder tablets (0.6-0.9 g powder), where systolic and diastolic blood pressures decreased by 6.3% and 6.9%, respectively, in the controlled studies in 1-6 months *(72)*. The only studies that showed no effect were those with persons who had normal or only slightly elevated blood pressures. A meta-analysis of eight of the clinical studies concluded that garlic may be of value for persons with mild hypertension *(124)*. Over 30 animal studies have also shown positive effects *(72)*. One mechanism for the effect appears to be stimulation of nitric oxide synthetase *(125, 126)*. There are no good indications of the active compounds. Allicin is not involved since selective removal of the thiosulfinates with organic solvents *(127-129)* or inactivation of alliinase by cooking *(130)* does not change the hypotensive activity in animals. Proteins are also not involved since only the external dialysate has activity *(131)*. The only indications for possible active compounds come from in vitro studies, one of which showed that the γ-glutamylcysteines can inhibit angiotensin converting enzyme *(132)*, a hormone that increases blood pressure; and a second which showed that the fructans of garlic can inhibit adenosine deaminase in isolated cells *(21)*, an effect that could increase levels of adenosine, a compound that acts at purinergic receptors to relax and dilate blood vessel smooth muscles.

Anticancer Effects. The best and almost only evidence for the anticancer effects of garlic in humans comes from eleven epidemiological studies in six countries (Table VIII). They have demonstrated a consistent correlation between garlic clove consumption and decreased risk of gastrointestinal cancers. Indeed, 79% of the 34 epidemiological studies on allium vegetables (garlic, onion, leek) and cancer (mainly gastrointestinal) have shown a protective effect, a similar percentage as for green vegetables and cruciferous vegetables *(144)*. A study from the Shandong Province of China showed that daily consumption of 20 g (4-6 cloves) of raw garlic was associated with a 92% decrease in stomach cancer compared to consuming less than 1 g daily *(134)*. The effect appears to be due to the killing of nitrate-reducing bacteria (decreased nitrosamines) by the thiosulfinates *(145, 146)*.

An important study for Americans is the Iowa Women's Health Study, a 5-year cohort study which monitored the intake of 127 foods by 41,837 Iowa women, ages 55-69 *(142)*. Garlic was the only food which showed a statistically significant association with decreased colon cancer risk, as demonstrated by a 35% lower risk for consumption of only one or more servings per week. The results of this study have been supported by a second U.S. study with 488 case-controlled Californians, which showed a very significant association between consumption of garlic (3 or more servings/week) and a decrease (37%) in the occurrence of pre-cancerous colorectal polyps *(143)*.

Table VIII. Epidemiological Studies on Garlic Clove Consumption and Cancer

Location/Ref.	Cancer type	Cases/controls	Food	Odds ratio[a]
Argentina *(133)*	colon	110/220	garlic & onions	0.3
China *(134)*	stomach	>100,000 each	raw garlic (20 g/d)	0.08
China *(135)*	stomach	564/1131	garlic (>4 g/d)	0.7
China *(136)*	larynx	201/414	garlic	0.55
Italy *(137)*	stomach	1016/1159	cooked garlic	0.5
Italy *(138)*	stomach	1000/1150	garlic & onions	0.8
Iran *(139)*	esophagus	344/688	pickled garlic	0.6
Switzerland *(140)*	breast	107/318	garlic	0.65
Switzerland *(141)*	endometrium	274/572	garlic	0.6
Iowa women *(142)*	colon	41,837 (cohort)	garlic (≥1 serving/wk)	0.65
California *(143)*	colon polyps	488/488	garlic (≥3 serving/wk)	0.63

[a] Values less than one indicate a protective association; all are statistically significant.

Epidemiological studies in the Netherlands for garlic supplements have failed to find a protective association for colorectal, stomach, or breast cancers *(147-149)*; however, these studies have little relevance since neither brands nor types of supplements were distinguished on the questionnaire and since, as shown in Table V, garlic supplements vary greatly in their composition (much greater than do garlic cloves) depending on both the type of supplement and the brand, even for European brands *(50)*. Furthermore, it is doubtful that supplements are consumed consistently enough over as long of time as are garlic cloves to see an anticancer association.

Although no intervention trials with people have yet been conducted, a large number of animal (carcinogen-induced tumors and DNA aberrations) and in vitro anticancer studies have been conducted with garlic cloves, garlic oils, and aged garlic extract, most of which have shown very positive effects, although large doses have usually been used. Some recent animal studies with lower doses of crushed garlic *(150, 151)*, garlic juice *(152)*, and garlic powder (0.1% of dry diet) *(153)* have shown good activity at the equivalent of 1.7, 28, 10, and 0.6 g garlic/70 kg person, respectively. The study with garlic powder fed to rats *(153)* indicates a saturation effect around 0.2% of the diet since only a small improvement was seen at 0.5% with no further improvement at 1%. Furthermore, a recent ex vivo study with healthy men showed that consumption of 3 g of raw garlic for 8 days resulted in significantly decreased benzopyrene-induced lymphocyte DNA-adducts *(154)*.

Both allicin and other unidentified compounds in garlic appear to be about equally responsible for the anticancer effects. There are a large number of animal studies which have shown positive effects using pure allicin-derived diallyl sulfides *(72)*. Although most of these studies have used very high concentrations (100-200 mg/kg body wt.), some recent studies with diallyl disulfide at 4-10 mg/kg *(155, 156)* and steam-distilled garlic oil at 2 mg/kg *(151)* have significantly prevented induced tumor growth. In fact, diallyl disulfide was as effective as an equal amount of 5-fluorouracil, the chemotherapy drug of choice *(156)*. The *S*-allyl group is essential to the effect since neither dipropyl sulfides nor *O*-allyl compounds have activity *(157-159)*. Furthermore, diallyl trisulfide is more effective than the disulfide *(160, 161)*, which is more effective than the monosulfide *(162, 163)*. The weak trisulfide bond is also a critical to the mechanism of action for the calicheamicins and esperamicins, natural compounds containing a methyl allylic trisulfide that are among the most powerful antitumor agents known *(164)*. The importance of the allyl group being attached to at least two sulfur atoms (dithioallyl compounds) is also indicated by the much greater antiproliferative effects on cancer cells by allicin-derived *S*-allylmercaptocysteine (allyl-SS-cys) than by *S*-allylcysteine (allyl-S-cys) *(165, 166)*.

On the other hand, in most of the countries where the epidemiological studies were conducted, garlic is mainly eaten in cooked form (alliinase inhibited), indicating that garlic can significantly decrease intestinal cancers in the near absence of allicin or sulfides. Furthermore, positive effects with aged extracts (2-4% of dry diet), which do not form allicin, also indicate important activity by compounds unrelated to allicin; however, the identity of these compounds is unknown. *S*-Allylcysteine has been proposed as a possibility, but its activity as a pure compound accounts for only 0.3-3% of the activity of the extract *(162, 165, 167)* and less than 0.1% of the activity of cloves *(168)*. Since garlic contains much more γ-glutamyl-*S*-allylcysteine than *S*-allylcysteine, it is possible that the former may account for significant activity, particularly since it is probably hydrolyzed to the latter in the kidney. Saponins have also been proposed to account for some of the effects of garlic *(169)*, but their abundance in garlic (1 mg/g) is too small, especially when considering the low amount of garlic normally eaten (1-3 g/d). In contrast, a commonly consumed amount of kidney beans (100 g/d) contains 1400 mg of saponins *(170)*.

Antioxidant Effects. Since antioxidants are important natural means of combating both cancer and atherosclerosis, numerous (42) in vitro studies plus 15 animal and 4 human studies have been conducted with garlic, allicin-derived garlic oils, and allicin *(72)*. They have consistently shown good antioxidant activity as demonstrated by decreased lipid peroxidation, increased free radical scavenging (decreased hydroxyl radicals), and increased glutathione. A recent mouse study showed that diallyl trisulfide—which is metabolized similar to allicin—is much more effective than diallyl sulfide toward increasing tissue glutathione and glutathione-*S*-transferase activity *(159)*. Of the four human studies conducted so far (600-900 mg of alliin/allicin-standardized garlic powder per day as tablets), three have shown positive effects on decreasing oxidation of LDL-cholesterol *(171-173)*, while one has not *(73)*. Even though pure allicin can be an oxidizing compound at sufficient concentrations in vitro, the results of 12 animal studies with allicin-yielding garlic powder and allicin-derived garlic oils indicate that allicin is probably responsible for most of the in vivo antioxidant effects of garlic at low doses (0.1-0.5 mg/kg body wt.; equivalent to 1.4 to 8 g garlic for a 70 kg person), while other compounds (alliin, γ-glutamylcysteines, *S*-allylcysteine) appear to have activity only at higher doses (10-50 mg compound/kg body wt.).

Antimicrobial Effects. As one of the most common historical and modern uses of crushed garlic, a large number of antibacterial and antifungal studies (about 250, of which 27 have been human studies), as well as several antiviral (23; 7 human) and antiprotozoal (11) studies, have been reported *(72)*. Crushed garlic or its juice or aqueous extract are effective against a wide range of both Gram positive and Gram negative bacteria, including *Escherichia, Salmonella, Staphylococcus, Streptococcus, Klebsiella, Proteus, Bacillus, Mycobacterium,* and *Clostridium*. Recent reports show that it is also effective against *Helicobacter pylori*, the cause of gastric ulcers *(174, 175)*. Furthermore, it has been shown that crushed garlic is considerably more effective against many of these pathogenic bacteria than against normal flora bacteria, such as *Lactobacillus, Enterococcus,* and *Pediococcus (176, 177)*. A very important aspect of the antibacterial activity garlic is the apparent inability of bacteria to develop resistance to allicin and crushed garlic *(178-181)*. For example, it is almost a thousand times easier for *Staphyloccocus aureus* to develop resistance to penicillin than to allicin *(180)*.

In 1944, Cavallito and Bailey *(182)* were the first to discover that allicin was the main antibacterial agent of garlic. Since then, the thiosulfinate fraction of garlic, which is about 75% allicin, has been shown many times to be responsible for the antimicrobial effects, since selective removal of the thiosulfinates or inhibition of their formation eliminates all of the activity. Allicin-derived compounds present in garlic oils (especially diallyl trisulfide and ajoene) also have antimicrobial activity, although less than for allicin *(72)*. In vitro, allicin is about equally as effective (MIC of 50 μM) as nystatin and ketoconazole against *Candida* fungi *(183)* and more effective than ampicillin or kanamycin against a broad range of pathogenic bacteria *(184)*. Furthermore, Cavallito and Bailey *(182)* showed that while allicin has only 1% of the activity of penicillin against *Staphyloccocus aureus*, it is much more effective than penicillin against Gram negative bacteria. Recently, allicin has been shown in vitro to strongly reduce the pathological effects (50% decrease at 10 μM) of *Entamoeba histolytica*, a major world-wide cause of amoebic dysentery, apparently by inhibition of cysteine proteinases *(185)*. It has long been known that a main mechanism by which allicin exerts its antimicrobial effects is by its rapid reaction with the SH-group of cysteine *(186)* (see Figure 2) in enzymes containing cysteine at their active sites *(187)*.

Most of the antimicrobial studies with crushed garlic or the thiosulfinates have been conducted in vitro. Because most of the antimicrobial uses of garlic are for infections of the skin or intestinal tract, the in vitro studies are probably a fairly good estimate of the in vivo effects. In fact, several in vivo topical and gastrointestinal studies with animals (10) and humans (6) verify the in vitro effects *(72)*. Furthermore, several studies in animals *(188-191)* and people *(192-196)* demonstrate that orally consumed crushed garlic and allicin-related compounds also have systemic antimicrobial effects in the lungs, kidney, blood, brain, and cerebrospinal fluid.

Immune Effects. A small number of human (3), animal (6), and in vitro (5) studies indicate that low doses of garlic products can stimulate the immune system, as shown by increased numbers of lymphocytes, increased phagocytosis, increased natural killer cell activity, and increased antibody titers *(72)*. The human studies have used (a) daily doses of 0.6 g allicin-standardized garlic powder to show increased phago-cytosis and lymphocyte counts *(197)*, (b) 1.8 g of an aged extract to show increased natural killer cell activity *(198)*, and (c) 5-10 g of an undefined garlic extract to show increased natural killer cell activity in AIDS patients *(199)*. Both allicin and other unknown compounds appear to be responsible for the effects, since allicin, allicin-derived sulfides, allicin-depleted garlic extracts, and a protein fraction all have activity *(72)*.

Hypoglycemic Effects. A number of animal studies (25) and a few human studies (3) indicate that garlic can modestly reduce blood glucose and increase insulin levels *(72)*. Although most studies have used large doses for a single day, a recent controlled clinical trial found significant blood glucose reduction (13%) with 0.8 g allicin-standardized garlic powder, the equivalent of 2.5 g fresh garlic, taken for four weeks *(118)*. Allicin appears to be responsible for most of the effect since allicin itself is nearly as effective as the standard drug, tolbutamide *(200)*; whereas, removal of alliin from garlic by spray-drying resulted in no effect *(85)*. Fresh garlic has not yet been compared to cooked (alliinase-inhibited) garlic.

Effects on memory loss and aging. There have been several recent reports with senescence-accelerated mice showing that lifelong consumption of an aged extract of garlic at 2% of the diet (dry wt.) improves learning memory retention, gives a better aging index, and decreases brain shrinkage *(201-204)*. Neither fresh nor cooked garlic have yet been tested in memory or aging studies, although it is expected that similar effects, especially for cooked garlic, may be found at considerably lower doses due to its higher content of garlic compounds (see Table V).

Toxicology

It is common experience that both fresh and cooked garlic are generally well-tolerated when consumed in reasonable amounts and with a meal. Several human studies have shown that long-term consumption of 10-15 grams (2-5 cloves) of fresh garlic daily without noticeable side effects, other than odor on breath and skin *(205)*. Furthermore, in some areas of China, it is common to consume an average of 20 grams of raw garlic daily *(134)*. A recommended safe level for consumption of fresh garlic is 10 grams (2-3 cloves or 3.3 grams dry weight) per day, eaten with meals. However, when consumed without other food or in high amounts, fresh garlic will irritate the throat and stomach, which effect is due mainly to allicin. Other side effects include skin allergies on the left hand from frequent slicing, skin burns when crushed cloves are placed on the skin for six or more hours (shorter times should be used for treating skin infections), and rare cases of asthma attacks from occupational exposure to garlic dust. Since cooking prevents most of the allicin formation, considerably higher amounts of cooked garlic can be safely consumed. Allicin has an LD50 of 60 mg/kg bw (intravenously), corresponding to 1200 grams fresh garlic for a 70 kg person, as tested in mice (Merck Index), which represents a toxicity level similar to that of ferrous sulfate (a common form of iron supplementation), but about ten times less toxicity than for selenium or Vitamin D. Allicin has been reported to oxidize hemoglobin *(206);* however, such an effect has only been found in vitro near a concentration that is 25,000 times the maximum possible level attainable by consumption of a 3 g clove *(207).* As spices, both garlic and allicin-derived garlic oil have GRAS (generally recognized as safe) status in the U.S. (Federal Register 39, 1994, 34213-34215). The toxicology of garlic has been recently reviewed by Koch *(205).*

Summary of the Evidence for the Active Compounds of Garlic

So far, the thiosulfinates are the only class of compounds with reasonably proven activity at levels representing normal amounts of garlic clove consumption, 2-4 g/day. Allicin is the most important of the thiosulfinates since it is 70-80% of the total and since it has somewhat better antithrombotic and antimicrobial activities than the other thiosulfinates *(53, 182).* Several lines of evidence strongly indicate that the thiosulfinates are responsible for most of the antimicrobial and hypolipidemic/ hypocholesterolemic effects of garlic, while they also appear to be responsible for most of the antithrombotic, antioxidant, and hypoglycemic effects as well as for significant portions of the anticancer and immune-stimulating effects, but not for the hypotensive effects (summarized in Table IX).

Table IX. Summary of the Main Compounds Essential to the Pharmacological Effects of Garlic Cloves at Normal Levels of Consumption

Effect	Good evidence	Some evidence
Antimicrobial	allicin/thiosulfinates	
Hypolipidemic	allicin/thiosulfinates	
Hypotensive	unknown (not thiosulfinates)	γ-glutamylcysteines, fructans
Antithrombotic	allicin/thiosulfinates	
Fibrinolysis	allicin/thiosulfinates	cycloalliin
Antioxidant	allicin/thiosulfinates	
Anticancer	unknown and thiosulfinates	γ-glutamylcysteines
Immune effects	unknown and thiosulfinates	

This does not mean that all of the effects of garlic are due solely to the thiosulfinates, but no other compound has yet been identified with significant activity at levels present in whole or crushed garlic. Likewise, in garlic products, only the alliin-derived or allicin-derived compounds have been shown to account for most of the activity of these products. Certainly, other active compounds do exist, particularly for the hypotensive, anticancer, and immune effects of garlic, but they have not yet been identified at levels that would account for the activity. Alliin, the parent compound of the thiosulfinates, has no significant activity for any known effect. *S*-Allylcysteine, adenosine, saponins, and a protein fraction have all been proposed to account for some of garlic's activities, but the levels of these compounds are much too small to account for a significant portion of the activity. Cycloalliin may have important activity for both garlic and onions, but it has only been examined for its fibrinolysis effect. The γ-glutamylcysteines and fructans are much more abundant and are good possibilities for future research, although they have as yet only been examined in a few in vitro studies. Pharmacological studies with fractions of garlic of known composition have rarely been conducted for its hypotensive, anticancer, and immune effects, but they will be essential to discovering the identity of garlic's still-unknown active compounds. This is the challenge for the future!

Acknowledgments. The author would like to thank Prof. Eric Block of the State University of New York at Albany and Prof. Ewald Sprecher of the University of Hamburg, Germany, for critically reviewing this manuscript, Dr. C.J. Cavallito of Chapel Hill, NC—the discoverer of allicin in 1944—for reading it, and Z. Jonathan Wang for nearly a decade of excellent technical assistance and graphic work.

Literature Cited

1. Gruenwald, J. *HerbalGram* **1995**, *34*, 60-65.
2. Hahn, G. In: *Garlic: the science and therapeutic application of Allium sativum L. and related species*; Koch, H. P.; Lawson, L. D. Eds. Williams & Wilkins: Baltimore, 1996; pp 25-36.
3. Brevoort, P. *HerbalGram* **1996**, *36*, 49-57. Johnston, B.A. *HerbalGram* **1997**,*40*, 52.
4. Koch, H. P.; Lawson, L. D. *Garlic: the science and therapeutic application of Allium sativum L. and related species*; Williams & Wilkins: Baltimore, 1996. [Publisher out of stock. Available from the American Botanical Council, Austin TX, tel. 512-331-8868.]
5. Lawson, L. D. *HerbalGram* **1996**, *38*, 65-67.
6. Block, E. *Angew. Chem. Int. Ed. Engl.* **1992**, *31*, 1135-1178.
7. Block, E.; Calvey, E. M. In: *Sulfur Compounds in Foods*; Mussinan, C.J.; Keelan, M.E. Eds.; ACS Symp. Ser. 564; American Chemical Soc.: Washington, DC, 1994; pp 63-69.
8. Block, E.; Cai, X. J.; Uden, P. C.; Zhang, X.; Quimby, B. D.; Sullivan, J. J. *Pure Appl. Chem.* **1996**, *68*, 937-944.
9. Lawson, L. D. In: *Human Medicinal Agents from Plants*; Kinghorn, A. D.; Balandrin, M. F.Eds.; ACS Symp. Ser. 534; Am. Chemical Soc.: Washington, DC, 1993; pp 306-330.
10. Sendl, A. *Phytomedicine* **1995**, *4*, 323-339.
11. Reuter, H. D. *Phytomedicine* **1995**, *2*, 73-91.
12. Srivastava, K. C.; Bordia, A.; Verma, S. K. *S. Afr. J. Sci.* **1995**, *91*, 68-77.
13. Agarwal, K. C. *Medicinal Research Reviews* **1996**, *16*, 111-124.
14. Bradley, P. *Garlic*. In: *British Herbal Compendium*; Peter, B. Ed. 1992; pp 105-108.
15. Leung, A. Y.; Foster, S. *Encyclopedia of common natural ingredients used in food drugs, and cosmetics*; John Willey & Sons, Inc. 1996; pp 260-264.
16. Weiss, R. F. *Herbal Medicine*; Beaconsfield Publishers: Beaconsfield, England, 1988; pp 170-175.
17. Fulder, S. *The Garlic Book*; Avery Publishing: Garden City Park, NY, 1997.

18. Fulder, S.; Blackwood, J. *Garlic - Nature's Original Remedy*; Healing Arts Press: Rochester, Vermont, 1991.
19. Bergner, P. *The Healing Power of Garlic*; Prima Publishing: Rocklin, CA, 1996.
20. Darbyshire, B.; Henry, R. J. *New Phytol.* **1981**, *87*, 249-256.
21. Koch, H. P.; Jäger, W.; Groh, U.; Hovie, J. E.; Plank, G.; Sedlak, U.; Praznik, W. *Phytother. Res.* **1993**, *7*, 387-389.
22. Lawson, L. D. In: *Garlic: the science and therapeutic application of Allium sativum L.*; Koch, H. P.; Lawson, L. D. Eds. Williams & Wilkins: Baltimore, 1996; pp 37-108. [see ref. *4* for availability]
23. Block, E. *Adv. Exper. Med. Biol.* **1996**, *401*, 155-169.
24. Ip, C.; Lisk, D. J.; Stoewsand, G. S. *Nutr. Cancer* **1992**, *17*, 279-286.
25. Ip, C.; Lisk, D. J. *Adv. Exper. Med. Biol.* **1996**, *401*, 179-187.
26. Cai, X. J.; Block, E.; Uden, P. C.; Zhang, X.; Quimby, B. D.; Sullivan, J. J. *J. Agric. Food Chem.* **1995**, *43*, 1754-1757.
27. Ge, H.; Cai, X. J.; Tyson, J. F.; Uden, P. C.; Denoyer, E. R.; Block, E. *Analytical Communications* **1996**, *33*, 279-281.
28. Lawson, L. D.; Wang, Z. Y. J.; Hughes, B. G. *J. Nat. Prod.* **1991**, *54*, 436-444.
29. Ueda, Y.; Kawajiri, H.; Miyamura, N.; Miyajima, R. *Nippon Shokuhin Kogyo Gakkaishi (J. Jpn. Soc. Food Sci. Technol.)* **1991**, *38*, 429-434.
30. Matsuura, H.; Inagaki, M.; Maeshige, K.; Ide, N.; Kajimura, Y.; Itakura, Y. *Plant Med.* **1996**, *62*, 70-71.
31. Lawson, L. D.; Wood, S. G.; Hughes, B. G. *Planta Med.* **1991**, *57*, 263-270.
32. Ellmore, G. S.; Feldberg, R. S. *Am. J. Bot.* **1994**, *81*, 89-94.
33. Wen, G. Y.; Mato, A.; Malik, M. N.; Jenkins, E. C.; Sheikh, A. M.; Kim, K. S. *J. Cell. Biochem.* **1995**, *58*, 481-489.
34. Tchernychev, B.; Rabinkov, A.; Mirelman, D.; Wilchek, M. *Immunol. Lett.* **1995**, *47*, 53-57.
35. Stoll, A.; Seebeck, E. *Adv. Enzymol.* **1951**, *11*, 377-400.
36. Lawson, L. D.; Hughes, B. G. *Planta Med.* **1992**, *58*, 345-350.
37. Blania, G.; Spangenberg, B. *Planta Med.* **1991**, *57*, 371-375.
38. Van Damme, E. J. M.; Smeets, K.; Torrekens, S.; Van Leuven, F.; Peumans, W. J. *Eur. J. Biochem.* **1992**, *209*, 751-757.
39. Bayer, T.; Wagner, H.; Block, E.; Grisoni, S.; Zhao, S.H.; Neszmelyi, A. *J. Am. Chem. Soc.* **1989**, *111*, 3085-3086.
40. Block, E.; Naganathan, S.; Putnam, D.; Zhao, S.H. *J. Agric. Food Chem.* **1992**, *40*, 2418-2430.
41. Block, E.; Ahmad, S.; Catalfamo, J. L.; Jain, M. K.; Apitz-Castro, R. *J. Am. Chem. Soc.* **1986**, *108*, 7045-7055.
42. Ferary, S.; Auger, J. *J. Chromatogr. A* **1996**, *750*, 63-74.
43. Cai, X. J.; Block, E.; Uden, P. C.; Quimby, B. D.; Sullivan, J. J. *J. Agric. Food Chem.* **1995**,*43*, 1751-1753.
44. Laakso, I.; Seppänen-Laakso, T.; Hiltunen, R.; Müller, B.; Jansen, H.; Knobloch, K. *Planta Med.* **1989**, *55*, 257-261.
45. Blankenhorn, M. A.; Richards, C. F. *J. Am. Med. Assoc* **1936**, *107*, 409-410.
46. Taucher, J.; Hansel, A.; Jordan, A.; Lindinger, W. *J.Agric. Food Chem.* **1996**, *44*, 3778-82.
47. Ruiz, R.; Hartman, T. G.; Karmas, K.; Lech, J.; Rosen, R. T. In: *Food Phytochemicals for Cancer Prevention I*; Huang, M.T.; Osawa, T.; Ho, C.T.; Rosen, R.T. Eds.; ACS Symp. Ser. 546; American Chemical Soc.: Washington, DC, 1994; pp 102-119.
48. Mennella, J. A.; Beauchamp, G. K. *Pediatrics* **1991**, *88*, 737-744.
49. Mennella, J. A.; Beauchamp, G. K. *Pediatr. Res.* **1993**, *34*, 805-808.
50. Lawson, L. D.; Wang, Z. J.; Hughes, B. G. *Planta Med.* **1991**, *57*, 363-370.
51. Voigt, M.; Wolf, E. *Dtsch. Apoth. Ztg.* **1986**, *126*, 591-593.

206

52. Iberl, B.; Winkler, G.; Knobloch, K. *Planta Med.* **1990**, *56*, 202-211.
53. Lawson, L. D.; Ransom, D. K.; Hughes, B. G. *Thromb. Res.* **1992**, *65*, 141-156.
54. Hirao, Y.; Sumioka, I.; Nakagami, S.; Yamamoto, M.; Hatono, S.; Yoshida, S.; Fuwa, T.; Nakagawa, S. *Phytother. Res.* **1987**, *1*, 161-164.
55. Nakagawa, S.; Kasuga, S.; Matsuura, H. *Phytother. Res.* **1989**, *3*, 50-53.
56. Lawson, L. D.; Wang, Z. Y. *J. J. Toxicology* **1995**, *14*, 214
57. Pentz, R.; Siegers, C. P. In: *Garlic: the science and therapeutic application of Allium sativum L.*; Koch, H. P.; Lawson, L. D. Eds. Williams & Wilkins: Baltimore, 1996; pp 109-134. [see ref. *4* for availability]
58. Han, J.; Lawson, L.; Han, G.; Han, P. *Anal. Biochem.* **1995**, *225*, 157-160.
59. Müller, B. *Planta Med.* **1990**, *56*, 589-590.
60. Müller, B.; Ruhnke, A. *Dtsch. Apoth. Ztg.* **1993**, *133*, 2177-2187.
61. Ziegler, S. J.; Sticher, O. *Planta Med.* **1989**, *55*, 372-378.
62 Mennella, J. A.; Johnson, A.; Beauchamp, G. K. *Chemical Senses* **1995** *20*, 207-209.
63. Lachmann, G.; Lorenz, D.; Radeck, W.; Steiper, M. *Arzneim. Forsch.* **1994**, *44*, 734-743.
64. Bordia, T.; Mohammed, N.; Thomson, M.; Ali, M. *Prostaglandins Leukotrienes Essent. Fatty Acids* **1996**, *54*, 183-186.
65. Lawson, L. D.; Wang, Z. J. *Planta Med.* **1993**, *59*, A688
66. Brady, J.F.; Ishizaki, H.; Fukuto, J.M.; Lin, M.C.; Fadel, A.; Gapac, J.M.; Yang, C.S. *Chem. Res. Toxicol.* **1991**, *4*, 642-647.
67. Pentz, R.; Guo, Z.; Siegers, C. P. *Med. Welt* **1991**, *42*, 46-47.
68. Egen-Schwind, C.; Eckard, R.; Kemper, F. H. *Planta Med.* **1992**, *58*, 301-305.
69. Jandke, J.; Spiteller, G. *J. Chromatogr.* **1987**, *421*, 1-8.
70. de Rooij, B. M.; Bhoogaard, P. J.; Rijksen, D. A.; Commandeaur, J. N. M.; Vermeulen, N. P. E. *Arch. Toxicol.* **1996**, *70*, 635-639.
71. Nagae, S.; Ushijima, M.; Hatano, S.; Imai, J.; Kasuga, S.; Matsuura, H.; Itakura, Y.; Higashi, Y. *Planta Med.* **1994**, *60*, 214-217.
72. Reuter, H. D.; Koch, H. P.; Lawson, L. D. In: *Garlic: the science and therapeutic application of Allium sativum L. and related species*; Koch, H. P.; Lawson, L. D. Eds. Williams & Wilkins: Baltimore, 1996; pp 135-212. [see ref. *4* for availability]
73. Simons, L. A.; Balasubramaniam, S.; Konigsmark, M. *Atherosclerosis* **1995**, *113*, 219-225.
74. Mansell, P.; Leatherdale, B.; Lloyd, E. J.; Reckless, J. P. D. *Zeitschrift fur Phytotherapie Abstractband* **1995**, *16*, 16.
75. Adler, A. J.; Holub, B. J. *Am. J. Clin. Nutr.* **1997**, *65*, 445-450.
76. Melvin, K. R. *Zeitschrift fur Phytotherapie Abstractband* **1995**, *16*, 17-18.
77. Saradeth, T.; Seidl, S.; Resch, K.-L.; Ernst, E. *Phytomedicine* **1994**, *1*, 183-185.
78. Neil, H. A. W.; Silagy, C. A.; Lancaster, T.; Hodgeman, J.; Vos, K.; Moore, J. W.; Jones, L.; Cahill, J.; Fowler, G. *J. Royal Coll. Physic. London* **1996**, *30*, 329-334.
79. Warshafsky, S.; Kamer, R. S.; Sivak, S. L. *Ann. Intern. Med.* **1993**, *119*, 599-605.
80. Silagy, C.; Neil, A. *J. R. Coll. Physicians London* **1994**, *28*, 39-45.
81. Holzgartner, H.; Schmidt, U.; Kuhn, U. *Arzneim. Forsch./Drug Res.* **1992**, *42*, 1473-1477.
82. Newman, T.B.; Hulley, S.B. *J. Am. Med. Assoc.* **1996**, *275*, 55-69.
83. Scandinavian Simvastatin Survival Study Group *Lancet* **1994**, *344*, 1383-1389.
84. Plengvidhya, C.; Chinayon, S.; Sitprija, S.; Pasatrat, S.; Tankeyoon, M. *J. Med. Assoc. Thailand* **1988**, *71*, 248-252.
85. Sitprija, S.; Plengvidhya, C.; Kangkaya, V.; Bhuvapanich, S.; Tunkayoon, M. *J. Med. Assoc. Thailand* **1987**, *70*, 223-227.
86 Steiner, M.; Khan, A. H.; Holbert, D.; Lin, R. I. S. *Am. J. Clin. Nutr.* **1996**, *64*, 866-870.
87. Lau, B. H. S.; Lam, F.; Wang-Cheng, R. *Nutr. Res.* **1987**, *7*, 139-149.
88. Lawson, L. D. *Quarterly Review of Natural Medicine* **1997** (Spring), 11-12.
89. Eilat, S.; Oestraicher, Y.; Rabinkov, A.; Ohad, D.; Mirelman, D.; Battler, A.; Eldar, M.; Vered, Z. *Coron. Artery Dis.* **1995**, *6*, 985-990.

90. Augusti, K. T.; Mathew, P. T. *Experientia* **1974**, *30*, 468-470.
91. Billau, H. *Dissertation,Univ. Giessen,Germany* **1961**,
92. Silber, W. *Klin. Wochenschr.* **1933**, *12*, 509
93. Kamanna, V. S.; Chandrasekhara, N. *Indian J. Med. Res.* **1984**, *79*, 580-583.
94. Schiewe, F. P.; Hein, T. *Zeitschrift fur Phytotherapie* **1995**, *16*, 343
95. Barrie, S. A.; Wright, J. V.; Pizzorno, J. E. *J. Orthomol. Med.* **1987**, *2*, 15-21.
96. Bordia, A. *Apoth. Magazin.* **1986**, *4*, 128-131.
97. Bordia, A. *Am. J. Clin. Nutr.* **1981**, *34*, 2100-2103.
98. Bordia, A.; Sharma, K. D.; Parmar, Y. K.; Verma, S. K. *Indian Heart J.* **1982**, *34*, 86-88.
99. Bordia, A. *Dtsch. Apoth. Ztg.* **1989**, *129*, 16-17.
100. Bordia, A. *Cardiol. Pract.* **1989**, *7*, 14
101. Bordia, A.; Bansal, H. C.; Arora, S. K.; Singh, S. V. *Atherosclerosis* **1975**, *21*, 15-19.
102. Das, S. N.; Pramanik, A. K.; Mitra, S. K.; Mukherjee, B. N. *Indian Vet. J.* **1982**, *59*, 937-8.
103. Kaul, P. L.; Prasad, M. C. *Indian Vet. J.* **1990**, *67*, 1112-1115.
104. Bordia, A.; Verma, S. K. *Artery* **1980**, *7*, 428-437.
105. Lata, S.; Saxena, K. K.; Bhasin, V.; Saxena, R. S.; Kumar, A.; Srivastava, V. K. *J. Postgrad. Med.* **1991**, *37*, 132-135.
106. Gebhardt, R. *Lipids* **1993**, *28*, 613-619.
107. Gebhardt, R.; Beck, H.; Wagner, K. G. *Biochim. Biophys. Acta* **1994**, *1213*, 57-62.
108. Gebhardt, R. *Phytomedicine* **1995**, *2*, 29-34.
109. Gebhardt, R.; Beck, H. *Lipids* **1996**, *31*, 1269-1276.
110. Gebhardt, R. *Nutrition* **1997**, *13*, 379-380.
111. Merat, A.; Fallahzadeh, M. *Iranian J. Med. Sci.* **1996**, *21*, 141-146.
112. Boullin, D. J. *Lancet* **1981**, *1*, 776-777.
113. Kiesewetter, H.; Jung, F.; Jung, E. M.; Mrowietz, C.; Koscielny, J.; Wenzel, E. *Eur. J. Clin. Pharmacol.* **1993**, *45*, 333-336.
114. Bordia, A. *Atherosclerosis* **1978**, *30*, 355-360.
115. Zhejiang *J. Trad. Chin. Med.* **1986**, *6*, 117-120.
116. Bordia, A.; Verma, S. K.; Srivastava, K. C. *Prost. Leuko. Ess. Fatty Acids* **1996**, *55*, 201-5.
117. Jung, E. M.; Jung, F.; Mrowietz, C.; Kiesewetter, H.; Pindur, G.; Wenzel, E. *Arzneim. Forsch.* **1991**, *41*, 626-630.
118. Kiesewetter, H.; Jung, F.; Pindur, G.; Jung, E. M.; Mrowietz, C.; Wenzel, E. *Int. J. Clin. Pharmacol. Ther. Toxicol.* **1991**, *29*, 151-155.
119. Burnham, B. E. *Plast. Reconstr. Surg.* **1995**, *95*, 213
120. Petry, J. J. *Plastic Recon. Surg.* **1995**, *96*, 483-484.
121. Chutani, S. K.; Bordia, A. *Atherosclerosis* **1981**, *38*, 417-421.
122. Augusti, K.T.; Benaim, M.E.; Dewar, H.A.; Virden, R. *Atherosclerosis* **1975**, *21*, 409-416.
123. Agarwal, R.K.; Dewar, H.A.; Newell, D.J.; Das, B. *Atherosclerosis* **1977**, *27*, 347-351.
124. Silagy, C.; Neil, H. A. W. *J. Hypertension* **1994**, *12*, 463-468.
125. Das, I.; Khan, N. S.; Sooranna, S. R. *Biochem. Soc. Trans.* **1995**, *23*, S136
126. Das, I.; Khan, N. S.; Sooranna, S. R. *Curr. Med. Res. Opinion* **1995**, *13*, 257-263.
127. Torrescasana, E. U. d. *Rev. Esp. Fisiol. [Barcelona]* **1946**, *2*, 6-31.
128. Sial, A. Y.; Ahmad, S. I. *J. Pak. Med. Assoc.* **1982**, *32*, 237-239.
129. Rashid, A.; Khan, H. H. *J. Pak. Med. Assoc.* **1985**, *35*, 357-362.
130. Rao, H. T. R.; Humar, U.; Jayrajan, P.; Devi, G.; Shakunthala, V. T.; Srinath, U.; Rai, A. Y.; Basavaraju, M.; Prasad, V. S. *Indian J. Physiol. Pharmacol.* **1981**, *25*, 303
131. Martin, N. *J. Ethnopharmacol.* **1992**, *37*, 145-149.
132. Elbl, G. Ph.D. Dissertation, University of Munich, Germany, 1991.
133. Iscovich, J. M.; L'Abbe, K. A.; Castelleto, R.; Calzona, A.; Bernedo, A.; Chopita, N. A.; Jmelnitzsky, A. C.; Kaldor, J. *Int. J. Cancer* **1992**, *51*, 851-857.
134. Mei, X.; Wang, M. L.; Xu, H. X.; Pan, X. Y.; Gao, C. Y.; Han, N.; Fu, M. Y. *Acta Nutr. Sin.* **1982**, *4*, 53-56.

135. You, W. C.; Blot, W. J.; Chang, Y. S.; Ershow, A.; Yang, Z. T.; An, Q.; Henderson, B. E.; Fraumeni, J. F.,Jr.; Wang, T. G. *J. Natl. Cancer Inst.* **1989**, *81*, 162-164.
136. Zheng, W.; Blot, W. J.; Shu, X. O.; Gao, Y. T.; Ji, B. T.; Ziegler, R. G.; Fraumeni, J. F.,Jr. *Am. J. Epidemiol.* **1992**, *136*, 178-191.
137. Buiatti, E.; Palli, D.; Decarli, A.; Amadori, D.; Avellini, C.; Bianchi, S.; Biserni, R.; Cipriani, F.; Cocco, P.; Giacosa, A.; Marubini, E.; Puntoni, R.; Vindigni, C.; Fraumeni, J.; Jr.; Blot, W. *Int. J. Cancer* **1989**, *44*, 611-616.
138. Cipriani, F.; Buiatti, E.; Palli, D. *Ital. J. Gastroenterol.* **1991**, *23*, 429-435.
139. Cook-Mozaffari, P. J.; Azordegan, F.; Day, N. E.; Ressicaud, A.; Sabai, C.; Aramesh, B. *Br. J. Cancer* **1979**, *39*, 293-309.
140. Levi, F.; LaVecchia, C.; Gulie, C.; Negri, E. *Nutr. Cancer* **1993**, *19*, 327-335.
141. Levi, F.; Franceschi, S.; Negri, E.; LaVecchia, C. *Cancer* **1993**, *71*, 3575-3581.
142. Steinmetz, K. A.; Kushi, L. H.; Bostick, R. M.; Folsom, A. R.; Potter, J. D. *Am. J. Epidemiol.* **1994**, *139*, 1-15.
143. Witte, J. S.; Longnecker, M. P.; Bird, C. L.; Lee, E. R.; Frankl, H. D.; Haile, R. W. *Am. J. Epidemiol.* **1996**, *144*, 1015-1025.
144. Steinmetz, K. A.; Potter, J. D. *J. Am. Dietet. Assoc.* **1996**, *96*, 1027-1039.
145. Mei, X.; Wang, M.; Li, T.; Gao, C.; Han, N.; Fu, M.; Lin, B.; Nie, H. *Yingyang Xuebao (Acta Nutr. Sinica)* **1985**, *7*, 173-177.
146. Mei, X.; Lin, X.; Liu, J.; Song, P.; Hu, J.; Liang, X. *Acta Nutr. Sin.* **1989**, *11*, 141-145.
147. Dorant, E.; Van den Brandt, P. A.; Goldbohm, R. A. *Carcinogenesis* **1996**, *17*, 477-484.
148. Dorant, E.; Van den Brandt, P. A.; Goldbohm, R. A. *Gastroenterology* **1996**, *110*, 12-20.
149. Dorant, E.; Van den Brandt, P. A.; Goldbohm, R. A. *Breast Cancer Res. Treatment* **1995**, *33*, 163-170.
150. Das, T.; Choudhury, A. R.; Sharma, A.; Talukder, G. *Food Chem. Toxicol.* **1996**, *34*, 43-47.
151. El Mofty, M. M.; Sakr, S. A.; Essawy, A.; Gawad, H. S. A. *Nutr. Cancer* **1994**, *21*, 95-100.
152. Cheng, J.-Y.; Meng, C.-L.; Tzeng, C.-C.; Lin, J.-C. *World J. Surg.* **1995**, *19*, 621-626.
153. Polasa, K.; Krishnaswamy, K. *Cancer Lett.* **1997**, *114*, 185-186.
154. Hageman, G.; Krul, C.; van Herwijnen, M.; Schilderman, P.; Kleinjans, J. *Cancer Lett.* **1997**, *114*, 161-162.
155. Reddy, B. S.; Rao, C. V.; Rivenson, A.; Kelloff, G. *Cancer Res.* **1993**, *53*, 3493-3498.
156. Sundaram, S. G.; Milner, J. A. *J. Nutr.* **1996**, *126*, 1355-1361.
157. Sundaram, S. G.; Milner, J. A. *Biochim. Biophys. Acta* **1996**, *1315*, 15-20.
158. Hu, X.; Benson, P. J.; Srivastav, S. K.; Mack, L. M.; Xia, H.; Gupta, V. *Arch. Biochem. Biophys.* **1996**, *336*, 199-214.
159. Singh, S.V.; Mack, L.M.; Xia, H.; Srivastava, S.K.; Hu, X.; Benson, P.J.; Murthy, M.; McFadden, E.; Gupta, V.; Zaren, H.A. *Clin. Chem. Enzym. Comms.* **1997**, *7*, 287-297.
160. Sundaram, S. G.; Milner, J. A. *Cancer Lett.* **1993**, *74*, 85-90.
161. Sakamoto, K.; Lawson, L. D.; Milner, J. A. *FASEB J.* **1996**, *10*, A497.
162. Dion, M.E.; Agler, M., Milner, J.A. *Nutr. Cancer* **1997**, *28*, 1-6.
163. Haber-Mignard, D.; Suschetet, M.; Berges, R.; Astorg, P.; Siess, M.-H. *Nutrition and Cancer* **1996**, *25*, 61-70.
164. Smith, A.L.; Nicolaou, K.C. *J. Med. Chem.* **1996**, *39*, 2103-2117.
165. Li, G.; Qiao, C. H.; Lin, R. I.; Pinto, J.; Tiwari, R. K. *Oncology Reports* **1995**, *2*, 787-791.
166. Sigounas, G.; Hooker, J.; Anagnostou, A.; Steiner, M. *Nutr. Cancer* **1997**, *27*, 186-191.
167. Amagase, H.; Milner, J. A. *Carcinogenesis* **1993**, *14*, 1627-1631.
168. Hageman, G.J.; van Herwijnen, M.H.M.; Schilderman, P.A.E.L.; Rhijnsburger, E.H.; Moonen, J.C.; Kleinjans, J.C.S. *Nutr. Cancer* **1997**, *27*, 177-185.
169. Koch, H. P. *Dtsch. Apoth. Ztg.* **1993**, *133*, 3733-3743.
170. Fenwick, D. E.; Oakenfull, D. *J. Sci. Food Agric.* **1983**, *34*, 186-191.

171. Grune, T.; Scherat, T.; Behrend, H.; Conradi, E.; Brenke, R.; Siems, W. *Phytomedicine* **1996**, *2*, 205-207.

172. Harris, W. S.; Windsor, S. L.; Lickteig, J. Z. *Phytotherapie Abstractband* **1995**, *16*, 15.

173. Phelps, S.; Harris, W. S. *Lipids* **1993**, *28*, 475-477.

174. Cellini, L.; Campli, E. D.; Masulli, M.; Bartolomeo, S. D.; Allocati, N. *FEMS Immunology and Medical Microbiology* **1996**, *13*, 273-277.

175. Sivam, G. P.; Lampe, J. W.; Ulness, B.; Swanzy, S. R.; Potter, J. D. *Nutr. Cancer* **1997**, *27*, 118-121.

176. Rees, L. P.; Minney, S. F.; Plummer, N. T.; Slater, J. H.; Skyrme, D. A. *World J. Microbiol. Biotechnol.* **1993**, *9*, 303-307.

177. Weiss, E. *Med. Rec.* **1941**, *153*, 404-408.

178. Singh, K. V.; Shukla, N. P. *Fitoterapia* **1984**, *55*, 313-315.

179. Dankert, J.; Tromp, T. F. J.; De Vries, H.; Klasen, H. J. *Zentralbl. Bakteriol. Parasitenkd. Infektionskrankh. Hyg. ,Part I* **1979**, *A245*, 229-239.

180. Klimek, J. W.; Cavallito, C. J.; Bailey, J. H. *J. Bacteriol.* **1948**, *55*, 139-145.

181. Ahsan, M.; Chowdhurry, A.K.A.; Islam, S.N.; Ahmed, Z.U. *Phytotherapy Res.***1996**, *10*, 329-331.

182. Cavallito, C. J.; Bailey, J. H. *J. Am. Chem. Soc.* **1944**, *66*, 1950-1951.

183. Hughes, B. G.; Lawson, L. D. *Phytother. Res.* **1991**, *5*, 154-158.

184. Ahsan, M.; Islam, S.N. *Fitoterapia* **1996**, *67*, 374-376.

185. Ankri, S.; Miron, T.; Rabinkov, A.; Wilchek, M.; Mirelman, D. *Antimicrob. Agents Chemother.* **1997**, *41*, 2286-2288.

186. Cavallito, C. J.; Buck, J. S.; Suter, C. M. *J. Am. Chem. Soc.* **1944**, *66*, 1952-1954.

187. Wills, E. D. *Biochem. J.* **1956**, *63*, 514-520.

188. Jain, R. C. *Indian Drugs* **1993**, *30*, 73-75.

189. Kumar, A.; Kumar, S.; Yadav, M. P. *Indian J. Anim. Res.* **1981**, *15*, 93-97.

190. Perez, H. A.; De la Rosa, M.; Apitz, R. *Antimicrob. Agents Chemother.* **1994**, *38*, 337-339.

191. Louria, D. B.; Lavenhar, M.; Kaminski, T.; Eng, R. H. K. *J. Med. Vet. Mycol.* **1989**, *27*, 253-256.

192. Jezowa, L.; Rafinski, T.; Wrocinski, T. *Herba Pol.* **1966**, *12*, 3-13.

193. Caporaso, N.; Smith, S. M.; Eng, R. H. K. *Antimicrob. Agents Chemother.* **1983**, *23*, 700-702.

194. Hunan Medical College *Chin. Med. J.* **1980**, *93*, 123-126.

195. Davis, L. E.; Shen, J. K.; Cai, Y. *Antimicrob. Agents Chemother.* **1990**, *34*, 651-653.

196. Alkiewicz, J.; Lutomski, J. *Herba Pol.* **1992**, *38*, 79-83.

197. Brosche, T.; Platt, D. *Med. Welt* **1993**, *44*, 309-313.

198. Kandil, O.; Abdullah, T.; Tabuni, A. M.; Elkadi, A. *Arch. AIDS Res.* **1988**, *1*, 230-231.

199. Abdullah, T. H.; Kirkpatrick, D. V.; Carter, J. *Dtsch. Ztschr. Onkologie* **1989**, *21*, 52-53.

200. Mathew, P. T.; Augusti, K. T. *Indian J. Biochem. Biophys.* **1973**, *10*, 209-212.

201. Moriguchi, T.; Takashina, K.; Chu, P. J.; Saito, H.; Nishiyama, N. *Biol. Pharm. Bull.* **1994**, *17*, 1589-1594.

202. Moriguchi, T.; Saito, H.; Nishiyama, N. *Biol. Pharm. Bull.* **1996**, *19*, 305-307.

203. Moriguchi, T.; Saito, H.; Nishiyama, N. *Clin. Exper. Pharmacol. Physiol.* **1997**, *24*, 235-42.

204. Nishiyama, N.; Moriguchi, T.; Saito, H. *Experimental Gerontology* **1997**, *32*, 149-160.

205. Koch, H. P. In: *Garlic: the science and therapeutic application of Allium sativum L.*; Koch, H. P.; Lawson, L. D. Eds. Williams & Wilkins: Baltimore, 1996; pp 221-228. [see ref. *4*]

206. Freeman, F.; Kodera, Y. *J. Agric. Food Chem.* **1995**, *43*, 2332-2338.

207. Lawson, L.; Block, E. *J. Agric. Food Chem.* **1997**, *45*, 542.

Chapter 15

Neuroprotective Effects of *Ginkgo biloba* Extract

Barbara Ahlemeyer and Josef Krieglstein

Institut für Pharmakologie und Toxikologie, Fachbereich Pharmazie der Philipps-Universität Marburg, Ketzerbach 63, D–35032 Marburg, Germany

Extracts of the leaves of *Ginkgo biloba* (such as EGb 761 and LI 1370) are widely used for the treatment of dementia and cerebral insufficiency. This review will summarize experimental results concerning the protective effects of EGb 761 and its constituents in different models of cerebral ischemia. Furthermore, we present the supposed mechanisms of action and the current status of clinical trials.

For more than 40 years, the extract of the leaves of Ginkgo biloba (EGb 761) has been widely used therapeutically to increase peripheral and cerebral blood flow, as well as for the treatment of cerebral ischemia, brain disorders caused by old age, and of dementia due to neuronal degeneration or to vascular deficiency. In these age-related disorders, EGb 761 should enhance the intellectual capacity of the patients, that is, improve the learning, memory and cognitive faculties. Although it has been shown that organ systems other than the brain, e.g. the lung *(1, 2)* and the heart *(3, 4)*, could also benefit from the effects of EGb 761, this review will focus only on the effects of Ginkgo biloba extract on the brain function. First, we would like to present the neuroprotective effects of EGb 761 in different in vivo and in vitro models of cerebral ischemia and relate them to the different constituents of EGb 761. Thereafter, we would like to discuss in detail the effects of EGb 761 under normoxic as well as under ischemic conditions of the brain, including the effects of EGb 761 on cerebral blood flow, on cerebral glucose and lipid metabolism, and on the neurotransmitter systems. The last part of this review will summarize the present status of clinical trials on the commercial extracts of Ginkgo biloba (EGb 761 and LI 1370).

The Constituents of Ginkgo Biloba Extract (EGb 761)

This extract of Ginkgo biloba is prepared by highly extracting the leaves and removing the constituents ginkgole acid and the biflavones. The purified extract, EGb 761, is standardized to 24 % flavonoids. In addition, it contains approximately 6% terpene lactones *(5)*. The flavonoid fraction consists of flavone glycosides and proanthocyanidines; whereas, the terpene lactones (non-flavonoid fraction) are composed of 2.9% bilobalide and 3.1% ginkgolides A, B, C and J (Figure 1). In addition, several organic acids, e.g. kynurenic acid and hydroxykynurenic acid, are also found. Most of the extracts of ginkgo biloba on the market are standardized to 24% flavone glycosides and 6% terpene lactones. There is convincing evidence that the terpene lactones are the effective constituents of EGb 761.

Ginkgolides

	R^1	R^2
Ginkgolide A	H	H
Ginkgolide B	OH	H
Ginkgolide C	OH	OH
Ginkgolide J	H	OH

Bilobalide

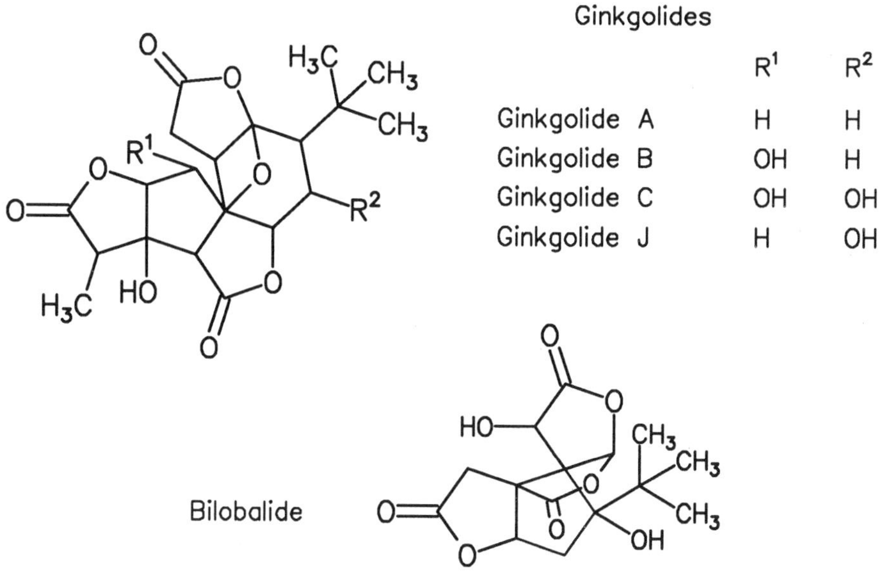

Figure 1: Chemical structures of the ginkgolides A, B, C and J and bilobalide.

The Neuroprotective Effects of EGb 761 and its Constituents in Different Models of Cerebral Ischemia and Hypoxia.

Previous studies of Karcher et al. *(6)* showed that animals exposed to hypobaric hypoxia survived 3-4 min and that the survival time was prolonged up to 25 min in animals treated with EGb 761. Based on these findings, Oberpichler et al. *(7)* measured again the survival of rats and mice after lethal hypoxic conditions (the animals were exposed to an atmosphere of 3-4% O_2) which was prolonged from 9±2 min to 21±4 min after a 30-min pretreatment with 100 mg/kg EGb 761. In this study only the non-flavonoid fraction of EGb 761 was effective. Further studies *(8)* were performed to evaluate which constituents of EGb 761 mediate the neuroprotective action. In mouse and rat models of permanent focal ischemia, administration of 5 mg/kg bilobalide, of 100 mg/kg ginkgolide B and of 50 mg/kg ginkgolide A (i.v.) 30

min before ischemia significantly reduced the infarct volume 2 days after focal ischemia. Bilobalide (10 mg/kg), but not ginkgolides A and B, was also effective when added immediately after focal ischemia. None of the drugs showed neuroprotection when given 60 min after focal ischemia. Ginkgolides C and J, as well as EGb 761 (200 mg/kg) were not effective (Table I).

Table I. Neuroprotective effects of EGb 761, bilobalide and the ginkgolides A and B which were administered directly after focal ischemia in mice and rats.

Drug	Dosage	Percentage of infarcted area (mm²) after 48 h		
		Control	Drug-treated	
EGb 761	200 mg/kg	29.1 ± 2.3 (14)	29.0 ± 4.4 (14)	n.s.
Ginkgolide A	50 mg/kg	29.5 ± 3.1 (12)	24.7 ± 2.3 (13)	***
Ginkgolide B	100 mg/kg	25.2 ± 3.4 (13)	19.9 ± 4.0 (13)	*
Bilobalide	10 mg/kg	130 ± 28 (15)	105 ± 20 (15)	**

Different from controls with * $p<0.05$, **$p<0.01$, ***$p<0.001$.

Under severe hypoxic conditions, again EGb 761 was less protective *(8)*. Global ischemia for 10 min in rats made it possible to discriminate the response between different brain regions, especially to observe the delayed neuronal death in the CA1 region of the hippocampus. Bilobalide (10 mg/kg s.c., 30 min before ischemia) and EGb 761 (100 mg/kg, i.v., 30 min before ischemia; 2 x 100 mg/kg i.v., 15 and 45 min after ischemia) failed to reduce the number of damaged neurons 7 days after ischemia. Although EGb 761 was able to attenuate postischemic hypoperfusion, the enhancement of local cerebral blood flow was not sufficient to cause neuroprotection and was probably abolished by the effect of EGb 761 to elevate plasma glucose levels. Similar results were described by Karkoutly *(9)*, who found no protective effect of 100 mg/kg EGb 761 after 10 min of global ischemia in rats either by adding it 30 min before ischemia or by a 10-day pretreatment. Interestingly, in a gerbil model of 5 min of global ischemia, a pretreatment of two weeks with EGb 761 (100 mg/kg, p.o.) reduced the percentage of damaged neurons in the hippocampus from 90% to 50% and improved the neurological disorders of the gerbils *(10)*.

The protective effects of EGb 761 and its constituents were also studied in primary cell cultures. Cytotoxic hypoxia was performed in pure neuronal cultures from chick embryo telencephalons by incubating the cells for 30 min with 1 mM NaCN and by measuring cell viability and the decrease in intracellular ATP concentration after three days of reoxygenation. EGb 761 (300 mg/l) failed to attenuate the loss in ATP concentration *(11)*, but bilobalide (0.1 μM) reduced the percentage of damaged neurons from 35±4 % to 25±3% *(12)*. In mixed cultures of neurons and astrocytes from neonatal rat hippocampus, bilobalide (10 μM) and ginkgolide B (1-100 μM) could protect the neurons from glutamate-induced toxicity *(8)*. The drugs were added 30 min before the addition of 1 mM glutamate and the cell viability was determined 18 h later (Table II).

Table II. The effects of bilobalide and ginkgolide B on cell viability after glutamate-induced excitotoxicity in primary hippocampal cultures.

Drug	Dosage	Cell viability (%) after 18 h	
		Control	Drug-treated
Ginkgolide B	1 μM	50 ± 2 (6)	60 ± 3 (6) ***
Ginkgolide B	100 μM	50 ± 2 (6)	70 ± 4 (6) ***
Bilobalide	10 μM	45 ± 5 (15)	58 ± 7 (6) **

Different from controls with * $p < 0.05$, ** $p < 0.01$, *** $p < 0.001$.

Taken together, these studies demonstrate that the EGb 761 constituents bilobalide and the ginkgolides A and B (50-100 mg/kg) significantly protected neurons after mild ischemia even when given shortly after the injury, but they failed to be effective under conditions of severe ischemia. With EGb 761, protective effects could only be observed in mild ischemia with long periods of pretreatment and only a tendency of neuroprotection after a 30-min pretreatment. Bearing in mind that 100 mg/kg EGb 761 contains only 3.1 mg/kg and 2.9 mg/kg ginkgolides and bilobalide, respectively, this is to be expected. Similar results were found in neuronal cell cultures.

Possible Mechanisms of the Neuroprotection of EGb 761 and its Constituents

Effect on Cerebral Blood Flow. The ability of EGb 761 to increase the blood flow in the whole body under normal conditions is well accepted. In the first characterization of EGb 761, an increase in the blood flow through peripheral vessels of guinea pig leg was shown *(13)*. In the brain, the infusion of 130 mg/kg EGb 761 in rats increased the local cerebral blood flow (LCBF) by 50-100% in nearly all brain regions *(14)*, as evaluated by [14][C]iodoantipyrine autoradiography. This increase in LCBF was shown to be related to the non-flavonoid fraction of EGb 761.

After global ischemia, the cerebral blood flow at first increases (postischemic hyperperfusion), but then it decreases for several hours to 40-60% of preischemic values (postischemic hypoperfusion) *(15)*. Under conditions of 10 min of global cerebral ischemia in rats, the administration of 2 x 100 mg/kg EGb 761, 15 and 45 min after ischemia, is able to reverse the hypoperfusion to nearly control levels, although the infarct volume was not reduced *(8)*. Le Poncin-Laffite et al. *(16)* have also found a less pronounced decrease of LCBF after mild ischemia induced by microembolization in gerbils by pretreatment with EGb 761 (100 mg/kg p.o.) for 3 weeks.

The increase in LCBF under normal as well as under ischemic conditions could be due to an increase in sympathetic activity and/or to vessel relaxation. There is some experimental evidence of the potency of EGb 761 to induce the release of catecholamines from their intracellular stores (17, 18), thereby acting as an indirect sympathomimetic agent. This hypothesis was supported by the findings that EGb 761 forced the contractile effect of noradrenaline in isolated vessels (17) and that cocaine, which stabilizes the cellular membrane, diminished the constrictive effect of EGb 761 by inhibiting the release of the catecholamines (18). In addition, an increase in the levels of biogenic monoamines in certain brain regions (19) and an inhibition of monoaminoxidase by EGb 761 (20) were found.

The relaxation of blood vessels by EGb 761 was discussed as being an indirect sympathomimetic effect but also due to an increased synthesis of prostacyclin (21). Furthermore, during ischemia and especially during reperfusion the amount of free radicals is enhanced, and, in this case, the radical scavenger capacity of EGb 761 (22) will inhibit formation of nitric dioxide radicals (NOO·) from nitric oxide (NO). As a consequence, sufficient amounts of NO activate the guanyl cyclase, leading to an increase in intracellular c-GMP concentration which results in vasodilation. In addition, an inhibitory effect of EGb 761 on c-GMP phosphodiesterase has been found which also increases the intracellular c-GMP concentration (23).

An additional point to discuss is the beneficial effect of EGb 761 on brain perfusion by its thrombolytic activity. After ischemia (and also during other brain diseases such as dementia) blood viscosity was found to be increased and this phenomenon could be partly blocked by high doses of EGb 761 (24). Experimental hints suggest that EGb 761 reduces the thrombocyte aggregation by increased synthesis of prostacyclin (21).

In summary, sympathic activation, in combination with vessel dilation, are the main reasons for enhanced brain tissue perfusion by EGb 761 under both normal and ischemic conditions. The increase in LCBF by EGb 761 could protect brain tissue after mild focal ischemia, but not after severe global ischemia.

Effect on Brain Energy Metabolism. Under normoxic conditions, there are regional differences in brain glucose utilization, which remained unchanged after injection of 100 mg/kg EGb 761 (25). Because EGb 761 increased plasma glucose level with no change in the rate of glucose consumption, an enhanced brain tissue level of glucose could be measured (6).

Under mild hypoxic conditions (12% O_2 for 2 hours), glucose consumption decreased, but severe hypoxia led to an increase in glucose consumption in different regions of the brain due to the non-economic use of glucose for ATP production by anaerobic glycolysis. As a consequence, the ATP level dropped in the whole brain tissue and the lactate concentration increased. Animals pretreated for 5 days with 100 mg/kg EGb 761 (p.o., daily) showed, after focal ischemia, approximately the same values of glucose utilization compared to normoxic controls (26). Also, under severe reduction of O_2 to 7%, EGb 761-pretreated animals showed an increased glucose consumption by 40-70% in gray matter and by 50-90% in white matter compared to ischemic animals (27).

In addition, EGb 761 can prolong the decrease in energy metabolism. During incomplete ischemia induced by microembolization, Le Poncin-Lafitte et al. *(16)* have found that pretreatment with 100 mg/kg EGb 761 (p.o., daily) for three weeks prevents ATP loss. Also, after severe ischemia, EGb 761-treated animals have higher ATP, creatine phosphate, and glucose levels than ischemic controls *(28)*. The increase in lactate observed during ischemia was also partly blocked *(28)*, suggesting that EGb 761 reduces utilization of anaerobic pathways. Furthermore, Spinnewyn *(10)* has shown that, in the presence of EGb 761, the mitochondria could more effectively use O_2 for ATP synthesis than the mitochondria of untreated ischemic animals. When discussing the effect of EGb 761 on energy metabolism, one has to keep in mind that EGb 761 enhances blood flow and thereby ameliorates brain perfusion and nutrition. Therefore, the effect of EGb 761 on energy metabolism could also be an indirect one.

In conclusion, the ability of EGb 761 to preserve energy and to inhibit the increase of cerebral lactate concentration during and after ischemia is an important factor in neuronal survival following transient ischemia.

Effects on the Adrenergic System. There is experimental evidence for a direct and indirect sympathomimetic action of EGb 761, as already described in the section *Effects on Cerebral Blood Flow.* Brunello et al. *(29)* showed that a single dose application of EGb 761 enhanced the release of noradrenaline in rat brain and that longer treatments for at least 27 days reduced the density of cerebral ß-receptors as an adaptive mechanism. However, Petkov et al. *(30)* have found a decrease in noradrenaline level by EGb 761, especially in the rat hippocampus, and Taylor *(31)* did not find any change in ß-receptor density, even after a 28-day treatment with EGb 761 (100 mg/kg, p.o.). Furthermore, an age-related decrease in hippocampal α_2-receptors could be prevented by treatment with EGb 761 *(32)*. Thus, the effect of EGb 761 on the noradrenaline level and on the density of adrenergic receptors in the brain needs further investigation.

Effects on the Dopaminergic System. There are some indications that EGb 761 influences the dopaminergic system. In a mouse model of N-methyl-4-phenyl-1,2,3,6-tetrahydropyridine (MPTP$^+$)-induced toxicity on dopaminergic neurons, EGb 761 inhibited the degeneration of the neurons in the striatum *(33)*. The authors suggested that inhibition of the re-uptake of dopamine and MPTP$^+$ is the protective mechanism of action of EGb 761. Indeed, they found that the uptake of dopamine in the synaptosomes could be inhibited in vitro by high doses of EGb 761 (IC_{50} 637 μg/ml), and they related the effect to the flavonoid fraction. However, at doses of 100 mg/kg EGb 761, this effect could not be involved in the neuroprotection. No effect of EGb 761 was found concerning the density of dopamine (D_2) receptors *(31)*.

Effects on the Cholinergic System. It is well known that the density of acetylcholine receptors in rat hippocampus decreases with age. This decline could be blocked by pretreatment with EGb 761 (100 mg/kg, p.o., daily) for four weeks and was found only in older animals with decreased acetylcholine receptor densities, but not in younger rats *(31)*.

Effects on the Serotoninergic System. Older rats show, in general, a decrease in serotonin (5-HT) levels and a reduced number of 5-HT_{1A} receptors *(33)*. The data of Petkov et al. *(30)* showed that an extract of ginkgo biloba, GK 501, increased 5-HT levels in nearly all rat brain structures, except for the pons, by treatment with 30 mg/kg GK 501 (p.o.) for 7 days. The authors performed their studies with old rats. In addition, it has been shown in synaptosomes from mouse cortex that EGb 761 could enhance 5-HT uptake and that the effect was related to the flavonoid fraction, especially to quercetin *(34)*.

Furthermore, Huguet et al. *(35)* have found that chronic treatment with 5 mg/kg EGb 761 (i.p.) for 3 weeks reversed the age-related decrease in 5-HT_{1A} receptor density. The authors suggest that the observed reverse effect on 5-HT_{1A} receptors is due to the ability of EGb 761 to inhibit lipid peroxidation. They hypothesize that EGb 761 has a restorative effect on the age-related decrease in receptor density by modulating receptor synthesis and by inhibiting membrane destruction of cerebral neurons which are apparently modified with aging *(36,37)*. They assumed that the effect of EGb 761, which also blocked the age-related decrease in acetylcholine- and α_2-receptors, was a general effect, independent of the receptor type. Furthermore, the decrease in neuronal membrane fluidity with age was also found to be attenuated by EGb 761, but independently from the improvement in changes in passive avoidance learning in the aging mouse *(38)*.

Radical Scavenging Activity. Several studies suggest an interaction between EGb 761 and free radicals. The free radicals are unstable products of oxygen metabolism, generated during the reperfusion phase after ischemic injury. They can rapidly interact with membrane unsaturated lipids, proteins or DNA. They also disturb calcium homeostasis and lead to the disruption of normal cell metabolism *(39)*. Free radicals play a role during the initial onset of ischemia, as well as in postischemic delayed neuronal death. In addition, the superoxide anion radical (O_2^-) inactivates NO and, thereby, inhibits postischemic vasodilatation. The hydroxyl radical (OH^-) inhibits the synthesis of prostacyclin, leading to vasoconstriction and to an increase in blood viscosity. The early studies of Pincemail and Deby *(40)* showed that the in vitro production of free radicals could be blocked by adding EGb 761 (0.125 - 0.5 mg/ml) to the reaction mixture. A detailed study revealed that the constituents ginkgolides A and B and bilobalide interact preferentially with superoxide anions, but the flavones also have a scavenging activity with a high affinity for hydroxyl radicals *(41)*. One explanation for the failure of flavones to protect neurons in the experimental models of cerebral ischemia may be their inability to pass the blood-brain barrier. Although Moreau et al. *(42)* used a [14][C]-labelled flavone and have found accumulation of radioactivity in the brain, the authors did not show whether the intact flavone reached the extracellular fluid of the brain.

PAF Antagonism. PAF (platelet activating factor) has been shown to play an important role during postischemic reperfusion. After ischemia, the activation of a Ca^{2+} dependent phospholipase A_1/A_2 leads to an excess of lyso-PAF (the precursor of PAF) and a release of free fatty acids into the cytoplasm. PAF is thought to be involved in the regulation of blood pressure, anaphylaxis, and inflammation *(43)* and in necrosis after ischemic injury *(44)*. Interestingly, treatment with 50 mg/kg ginkgolide B (a dose which is able to reduce the percentage of damaged neurons in a rat model of transient focal ischemia) markedly reduced the release of free fatty acids and the production of diacylglycerol during and after ischemia *(45)*. A direct antagonistic action of ginkgolide B on PAF receptors was also demonstrated, but at doses much higher than in the original extract *(46,47)*. The potency of the ginkgolides to antagonize PAF was demonstrated by their ability to inhibit PAF-induced thrombocyte aggregation *(46)* and to inhibit the PAF-induced increase in intracellular calcium concentration *(48,49)*.

Antiapoptotic Activity. Recent studies have shown an antiapoptotic activity of EGb 761 *(50)*. Apoptosis (programmed cell death) was induced in primary olfactory neurons by olfactory nerve sectioning and was determined by measuring the thickness of the epithelium 2-3 days after axotomy and by quantification of ^{32}P-end labelled genomic DNA as a hallmark of apoptosis. The reduction of epithelium thickness and the increase in DNA fragmentation was reversed in rats by a 10-day pretreatment with EGb 761 (50 mg/kg, p.o.) compared to the untreated control. The neuroprotective effect in animals, with continued treatment, remained even 1-2 months after axotomy. The authors suggested the free radical scavenging properties of EGb 761 as the main mechanism of the antiapoptotic activity. However, nerve sectioning led to apoptosis because nerve cells lost contact and were deprived of growth factors from neighboring cells. This opens the possibility that induction of neurotrophic factors is involved in the apoptotic effect of EGb 761. In addition, it has yet to be clarified which constituents of EGb 761 mediate the antiapoptotic effect.

Clinical Trials. Special extracts of the leaves of Ginkgo biloba (EGb 761 and LI 1370) are licensed in Germany for the treatment of cerebral dysfunction with symptoms like difficulties in memory, dizziness, tinnitus, headaches, and emotional instability with anxiety. It is also used for supportive treatment of hearing loss due to cervical syndrome and/or peripheral arterial circulatory disturbances with intact circulatory reserve (intermittent claudication).

More than forty clinical trials on treatment of cerebral insufficiency have been published; 10-12 of them are of good quality *(51,52)*. Nearly all trials showed positive effects of the Ginkgo biloba extracts compared with placebo on the symptoms mentioned above. However, the number of studies to be considered becomes much smaller if one applies the most modern standards. These up-dated assessment criteria are the operationalized diagnosis of dementia, the inclusion of mild and moderate, but not severe, cases and demonstration of the statistically significant effects at three efficacy levels (psychopathological, psychometric, and behavioral). Only two studies with Ginkgo biloba extracts fulfill the methodological criteria of the advisory committees A and B3 of the German Federal Health Office *(52)*.

The most relevant study on the efficacy of EGb 761 in outpatients suffering from dementia of the Alzheimer type of multi-infarct dementia was recently published by Kanowski et al. *(53)*. After a 4-week run-in period, 216 patients were included in the prospective, randomized, double-blind, placebo-controlled, multi-center study lasting for 24 weeks. The patients received a daily oral dose of 240 mg EGb 761 or placebo. Clinical global impressions for psychological assessment, the *Syndrom-Kurztest* for the assessment of the patient's attention and memory, and the *Nürnberger Alters-Beobachtungsskala* (NAB) for behavioral assessment of activities of daily life were chosen for evaluation. Clinical efficacy was assessed by means of a responder analysis. The study design fulfilled the methodological standards currently required for studies with nootropics (groups of neurotropics). Proof of efficacy was carried out by using validated measurement instruments. The results of the study suggests that EGb 761 is of clinical efficacy in the treatment of outpatients with dementia.

Thus, the efficacy of Ginkgo biloba has been demonstrated according to the strictest criteria: operationalized diagnosis of dementia, inclusion of mild-to-moderate cases, statistical proof of all three levels of efficacy, and global clinical evaluation.

No serious adverse effects have been noted in any trial. In rare cases, mild gastrointestinal complaints, headache, sleep disturbances, dizziness, and allergic skin reactions have been reported. Ginkgo biloba extracts seem to be well tolerated. Therefore, the data obtained so far show a clearly positive therapeutic risk-benefit relationship.

Conclusion. From various in vitro and in vivo experiments, it may be concluded that constituents of Ginkgo biloba have neuroprotective properties. PAF antagonism, free radical scavenging, interaction with neurotransmitters, and even induction of growth factors may be discussed as putative mechanisms of action. Clinical studies suggest the extract of Ginkgo biloba leaves to be of clinical efficacy in the treatment of mild and moderate dementia.

Literature Cited

1. Touvay, C.; Vilain, B.; Taylor, J.E.; Etienne, A.; Braquet, P. *Prog. Lipid Res.* **1986**, *25*, 277-288.
2. Berti, F.; Omini, C.; Rossoni, G.; Braquet, P. *Pharmacol. Res. Commun.* **1986**, *18*, 775-793.
3. Haramaki, N.; Aggarwal, S., Kawabata, T.; Droylefaix, M.T.T.; Packer, L. *Free Rad. Biol. Med.* **1994**, *16*, 789-794.
4. Guillon, J.M.; Rochette, L.; Baranès, J. *Presse Méd.* **1986**, *15*, 1516-1519.
5. Weinges, K.; Baehr, W.; Kloss, P. *Drug Res.* **1985**, *1459*, 35-41.
6. Karcher, L.; Zagermann, P.; Krieglstein, J. *Naunyn-Schmiedeberg´s Arch. Pharmacol.* **1984**, *327*, 31-35.
7. Oberpichler, H.; Beck, T.; Abdel-Rahman, M.M.; Bielenberg, G.W.; Krieglstein, J. *Pharmacol. Res. Commun.* **1988**, *20*, 349-368.
8. Krieglstein, J.; Ausmeier, F.; El-Abhar, H.; Lippert, K.; Welsch, M.; Rupalla, K.; Henrich-Noack, P. *Eur. J. Pharm. Sci.* **1995**, *3*, 39-48.
9. Karkoutly, C. Inauguraldissertation, Marburg **1990**.
10. Spinnewyn, B. In *Advances in Ginkgo biloba Extract Research. Effects of Ginkgo biloba Extract (EGb 761) on the Central Nervous System.* Christen, Y., Constentin, J., Lacour, M. Eds.; Elsevier: Paris, Franc, 1992, *Vol. 2*; pp 113-118.
11. Ahlemeyer, B.; Krieglstein, J. *Life Sci.* **1989**, *45*, 835-842.
12. Prehn, J.H.M.; Oberpichler, H.; Roßberg, C.; Mennel, H.; Krieglstein, J. *J. Cereb. Blood Flow Metabol.* **1991**, *11* Suppl. 2, S722.

13. Peter, H.; Fisel, J.; Weisser, W. *Drug Res.* **1966**, *16*, 719-725.
14. Krieglstein, J.; Beck, T.; Seibert, A. *Life Sci.* **1986**, *39*, 2327-2334.
15. Ginsberg, M.D.; Busto, R. *Stroke* **1989**, *20*, 1627-1642.
16. Le Poncin-Lafitte, M.; Rapin, J.; Rapin, J.R. *Arch. Int. Pharmacodyn.* **1980**, *243*, 236-244.
17. Auguet, M.; DeFeudis, F.V.; Clostre, F. *Gen. Pharmac.* **1982**, *13*, 169-171.
18. Auguet, M.; DeFeudis, F.V.; Clostre, F.; Deghenghi, R. *Gen. Pharmac.* **1982**, *13*, 225-230.
19. Morier-Teissier, E.; Helgoualch, A.; Drieu, K.; Rips, R. *Biogen. Amines* **1987**, *4*, 351-358.
20. White, H.L.; Scates, P.W.; Cooper, B.R. *Life Sci.* **1996**, *58*, 1315-1321.
21. Chatterjee, S.S. In *Effects of Ginkgo biloba on Organic Cerebral Impairments*; Agnoli, A., Rapin, J.R., Scapagnini, V., Weitbrecht, W.V. Eds.; John Libbey: London, United Kingdom, **1985**, pp 5-14 .
22. Ferradini, C.; Droy-Lefaix, M.T.; Christen, Y. In *Advances in Ginkgo biloba Extract Research. Effects of Ginkgo biloba Extract (EGb 761) on the Central Nervous System.* Christen, Y., Constentin, J., Lacour, M. Eds.; Elsevier: Paris, France, **1992**.
23. Ruckstuhl, M.; Beretz, A.; Anton, R.; Landry, Y. *Biochem. Pharmacol.* **1979**, *28*, 535-538.
24. Macovschi, O.; Prigent, A.F.; Nemoz, G.; Pacheco, H. *J. Neurochem.* **1996**, *49*, 107-114.
25. Witte, S.; Anadere, I.; Chmiel, H. *Clin. Hemorheol.* **1983**, *3*, 291-297.
26. Rapin, J.R.; Le Poncin-Lafitte, M. *Presse Méd.* **1986**, *15*, 1494-1497.
27. Beck, T.; Krieglstein, J. *Am. J. Physiol.* **1987**, *252*, H504-512.
28. Beck, T. Inauguraldissertation, Marburg **1985**.
29. Brunello, N.; Racagni, G.; Clostre, J.F.; Drieu, K.; Braquet, P. *Pharm. Res. Comm.* **1985**, *17*, 1063-1072.
30. Petkov, V.D.; Hadjiivanova, C.; Petkov, V.V.; Milanov, S.; Visheva, N.; Boyadjieva, N. *Phyto. Res.* **1993**, *7*, 139-145.
31. Taylor, J.E. *Presse Méd.* **1985**, *15*, 1491-1493.
32. Huguet, F.; Tarrade, T. *J. Pharm. Pharmacol.* **1992**, *44*, 24-27.
33. Ramassamy, C.; Clostre, F.; Christen, Y.; Constentin, J. *J. Pharmacol.* **1990**, *42*, 785-789.
34. Ramassamy, C.; Christen, Y.; Clostre, F.; Constentin, J. *J. Pharm. Pharmacol.* **1992**, *44*, 943-945.
35. Huguet, F.; Drieu, K.; Piriou, A. *J. Pharm. Pharmacol.* **1994**, *46*, 316-318.
36. Heron, D.S.; Schinitzky, M.; Hershkowitz, M.; Samual, D. *Proc. Natl. Acad. Sci. USA* **1980**, *77*, 7463-7467.
37. Greenberg, L.H. *Fed. Proc.* **1986**, *45*, 55-59.
38. Stoll, S.; Scheuer, K., Pohl, O.; Müller, W.E. *Pharmacopsychiat.* **1996**, *29*, 144-149.
39. Halliwell, B. *J. Neurochem.* **1992**, *59*, 1609-1623.
40. Pincemail, J.; Deby, C. *Presse Méd.* **1986**, *15*, 1475-1479.
41. Marcocci, L.; Packer, L.; Droy-Lefaix, M.T.; Sekaki, A.; Gardès-Albert, M. *Methods Enzymol.* **1994**, *234*, 462-475.
42. Moreau, J.P.; Eck, C.R.; McCabe, J.; Skinner, S. In *Rökan (Ginkgo biloba). Recent Results in Pharmacology and Clinic*; Fünfgeld, E.W. Ed.; Springer: Berlin, Germany **1988**, pp. 37-45.
43. Braquet, P.; Touqui, L.; Shen, T.Y.; Vargaftig, B.B. *Pharmacol. Rev.* **1987**, *39*, 97-145.
44. Braquet, P.; Paubert-Braquet, M.; Koltai, M.; Bourgain, R.; Bussolino, F.; Hosford, D. *Trends in Pharmacol. Sci.* **1989**, *10*, 23-30.

45. Panetta, T.; Marcheselli, V.L.; Braquet, P.; Spinnewyn, B.; Bazan, N.G. *Biochem. Biophys. Res. Comm.* **1987**, *149*, 580-587.

46. Nunez, D.; Chignard, M.; Korth, R.; Le Couedic, J.P.; Norel, X.; Spinnewyn, B.; Braquet, P.; Benveniste, J. *Eur. J. Pharmacol.* **1986**, *123*, 197-205.

47. Korth, R.; Nunez, D.; Bidault, J.; Benveniste, J. *Eur. J. Pharmacol.* **1988**, *152*, 101-110.

48. Baroggi, N.; Cachia, H.; Etienne, A.; Braquet, P. *Prostaglandins* **1985**, *30*, 700.

49. Hirafuji, M.; Maeyama, K.; Watanabe, T.; Ogura, Y. *Biochem. Biophys. Res. Comm.* **1988**, *154*, 910-917.

50. Didier, A.; Rouiller, D.; Coronas, V.; Jourdan, F.; Droy-Lefaix, M.T. In *Advances in Ginkgo biloba Extract Research. Effects of Ginkgo biloba Extract (EGb 761) on Neuronal Plasticity;* Christen, Y.; Droy-Lefaix, M.T.; Macias-Nunez, J.F. Ed.; Elsevier: Paris, France, **1996**, *Vol. 5*, pp 45-52.

51. Kleijnen, J; Knipschild, P. *Lancet* **1992**, *340*, 1136-1139.

52. Letzel, H.; Haan, J.; Feil, W.B. J. *Drug Dev. Clin. Pract.* **1996**, *8*, 77-94.

53. Kanowski, S.; Herrmann, W.M.; Stephan, K.; Wierich, W.; Hörr, R. *Pharmacopsychiatry* **1996**, *29*, 47-56.

Chapter 16

Biochemical, Pharmaceutical, and Medical Perspectives of Ginseng

O. Sticher

Department of Pharmacy, Swiss Federal Institute of Technology (ETH) Zurich, CH-8057 Zürich, Switzerland

Ginseng, the root of *Panax ginseng (Araliaceae)*, has been used in oriental medicine since ancient times as a stimulant, tonic, diuretic, and stomachic agent. In Europe, ginseng phytomedicines are used for a wide range of pharmacological activities. These preparations can increase physical and mental performance, as well as resistance to stress and disease and prevent exhaustion. An update is given on the chemistry, biological effects, molecular mechanisms, and possible therapeutic uses, as well as on quality control, of *Panax ginseng* and of phytomedicines based on root extracts of this plant. Special emphasis is placed on the structures of the ginsenosides—monodesmosidic and bisdesmosidic saponins—which are considered to be the main active compounds, as well as on their quantitative determination in ginseng roots and phytomedicines.

Ginseng, the root of *Panax ginseng* C.A. Meyer, has been used in traditional medicine in Asian countries since ancient times as a stimulant, tonic, diuretic, and stomachic agent. In Europe, ginseng preparations are among the leading phytomedicines. They are drugs of the OTC market and are used for a wide range of pharmacological activities, including improvement of physical and mental performance, better resistance to stress and disease, as well as prevention of exhaustion. The annual retail sales in 1994 (Figure 1) were approximately 73 million DEM (approximately $ 50 million). Of this market, 49% is in Germany, followed by Spain (14%), France (13%), and Switzerland (11%) (*1*).

Panax Species

During the Han dynasty (206 BC-24 AD), a medical text was written about a plant of the Araliaceae family called sheng. Later, this plant was known as "Jen Sheng, the root of heaven" and "ginseng" has been the subject of writings and research in almost every country of the world. In 1883, the genus Panax was added to its name, "pan" meaning all and "axos" meaning cure - the word "jen-sheng" or "ginseng" means man-herb because of its man-shaped root. Since the word *Panax ginseng* virtually means "the all-healing man-herb" it is little wonder that it has been regarded as something of a panacea - particularly in the East (*2*).

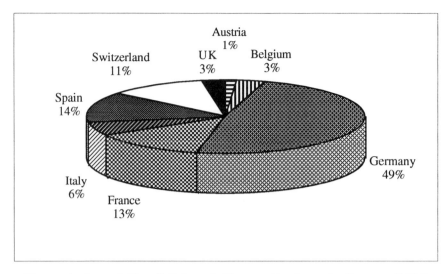

Figure 1. Annual Retail Sales of Ginseng Products in Europe (1994)

Several Panax species other than ginseng occur in the Northern hemisphere, from the Eastern Himalayas onward through China and Japan to North America. They can be divided into two groups, depending on the shape of the underground parts. Group A contains carrot-like roots and group B long, horizontal rhizomes (*3*) (Tables I and II).

Table I. Panax Species: Group A

Species	*Distribution*	*Cultivation*
1. *P. ginseng* C.A. Meyer [Korean Ginseng]	Korea, North eastern China, Far eastern Siberia	Korea, China, Japan
2. *P. quinquefolium* L. [American Ginseng]	North America	North America
3. *P. trifolius* L. [Dwarf Ginseng]	North America	—
4. *P. notoginseng* (Burk.) F.H. Chen [Sanchi Ginseng]	Unclear	Yunnan (China)
5. *P. pseudo-ginseng* Wall. subsp. *pseudo-ginseng* Hara	Eastern Himalayas	—

Originally, *P. ginseng* C.A. Meyer or Korean ginseng grew in northeastern China, Korea, and far eastern Siberia, but now the wild plant is very rare. Today, almost all the ginseng roots on the market were cultivated in China, Korea, and Japan.

Table II. Panax Species: Group B

Species	Distribution	Cultivation
1. *P. japonicus* C.A. Meyer = *P. pseudo-ginseng* Wall. subsp. *japonicus* (Meyer) Hara [Chikusetsu-ninjin (Japan); Zhujie-shen (China)]	Japan, Southern China	—
2. *P. japonicus* C.A. Meyer var. *major* (Burk.) C.Y. Wu et K.M. Feng (= *P. pseudo-ginseng* Wall. var. *major* (Burk.) Li [Zu-tzi-shen (China)]	Southern China	—
3. *P. zingiberensis* C.Y. Wu et K.M. Feng	Southern China	—
4. *P. pseudo-ginseng* Wall. subsp. *himalaicus* Hara with var. *angustifolius* (Burk.) Li and var. *bipinnatifidus* (Seem.) Li [Himalayan Ginseng]	Eastern Himalayas	—

Some other congeners of ginseng are also used as medicines. American ginseng, the root of *P. quinquefolium* L., which initially grew wild in the northeastern United States, is now cultivated for the export to the Hong Kong market. Chinese people use it in almost the same manner as Korean ginseng, but for slightly different purposes. Sanchi ginseng, the root of *P. notoginseng* (Burk.) F.H. Chen, is mainly cultivated in the Yunnan province in China and has been used in Chinese medicine for similar purposes as Korean ginseng. Japanese Chikusetsu ginseng is the rhizome of wild growing *P. japonicus* C.A. Meyer. It has been used as a medicine in Japan in place of Korean ginseng. *P. pseudo-ginseng* Wall. subsp. *himalaicus* Hara grows wild in Nepal and in the east Himalayan district. Some other species of *Panax*, such as *P. japonicus* C.A. Meyer var. *major* (Burk.) C.Y. Wu et K.M. Feng are also used partly in traditional medicine in China and Vietnam (*4*).

There are several other *Panax* species not listed in Tables I and II; for example, *P. japonicus* C.A. Meyer var. *stipuleanatus* H.T. Tsai et K.M. Feng is found in southern China (*3*). Recently, a new species of the genus Panax was discovered, *P. vietnamensis* Ha et Grush., Vietnamese Ginseng, used as a secret medicinal plant in parts of Vietnam (*5*). In the following, only Korean or *Panax ginseng* will be discussed in detail.

Botany

Ginseng is an aromatic herb with a perennial short underground stem (rhizome) associated with a fleshy white root (Figure 2). The aerial part consists of a single stem about 30-60 cm high (in cultivation 60-70 cm) which dies annually. Ginseng plants begin to flower at the age of three or four years. A flowering plant may bear from 3 to 6 palmately compound leaves, each with five leaflets, and with a peduncle terminated by a simple umbel at the center with 4 to 40 flowers according to the age of the plant and its growing conditions. Ginseng flowers in June or July. The fruit is pea-sized, green at first, and red at maturity, containing 2 or 3 white seeds. The root system of ginseng consists of the primary root and its branches, and of some adventitious roots developed from the rhizome (*6*).

Ginseng Plantations and Processing

Due to continuous use over thousands of years, the natural supply of ginseng roots was exhausted a long time ago. Most of the ginseng preparations now available are derived from plants cultivated in Korea, China or Japan. Naturally growing ginseng can be found in the cool temperate zone in rich, damp, but not wet or muddy soil, prefering the shade of hardwood forests. In cultivation, good drainage and artificial shade must be provided. Ginseng is propagated from seeds. For this purpose, ripe fruits of 4 to 5 year-old plants are collected in September. In a nursery bed, horizontal drills, 4-6 cm apart, are made and 25-30 seeds are sowed in each drill and covered with soil (approximately 3 cm). Ginseng seeds require 18-20 months to germinate. Seedlings may be transplanted to permanent beds when they are one or two years old. Wind, rain, and direct sun are harmful to ginseng plants. Therefore, frames for artificial shade must be built up over the beds and the ground must be kept moist by watering. Ginseng is a root crop. Preventing the plants from flowering helps them to produce larger and better roots. It takes 4 to 6 years from seedling to harvest. Ginseng is harvested between August and October, when the aerial portion turns yellow. Careful handling of the roots is very important. The branching main roots are 8 to 20 cm long and about 2 cm thick. The thin ends of main and secondary roots are traditionally removed so that the human form can be better brought into evidence (6).

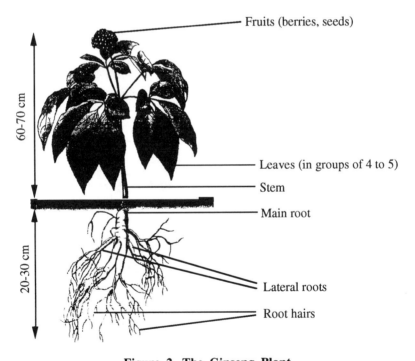

Figure 2. The Ginseng Plant

Ginseng is processed as two forms, white ginseng and red ginseng. The former is the dried root, whose peripheral skin is frequently peeled off, and the latter is the steamed root, showing a caramel-like color and resisting invasion of fungi and worms (*4*). In the European phytomedicines market, mainly white ginseng (including lateral roots and root hairs) is used as crude material, while in Asian traditional medicine red ginseng is preferred.

Ginseng Monographs in European Pharmacopeias

Ginseng is specified among others, in the German, Swiss, Austrian, and French pharmacopeias. The Swiss pharmacopeia demands a content of total ginsenosides, calculated as ginsenoside Rg_1, which should be not less than 2.0%. According to the German pharmacopeia, the total ginsenoside content should be not less than 1.5%. Both pharmacopeias use for the quantification a spectrophotometric method. On the other hand, a draft for the European pharmacopeia demands the content of ginsenosides Rg_1 and Rb_1 to be not less than 0.2% based on an HPLC method.

Constituents

Since the beginning of this century, the constituents of the ginseng roots have been investigated, and a series of classes of compounds have been isolated, such as triterpene saponins, essential oil containing polyacetylenes and sesquiterpenes, polysaccharides, peptidoglycans, and nitrogen containing compounds, as well as various ubiquitous compounds like fatty acids, carbohydrates, phenolic compounds, and others (*7*).

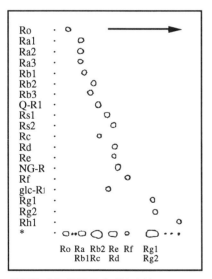

CHCl₃-MeOH-H₂O (14:6:1)

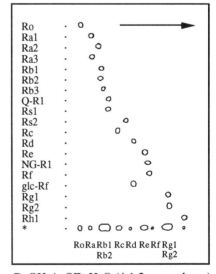

BuOH-AcOEt-H₂O (4:1:2, upper layer)

Figure 3. Thin Layer Chromatograms of the Saponins of *P. ginseng* Roots ((*3*); Plate: Silica gel 100 F₂₅₄ (Merck); * Crude saponin fraction of *Panax ginseng* roots)

226

The characteristic chemical constituents of ginseng, considered to contribute decisively to its pharmacological effects have been investigated extensively since 1955. They are triterpene saponins and named ginsenosides Rx according to their mobility on TLC plates: Their polarity decreases from index "a" to "h" (Figure 3). This property is, of course, a function of the number of monosaccharide residues in the sugar chain (8). Acid hydrolysis of the genuine glycosides originally yielded two aglycones, panaxadiol and panaxatriol, now known to be artifacts produced by the action of acid during the process of hydrolysis (4). Applying various chemical and spectroscopic methods, the genuine aglycones were found to be protopanaxadiol and protopanaxatriol both having a dammarane skeleton (3, 4). A tertiary hydroxyl group attached to C-20 participates in ring closure with a double bond in the side chain on acid treatment of protopanaxadiol and protopanaxatriol (Scheme 1).

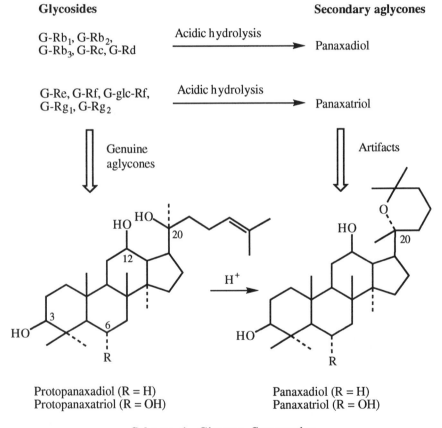

Glycosides   **Secondary aglycones**

G-Rb$_1$, G-Rb$_2$,   Acidic hydrolysis
G-Rb$_3$, G-Rc, G-Rd   ──────────────▶   Panaxadiol

G-Re, G-Rf, G-glc-Rf,   Acidic hydrolysis
G-Rg$_1$, G-Rg$_2$ ──────────────▶   Panaxatriol

Genuine aglycones   Artifacts

H$^+$

Protopanaxadiol (R = H)   Panaxadiol (R = H)
Protopanaxatriol (R = OH)   Panaxatriol (R = OH)

Scheme 1. Ginseng Sapogenins

So far, 31 ginsenosides have been isolated from the roots of white and red ginseng (3, 4, 9-15). They may be divided into three groups depending on their aglycones: protopanaxadiol- and protopanaxatriol-type ginsenosides, and oleanolic acid-type saponins (Schemes 2-4).

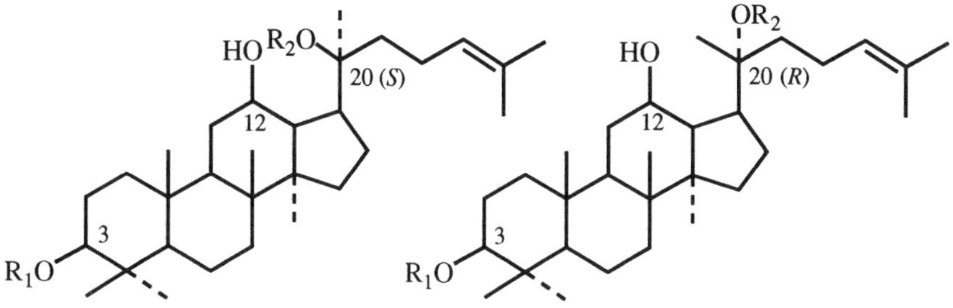

Ginsenoside	R_1	R_2	
Rb_1	-Glc^2-Glc	-Glc^6-Glc	• Other diols are: Ra_1, Ra_2, Ra_3, Rb_3, Rg_3, Rh_2, Rs_1, Rs_2, Q-Rh_1, NG-R_4, mRd
Rb_2	-Glc^2-Glc	-Glc^6-Ara(p)	
Rc	-Glc^2-Glc	-Glc^6-Ara(f)	
Rd	-Glc^2-Glc	-Glc	• 20 (S), except Rg_3 = 20 (S+R)
mRb_1	-Glc^2-Glc^6-Ma	-Glc^6-Glc	• Rh_2, Rs_2 and NG-R_4 occur only in red ginseng
mRb_2	-Glc^2-Glc^6-Ma	-Glc^6-Ara(p)	
mRc	-Glc^2-Glc^6-Ma	-Glc^6-Ara(f)	

Scheme 2. Ginseng Saponins of 20 (S) and 20 (R) Protopanaxadiol

All dammarane ginsenosides isolated from ginseng root (white ginseng) are derivatives of 20 (S) protopanaxadiol and 20 (S) protopanaxatriol, with the exceptions of ginsenoside Rg_3 (= 20 (R)) (9, 16), Rh_4 (15), and koryoginsenoside R_2 (12). According to Shibata et. al. (4), Rg_3 might be an artifact, since all other ginsenosides possess 20 (S) configuration. Rh_4 has an additional double bond between C-20 and 22 in the side chain instead of a 20-OH group. The compound is related e.g. to ginsenoside F_4, isolated from ginseng leaves (17). In koryoginsenoside R_2 the double bond is between C-22 and 23 instead of 24(25). Almost all the ginsenosides isolated from white ginseng are also found in red ginseng. However, some of the ginsenosides, such as 20 (R) ginsenoside Rg_2, 20 (R) ginsenoside Rh_1, ginsenosides Rh_2, Rs_1, Rs_2, Q-R_1, and NG-R_1, are characteristic saponins for red ginseng. The 20 (R) compounds are considered to be degradation products formed by heating and hydrolysis during the steaming process (4, 7). Therefore, the natural occurrence of either 20 (R) and 20 (S) ginsenoside Rg_3 needs to be verified. From recently published HPLC investigations, it is unclear if both occur in white and red ginseng (13) or only in red ginseng (14). Ginsenoside Ro is a minor compound and is the only oleanolic-type saponin identified in the root of *P. ginseng* (Scheme 4).

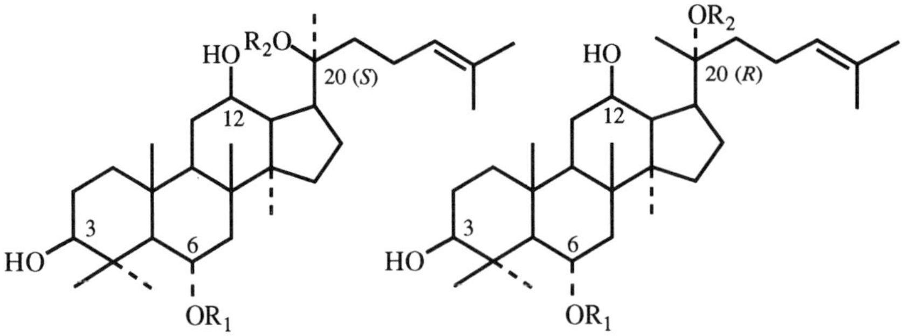

Ginsenoside	R_1	R_2	
Re	-Glc^2-Rha	-Glc	
Rf	-Glc^2-Glc	-H	20 (S), except Rg$_2$ and Rh$_1$ = 20 $(S+R)$.
glc-Rf	-Glc^2-Glc	-Glc	Rg$_2$ (R) and Rh$_1$ (R) may be produced
Rg$_1$	-Glc	-Glc	during production process. They are
Rg$_2$	-Glc^2-Rha	-H	characteristic for red ginseng
Rh$_1$	-Glc	-H	
NG-R$_1$	-Glc^2-Xyl	-Glc	

Scheme 3. Ginseng Saponins of 20 (S) and 20 (R) Protopanaxatriol

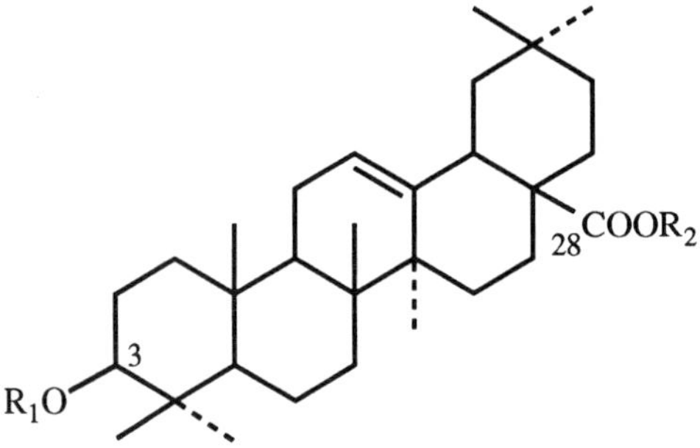

Ginsenoside	R_1	R_2
Ro	-GlcUA^2-Glc	-Glc

Scheme 4. Ginseng Saponin of Oleanolic Acid

Pharmacological Effects of Ginseng

Overviews of the pharmacological effects of ginseng extracts and preparations, as well as of ginsenoside-, polyacetylene- and polysaccharide fractions, of single ginsenosides and polyacetylenes, have been presented by several authors (e.g., *3-4, 7, 18-20*).

Reviewing the literature, the main activities of ginseng can be summarized as follows (*7*): general tonic, effects on immunological functions, effect on the cardiovascular system, effect on lipid metabolism, effect on alcohol metabolism, hypoglycemic activity, stimulating effect on pituitary-adrenocortical system, and antitumor activity. From earlier publications, only the main activities of ginsenosides Rg_1 and Rb_1 shall be mentioned here. Rg_1 stimulates the central nervous system and enhances protein, DNA and RNA synthesis, while Rb_1 has tranquillizing effects on the central nervous system and improves memory (for references see (*7*)).

Some recently published effects are antiplatelet activity (various ginsenosides— such as Ro, Rg_1, and Rg_2— and polyacetylenes (*21-23*), antioxidant activity (Rg_1, Rb_1) (*24-25*), antitumor activity (polyacetylens, polysaccharides, Rg_3, Rh_2) (*26-30*), cytoprotective activity (polysaccharides) (*31*), inhibition of calcium channels (Rf) (*32*), immunomodulatory activity (Rg_1, polysaccharides) (*33-39*), neuroprotection (ginsenosides, Rb_1) (*40*), and others. From recent publications about ginseng activities, inhibition of platelet aggregation, radical promotion, vasorelaxation, oxygen scavenging, lipid peroxidation, antioxidant, and neuroprotection are especially worth mentioning and have been briefly summarized as "anti-aging activity". Combined with the inhibitory effect on the proliferation of cancer cells and the effects on the immune system, the use of ginseng by east Asians to promote longevity is at least partially explained. Since, in general, the experimentally shown effects are observed in their totality only in the whole extract (e.g. *41-46*), no single constitutents are used individually as monopreparations. All commercially available products are extract preparations.

Based on the described activities, as well as on the traditional use in the Orient, ginseng preparations are mainly used as general tonics and geriatric remedies to regulate catabolic and anabolic processes of cellular metabolism, to increase the capacity of the organism to adapt to mental and physical stress, as well as to increase concentration and attention.

Only recently, was the role of compounds other than the ginsenosides demonstrated for of the ginseng extraxt. From the isolated polyacetylenes (*22-23, 26-29, 47*), mainly panaxytriol, panaxynol, and panaxydol show cytotoxic, antiplatelet and anti-inflammatory effects (Scheme 5).

In the case of the polysaccharides, the panaxans A-U (*48-53*), which are peptidoglycans, are responsible for the hypoglycemic activity of ginseng. The ginsenans PA, PB, S-IA, S-IIA and some other polysaccharides (*35-37*) show immunological activities, such as recticuloendothelial system (RES) potentiation, anti-complement effects, and alkaline phosphatase-inducing activity (*31, 54* and references cited therein). Various polysaccharides show cytoprotective activity (anti-ulcerogenic effects, treatment of gastroenteric diseases) and antitumor activity. The structures of the isolated polysaccharides are only partially known. Ginsenan PA, e.g., has a backbone chain mainly consisting of β-1,3-linked D-galactose (*35*).

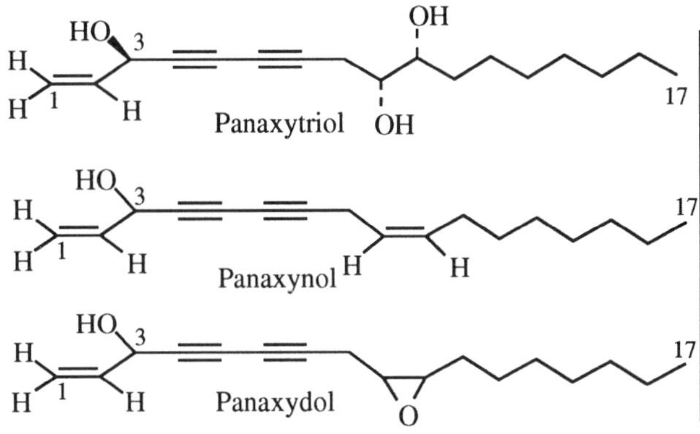

- Cytotoxic activity: e.g. Panaxytriol, panaxynol, panaxydol
- Antiplatelet activity: Panaxynol
- Anti-inflammatory activity: Panaxynol

Scheme 5. Polyacetylenes from Ginseng Roots

Pharmacokinetics and Metabolism of Ginsenosides

Until recently, little was known about the absorption, distribution, excretion and metabolism of ginseng saponins. In the last decade various investigations dealing with the pharmacokinetics and the metabolism of ginsenosides were published. Results were presented, e.g., on ginsenosides Rg_1, Rb_1, Rb_2, Re, and Rh_2. From these studies it can be concluded that the decomposition modes are different for protopanaxadiol and protopanaxatriol saponins.

As can be seen in Scheme 6, ginsenoside Rg_1 (protopanaxatriol-type) showed an extremely short half-life of 27 min after intravenous administration into minipigs. In contrast to Rg_1, the protopanaxadiol-type ginsenoside Rb_1 showed a half-life in the β-phase of 16 h. These results correlated with the pharmacokinetic results in rats and in rabbits. The high persistence of Rb_1 in serum and tissues was assumed to result from a high degree of plasma protein binding (*7, 55-57*).

Rg_1 was rapidly absorbed after oral administration (approximately 30% after one hour). The concentration of Rg_1 and metabolites was found to be high in blood, liver, bile, subcutis and conjunctiva, and epithelia of the oral cavity, oesophagus, and nasal cavity; whereas, the concentration was low in muscle and endocrine organs, and very low in the brain. Rg_1 was also rapidly metabolized. Intact Rg_1 was excreted in mouse urine and feces in very low amounts, while the concentration of metabolites was high. A total of five metabolites could be detected. Two of them were ginsenoside Rh_1 and 25-OH-Rh_1 (Scheme 7) (*7, 58-59*).

A different approach was chosen for ginsenoside Rh_2. The metabolism of Rh_2 was studied using intact cells (Scheme 8). In a medium containing 2% fetal calf serum and B16 melanoma cells, the uptake reached a maximum after 3-6 h (*60*). Rh_2 was deglycosylated to protopanaxadiol. Both, ginsenoside Rh_2 and protopanaxadiol were shown to inhibit the growth of B16 melanoma cells.

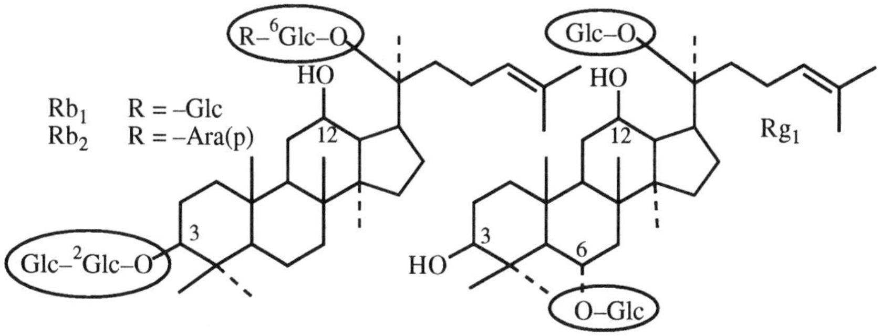

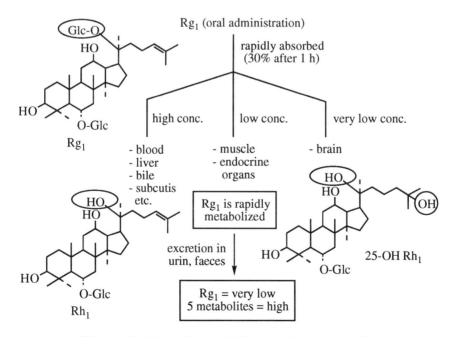

Half-life of protopanaxadiol-type and protopanaxatriol-type ginsenosides in serum

Ginsenoside	Animal		
	Rat	Rabbit	Minipig
Rb_1	14.5 h		16 h
Rb_2		8 h	
Rg_1	6.3 min	69.5 min	27 min

Rb_1 = high degree of plasma protein binding

Scheme 6. Pharmacokinetic Studies of Diol- and Triol-Type Ginsenosides

Scheme 7. Metabolism of Ginsenoside Rg_1 in Mice

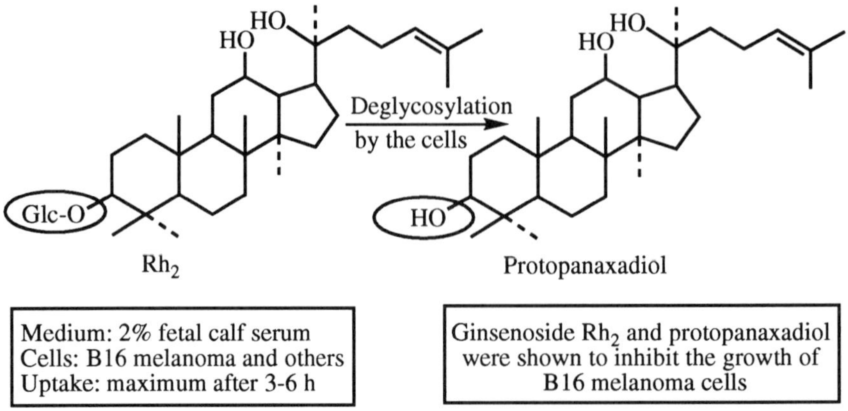
HO
HO
Deglycosylation
by the cells
HO
HO
Glc-O
HO
Rh2
Protopanaxadiol
Medium: 2% fetal calf serum
Cells: B16 melanoma and others
Uptake: maximum after 3-6 h
Ginsenoside Rh2 and protopanaxadiol
were shown to inhibit the growth of
B16 melanoma cells

Scheme 8. Metabolism of Ginsenoside Rh_2 by Intact Cells

Qualitative and Quantitative Analysis of Ginsenosides

In the last two decades, many attempts have been made to identify and quantify the ginsenosides in crude plant material as well as in commercial preparations. For the identification of ginsenosides the following methods have been used: thin layer chromatography (TLC) (*61-62*); high performance TLC (HPTLC) (*61-63*); TLC-densitometry (61-62); overpressured-layer chromatography (OPLC) (*64*); secondary ionization mass spectrometry (SIMS)-linked scanning analysis (*65*); liquid chromatography/mass spectrometry (LC/MS) equipped with a frit-fast atom bombardment interface (FRIT-FAB) (*66-67*); electrospray ion source (*68*); 252-cf plasma desorption ionization technique (*69*); arbitrarily-primed polymerase chain reaction (AP-PCR) and randomly amplified polmorphic DNA fingerprinting (RAPD) (*70-71*); high-performance liquid chromatography (HPLC) in either isocratic or gradient mode using various stationary phases in combination with UV detection (*63, 72-78*), RI detection (*79-81*), fluorescence detection (*82-83*), amperometric detection (*84*), or evaporative light scattering detection (ELSD) (*85*); HPLC and GLC of the aglycones protopanaxadiol and protopanaxatriol after acid or alkaline cleavage and derivatization (*86*); and micellar electrokinetic chromatography (MEKC) (*87*).

For the quantification (partly also for identification) of ginsenosides the following methods have been used: colorimetry (*88-90*); droplet counter-current chromatography (DCCC)-spectrophotometry (*91*); TLC-densitometry (*92-100*); GLC after acid cleavage and derivatization (*101-105*) or after alkaline cleavage (*106*), determination of panaxadiol and panaxatriol by supercritical fluid chromatography (SFC) (*107*); radioimmunoassay (Rg_1) (*108*); enzyme immunoassay (Rb_1) (*109*); HPLC after derivatization of the ginsenosides with benzoyl chloride (*110*); and HPLC in either isocratic or gradient mode using either a normal phase silica gel column (*111*), an amino bonded column (*111-113*), a reversed-phase octadecylsilyl column (RP-ODS) (*13-14, 114-125*), an octadecylsilyl porous glass column (MPG-ODS) (*126-127*), a borate ion exchange column (*111*), a carbohydrate column (*128*), or a hydroxyapatite column (*111*).

Quality Control of Ginseng Phytomedicines

Identification and quantitative determination of ginsenosides by HPLC, mainly using RP-ODS columns followed by UV detection, are now the favored methods, at least in quality control. In certain cases, especially for rapid identification of ginsenosides in plant material or for the determination in biological fluids, other methods, such as LC/MS or radioimmunoassay, are also preferable.

In the case of ginseng analysis by HPLC, sample clean-up has long been a limiting factor. Both, solvent extraction and solid phase extraction methods have been used. Now, sample clean-up using Sep Pak cartridges from Waters (Milford, MA) seems to be the most suitable extract purification procedure. In principal, all the columns used for solid phase extraction have the same goal and are based on the selective elution of the compounds of interest. Interfering compounds are washed through the columns or remain on the cartridge.

Most of the HPLC approaches used today in quality control go back to a method developed in our laboratory in the seventies (*115*). We used RP-ODS columns and acetonitrile-water mixtures as mobile phase. Using acetonitrile-water (30:70), the six ginsenosides Rd, Rb_2, Rc, Rb_1, Rg_2 and Rf, but not Rg_1/Re could be separated and quantitatively determined. The separation of Rg_1 and Re was possible using the same solvent in a ratio 18:82 (Figure 4).

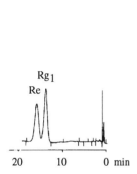

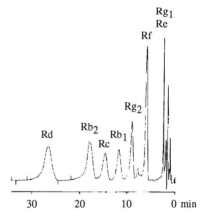

Column: μBondapak C18; 3.9 mm x 30 cm I.D.; eluent: acetonitrile-water (18:82); flow rate: 4 ml/min; detection: UV 203 nm

Column: μBondapak C18; 3.9 mm x 30 cm I.D.; eluent: acetonitrile-water (30:70); flow rate: 2 ml/min; detection: UV 203 nm

Figure 4. HPLC Chromatograms of a Mixture of Pure Ginsenosides Re, Rg1 (left chromatogram) and Rb1, Rb2, Rc, Rd, Re, Rf, Rg1, Rg2 (right chromatogram) (*115*)

Later, an improved HPLC method, also developed in our laboratory, was presented on the occasion of a ginseng meeting in Hong Kong using a programmed elution technique and a photodiode array detection system (*116*). As can be seen from

Table III, the analysis was performed using a Spherisorb ODS II C18 cartridge (3 μm; 10 cm x 4 mm I.D.) The analysis started with 100% eluent A (acetonitrile 20%-water 80%) and 0% eluent B (acetonitrile) under isocratic conditions. After 11 min, a linear rise was programmed until eluent A was 90% and eluent B 10%. After 22 min there was a linear rise until eluent A was 85% and eluent B 15%. Principially the same basic eluents as in the former isocratic analysis were used. However, the use of this programmed elution technique made it possible to separate all eight ginsenosides in a single analytical run (Figure 5). In the meantime a series of papers were published using HPLC and other separation techniques but this method with slight modifications is still the method of choice in quality control of ginseng preparations.

Table III. Ginseng Gradient (Eluent A: Acetonitrile 20%-Water 80%; Eluent B: Acetonitrile; Detection: UV 203 nm) (*116*)

Time (min)	Flow rate (ml)	Eluent A	Eluent B	Curve
0	1	100	0	
11	1	100	0	isocratic
12	1	90	10	linear rise
22	1	90	10	isocratic
23		85	15	linear rise
50	1	85	15	isocratic

HPLC separations, such as recently published by Samukawa et al. (*14*), enable the separation of additional ginsenosides. In this case 22 ginsenosides could be separated in a single run (Figure 6). Such methods are useful for the differentiation of, e.g., white from red ginseng and for the detection of degradation products, but for a quantitative determination they are not useful. As can be seen in this chromatogram malonyl ginsenosides occur only in white ginseng whereas red ginseng shows some additional peaks, such as peaks for Rh_1, $20(R)$ Rh_1, $20(R)$ Rg_2, $Q-R_1$, Rs_1, $20(S)$ and (R) Rg_3.

Conclusions

The chemical constituents of *P. ginseng* are known, and the available analytical methods satisfy the highest standards in quality control of ginseng preparations. Many studies are still being published, investigating the pharmacological effect of the total extract, and of single plant constituents, as well as the medical use of ginseng preparations. The metabolism of the chemical constituents of *P. ginseng* has also been investigated. Although much scientific work has been done in the field, there are still open questions regarding the active compounds, the relevance of the experimentally proven effects, and the therapeutic use of ginseng preparations. It is still unclear if root extracts of *P. ginseng* can be substituted by leaf extracts or possibly by extracts from other ginseng species which often contain similar active compounds. Ginseng preparations are mostly used as "general tonics" and "geriatric remedies", but it is very difficult to describe these fields in scientific terms and to present quantitative results.

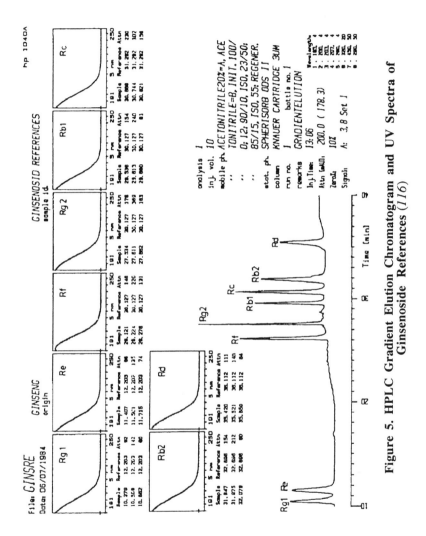

Figure 5. HPLC Gradient Elution Chromatogram and UV Spectra of Ginsenoside References (116)

236

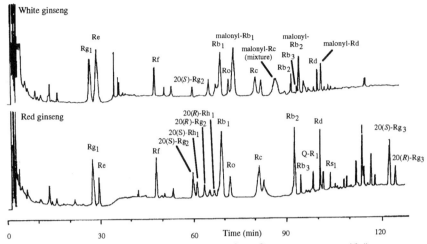

Figure 6. HPLC Gradient Elution Chromatogram (*14*)

In Asia, the general idea of the efficacy of ginseng is to restore Yang. Therefore ginseng preparations are probably used most successfully there. In Chinese medicine, the balance between Ying and Yang is essential for the physical and mental health. Chinese medicine uses a holistic approach to disease, taking into account different factors, while Western medicine diagnoses very analytically—for instance, establishing the cause of a disease as being of viral or bacterial or cancerous origin.

Acknowledgment

The author wishes to thank Annemarie Suter for editorial assistance.

Literature Cited

1. Soldati, F. *Personal communication* **1996**.
2. Owen, R.T. *Drugs of Today* **1981**, *17*, 343-351.
3. Tanaka, O.; Kasai, R. In *Progress in Chemistry of Natural Products*, Vol. 46; Herz, W.; Grisebach, H.; Kirby, G.W.; Tamm, Ch., Eds.; Springer-Verlag: New York, 1984; pp. 1-76.
4. Shibata, S.; Tanaka, O.; Shoji, J.; Saito, H. In *Economic and Medicinal Plant Research*, Vol. 1; Wagner, H.; Hikino, H.; Farnsworth, N.R., Eds.; Academic Press: London, 1985; pp. 217-284.
5. Nham, N.T.; De, P.V.; Luan, T.C.; Duc, N.M.; Shibata, S.; Tanaka, O.; Kasai, R. *J. Jpn. Bot.* **1995**, *70*, 1-10.
6. Hu, S.-Y. *American Journal of Chinese Medicine* **1977**, *5*, 1-23.
7. Tang, W.; Eisenbrand, G. *Chinese Drugs of Plant Origin*; Springer-Verlag: Berlin, Heidelberg, 1992; pp. 711-737.
8. Hostettmann, K.; Marston, A. *Saponins*; Cambridge University Press: Cambridge, 1995; p. 50.
9. Kasai, R.; Besso, H.; Tanaka, O.; Saruwatari, Y.-I.; Fuwa, T. *Chem. Pharm. Bull.* *1983*, *31*, 2120-2125.

237

10. Matsuura, H.; Kasai, R.; Tanaka, O.; Saruwatari, Y.-I.; Kunihiro, K.; Fuwa, T. *Chem. Pharm. Bull.* **1984**, *32*, 1188-1192.
11. Kitagawa, I.; Taniyama, T.; Yoshikawa, M.; Ikenishi, Y.; Nakagawa, Y. *Chem. Pharm. Bull.* **1989**, *37*, 2961-2970.
12. Kim, D.-S.; Chang, Y.-J.; Zedk, U.; Zhao, P.; Liu, Y.-Q.; Yang, C.-R. *Phytochemistry* **1995**, *40*, 1493-1497.
13. Samukawa, K.; Yamashita, H.; Matsuda, H.; Kubo, M. *Yakugaku Zasshi* **1995**, *115*, 241-249.
14. Samukawa, K.; Yamashita, H.; Matsuda, H.; Kubo, M. *Chem. Pharm. Bull.* **1995**, *43*, 137-141.
15. Baek, N.-I.; Kim, D.S.; Lee, Y.H.; Park, J.D.; Lee, C.B.; Kim, S.I. *Planta Med.* **1996**, *62*, 86-87.
16. Kaku, T.; Kawashima, Y. *Arzneim.-Forsch. / Drug Res.* **1980**, *30*, 936-943.
17. Zhang, S.; Takeda, T.; Zhu, T.; Chen, Y.; Yao, X.; Tanaka, O.; Ogihara, Y. *Planta Med.* **1990**, *56*, 298-300.
18. Liu, C.-X.; Xiao, P.-G. *J. Ethnopharmacol.* **1992**, *36*, 27-38.
19. Sonnenborn, U. *Dtsch. Apoth. Ztg.* **1987**, *127*, 433-441.
20. Sonnenborn, U.; Proppert, Y. *Zeitschrift für Phytotherapie* **1990**, *11*, 35-49.
21. Kimura, Y.; Okuda, H.; Arichi, S. *J. Pharm. Pharmacol.* **1988**, *40*, 838-843.
22. Teng, C.-M.; Kuo, S.-C.; Ko, F.-N.; Lee, J.-C.; Lee, L.-G.; Chen, S.-C.; Huang, T.-F. *Biochimica et Biophysica Acta* **1989**, *990*, 315-320.
23. Kuo, S.-C.; Teng, C.-M.; Lee, J.-C.; Ko, F.-N.; Chen, S.-C.; Wu, T.-S. *Planta Med.* **1990**, *56*, 164-167.
24. Deng, H.W.; Guan, Y.Y.; Kwan, C.Y. *Biochemical Archives* **1990**, *6*, 359-365.
25. Deng, H.; Zhang, J. *Chinese Medical Journal* **1991**, *104*, 395-398.
26. Ahn, B.-Z.; Kim, S.-I. *Arch. Pharm.* **1988**, *321*, 61-63.
27. Matsunaga, H.; Katano, M.; Yamamoto, H.; Mori, M.; Takata, K. *Chem. Pharm. Bull.* **1989**, *37*, 1279-1281.
28. Matsunaga, H.; Katano, M.; Yamamoto, H.; Fujito, H.; Mori, M.; Takata, K. *Chem. Pharm. Bull.* **1990**, *38*, 3480-3482.
29. Matsunaga, H.; Saita, T.; Nagumo, F.; Mori, M.; Katano, M. *Cancer Chemother. Pharmacol.* **1995**, *35*, 291-296.
30. Mochizuki, M.; Yoo, Y.C.; Matsuzawa, K.; Sato, K.; Saiki, I.; Tono-oka, S.; Samukawa, K.; Azuma, I. *Biol. Pharm. Bull.* **1995**, *18*, 1197-1202.
31. Sun, X.-B.; Matsumoto, T.; Kiyohara, H.; Hirano, M.; Yamada, H. *J. Ethnopharmacol.* **1991**, *31*, 101-107.
32. Nah, S.-Y.; Park, H.-J.; McCleskey, E.W. *Proc. Natl. Acad. Sci.* **1995**, *92*, 8739-8743.
33. Gao, Q.; Kiyohara, H.; Cyong, J.; Yamada, H. *Planta Med.* **1989**, *55*, 9-12.
34. Yamada, H.; Kiyohara, H. *Abstracts of Chinese Medicines* **1989**, *3*, 104-124.
35. Tomoda, M.; Takeda, K.; Shimizu, N.; Gonda, R.; Ohara, N.; Takada, K.; Hirabayashi, K. *Biol. Pharm. Bull.* **1993**, *16*, 22-25.
36. Tomoda, M.; Hirabayashi, K.; Shimizu, N.; Gonda, R.; Ohara, N.; Takada, K. *Biol. Pharm. Bull.* **1993**, *16*, 1087-1090.
37. Tomoda, M.; Hirabayashi, K.; Shimizu, N.; Gonda, R.; Ohara, N. *Biol. Pharm. Bull.* **1994**, *17*, 1287-1291.
38. Liu, J.; Wang, S.; Liu, H.; Yang, L.; Nan, G. *Mechanisms of Ageing and Development* **1995**, *83*, 43-53.
39. Yamada, H.; Otsuka, H.; Kiyohara, H. *Phytotherapy Reserach* **1995**, *9*, 264-269.
40. Wen, T.-C.; Yoshimura, H.; Matsuda, S.; Lim, J.-H.; Sakanaka, M. *Acta Neuropathol.* **1996**, *91*, 15-22.
41. Ni, X.-H.; Ohta, H.; Watanabe, H.; Matsumoto, K. *Phytotherapy Research* **1993**, *7*, 49-52.
42. Yamasaki, K.; Murakami, C.; Ohtani, K.; Kasai, R.; Kurokawa, T.; Ishibashi, S.; Soldati, F.; Stöckli, M.; Mulz, D. *Phytotherapy Research* **1993**, *7*, 200-202.
43. Liu, J.; Xiao, P.-G. *Phytotherapy Research* **1994**, *8*, 445-451.
44. Petkov, V.D.; Belcheva, S.; Konstantinova, E.; Kehayov, R.; Petkov, V.V.; Hadjiivanova, C. *Phytotherapy Research* **1994**, *8*, 470-477.
45. Nitta, H.; Matsumoto, K.; Shimizu, M.; Ni, X.-H.; Watanabe, H. *Biol. Pharm. Bull.* **1995**, *18*, 1286-1288.

238

46. Zhang, D.; Yasuda, T.; Yu, Y.; Zheng, P.; Kawabata, T.; Ma, Y.; Okada, S. *Free Radical Biology & Medicine* **1996**, *20*, 145-150.
47. Kobayashi, M.; Mahmud, T.; Umezome, T.; Kitagawa, I. *Chem. Pharm. Bull.* **1995**, *43*, 1595-1597.
48. Konno, C.; Sugiyama, K.; Kano, M.; Takahashi, M.; Hikino, H. *Planta Med.* **1984**, *50*, 434-436.
49. Tomoda, M.; Shimada, K.; Konno, C.; Sugiyama, K.; Hikino, H. *Planta Med.* **1984**, *50*, 436-438.
50. Tomoda, M.; Shimada, K.; Konno, C.; Hikino, H. *Phytochemistry* **1985**, *24*, 2431-2433.
51. Oshima, Y.; Konno, C.; Hikino, H. *J. Ethnopharmacol.* **1985**, *14*, 255-259.
52. Konno, C.; Hikino, H. *Int. J. Crude Drug Res.* **1987**, *25*, 53-56.
53. Konno, C.; Murakami, M.; Oshima, Y.; Hikino, H. *J. Ethnopharmacol.* **1985**, *14*, 69-74.
54. Yun, T.-K.; Choi, S.-C. *Cancer Epidemiology, Biomarkers & Prevention* **1995**, *4*, 401-408.
55. Odani, T.; Tanizawa, H.; Takino, Y. *Chem. Pharm. Bull.* **1983**, *31*, 292-298.
56. Odani, T.; Tanizawa, H.; Takino, Y. *Chem. Pharm. Bull.* **1983**, *31*, 1050-1066.
57. Odani, T.; Tanizawa, H.; Takino, Y. *Chem. Pharm. Bull.* **1983**, *31*, 3691-3697.
58. Jenny, E; Soldati, F. In *Advances in Chinese Medicinal Materials Research*; Chang, H.M.; Yeung, H.W.; Tso, W.-W.; Koo, A., Eds.; World Scientific Publ. Co.: Singapore, 1985; pp. 499-507.
59. Strömbom, J.; Sandberg, F. *Acta Pharmaceutica Suecica* **1985**, *22*, 113-122.
60. Ota, T.; Maeda, M.; Odashima, S. *J. Pharm. Sci.* **1991**, *80*, 1141-1146.
61. Xie, P.; Yan, Y. In *Proceedings of the Fourth International Symposium on Instrumental High Performance Thin-Layer Chromatography;* Traitler, H.; Studer, A., Eds.; Institute of Chromatography: Bad Duerkheim, F.R.G., 1987; pp. 401-413.
62. Xie, P.; Yan, Y. In *Proceedings of the Fourth International Symposium on Instrumental High Performance Thin-Layer Chromatography;* Traitler, H.; Studer, A., Eds.; Institute of Chromatography: Bad Duerkheim, F.R.G., 1987; pp. 415-426.
63. Sye, W.-F.; Tsai, P.-F. *J. Chin. Chem. Soc.* **1987**, *34*, 1-6.
64. Nyiredy, S.; Dallenbach-Toelke, K.; Zogg, G.C.; Sticher, O. *J. Chromatogr.* **1990**, *499*, 453-462.
65. Yamamoto, M.; Sugiyama, K.; Ichio, Y.; Yokota, M.; Maeda, Y.; Senda, N.; Shizukuishi, K. *Shoyakugaku Zasshi* **1992**, *46*, 394-398.
66. Hattori, M.; Kawata, Y.; Kakiuchi, N.; Matsuura, K.; Namba, T. *Shoyakugaku Zasshi* **1988**, *42*, 228-235.
67. Hattori, M.; Kawata, Y.; Kakiuchi, N.; Matsuura, K.; Tomimori, T.; Namba, T. *Chem. Pharm. Bull.* **1988**, *36*, 4467-4473.
68. van Breemen, R.B.; Huang, C.-R.; Lu, Z.-Z.; Rimando, A.; Fong, H.H.S.; Fitzloff, J.F. *Anal. Chem.* **1995**, *67*, 3985-3989.
69. Elkin, Y.N.; Makhankov V.V.; Uvarova, N.L.; Bondarenko, P.V.; Zubarev, R.A.; Knysh, A.N. *Acta Pharmacologica Sinica* **1993**, *14*, 97-100.
70. Cheung, K.-S.; Kwan, H.-S.; But, P.P.-H.; Shaw, P.-C. *J. Ethnopharmacol.* **1994**, *42*, 67-69.
71. Shaw, P.-C.; But, P.P.-H. *Planta Med.* **1995**, *61*, 466-469.
72. Pietta, P.; Mauri, P.; Rava, A. *J. Chromatogr.* **1986**, *356*, 212-219.
73. Yamaguchi, H.; Matsuura, H.; Kasai, R.; Mizutani, K.; Fujino, H.; Ohtani, K.; Fuwa, T.; Tanaka, O. *Chem. Pharm. Bull.* **1986**, *34*, 2859-2867.
74. Kanazawa, H.; Nagata, Y.; Matsushima, Y.; Tomoda, M.; Takai, N. *Chromatographia* **1987**, *24*, 517-519.
75. Kasai, R.; Yamaguchi, H.; Tanaka, O. *J. Chromatogr.* **1987**, *407*, 205-210.
76. Takai, N.; Kanazawa, H.; Matsushima, Y.; Nagata, Y.; Tomoda, M. *Seisan Kenkyu* **1989**, *41*, 773-776.
77. Kanazawa, H.; Nagata, Y.; Matsushima, Y.; Tomoda, M.; Takai, N. *J. Chromatogr.* **1990**, *507*, 327-332.

78. Kanazawa, H.; Nagata, Y.; Kurosaki, E.; Matsushima, Y.; Takai, N. *J. Chromatogr.* **1993**, *632*, 79-85.
79. Nagasawa, T.; Yokozawa, T.; Nishino, Y.; Oura, H. *Chem. Pharm. Bull.* **1980**, *28*, 2059-2064.
80. Kaizuka, H.; Takahashi, K. *J. Chromatogr.* **1983**, *258*, 135-146.
81. Lee, M.-K.; Lim, S.-U.; Park, H. *Korean J. Ginseng Sci.* **1988**, *12*, 164-172.
82. Kim, B.-Y.; Lee, M.Y.; Cho, K.H.; Park, J.H.; Park, M.K. *Arch. Pharm. Res.* **1992**, *15*, 328-332.
83. Park, M.K.; Kim, B.K.; Park, J.H.; Shin, Y.G.; Cho, K.H. *J. Liq. Chromatogr.* **1995**, *18*, 2077-2088.
84. Park, M.K.; Park, J.H.; Lee, M.Y.; Kim, S.J.; Park, I.J. *J. Liq. Chromatogr.* **1994**, *17*, 1171-1182.
85. Park, M.K.; Park, J.H.; Han, S.B.; Shin, Y.G.; Park, I.H. *J. Chromatogr. A* **1996**, *736*, 77-81.
86. Cui, J.; Carle, M.; Lund, E.; Björkhem, I.; Eneroth, P. *Analytical Biochemistry* **1993**, *210*, 411-417.
87. Iwagami, S.; Sawabe, Y.; Nakagawa, T. *Shoyakugaku Zasshi* **1992**, *46*, 339-347.
88. Hiai, S.; Oura, H.; Odaka, Y.; Nakajima, T. *Planta Med.* **1975**, *28*, 363-369.
89. Honerlagen, H.; Tretter, H.-R. *Dtsch. Apoth. Ztg.* **1979**, *119*, 1483-1486.
90. Liu, J.H.-C.; Staba, E.J. *J. Nat. Prod.* **1980**, *43*, 340-346.
91. Otsuka, H.; Morita, Y.; Ogihara, Y.; Shibata, S. *Planta Med.* **1977**, *32*, 9-17.
92. Wagner, H.; Wurmböck, A. *Dtsch. Apoth. Ztg.* **1978**, *118*, 1209-1213.
93. Sanada, S.; Shoji, J.; Shibata, S. *Yakugaku Zasshi* **1978**, *98*, 1048-1054.
94. Liberti, L.E.; der Marderosian, A. *J. Pharm. Sci.* **1978**, *67*, 1487-1489.
95. Saruwatari, Y.-I.; Besso, H.; Futamura, K.; Fuwa, T.; Tanaka, O. *Chem. Pharm. Bull.* **1979**, *27*, 147-151.
96. Kubo, M.; Tani, T.; Katsuki, T.; Ishizaki, K.; Arichi, S. *J. Nat. Prod.* **1980**, *43*, 278-284.
97. Vanhaelen, M.; Vanhaelen-Fastré, R. *J. Chromatogr.* **1984**, *312*, 497-503.
98. Schilke, O.; Hohaus, E.; Lentz, H; Kim, J.R. *Z. Naturforsch.* **1991**, *46b*, 829-834.
99. Xin, D.; Wen, W.; Yuqing, S. *Chinese J. Chromatogr.* **1994**, *12*, 173-174.
100. Gu, J.-Q.; Xu, X.-Y.; Ding, Y.; He, L.-Y.; Fang, Q.-C. *Phytochemical Analysis* **1996**, *7*, 152-155.
101. Sakamoto, I.; Morimoto, K.; Tanaka, O. *Yakugaku Zasshi* **1975**, *95*, 1456-1461.
102. Pleinard, J.F.; Delaveau, P.; Guernet, M. *Ann. pharm. franç.* **1977**, *35*, 465-473.
103. Brieskorn, C.H.; Mosandl, A. *Sci. Pharm.* **1978**, *46*, 106-116.
104. Tani, T.; Kubo, M.; Katsuki, T.; Higashino, M.; Hayashi, T.; Arichi, S. *J. Nat. Prod.* **1981**, *44*, 401-407.
105. Bombardelli, E.; Bonati, A.; Gabetta, B.; Martinelli, E.M. *J. Chromatogr.* **1980**, *196*, 121-132.
106. Cui, J. *Eur. J. Pharm. Sci.* **1995**, *3*, 77-85.
107. Li, Y.; Li, X.; Hong, L.; Liu, J.; Zhang, M.-Y. *Biomedical Chromatography* **1992**, *6*, 88-90.
108. Sankawa, U.; Sung, C.K.; Han, B.H.; Akiyama, T.; Kawashima, K. *Chem. Pharm. Bull.* **1982**, *30*, 1907-1910.
109. Kanaoka, M.; Kato, H.; Shimada, F.; Yano, S. *Chem. Pharm. Bull.* **1992**, *40*, 314-317.
110. Besso, H; Saruwatari, Y.; Futamura, K.; Kunihiro, K.; Fuwa, T.; Tanaka, O. *Planta Med.* **1979**, *37*, 226-233.
111. Yamaguchi, H.; Matsuura, H.; Kasai, R.; Tanaka, O.; Satake, M.; Kohda, H.; Izumi, H.; Nuno, M.; Katsuki, S.; Isoda, S.; Shoji, J.; Goto, K. *Chem. Pharm. Bull.* **1988**, *36*, 4177-4181.
112. Nishimoto, N.; Masaki, S.; Hayashi, S.; Takemoto, T.; Hayashi, T.; Tsuji, N. *Shoyakugaku Zasshi* **1986**, *40*, 345-351.
113. Yoshikawa, M.; Fukuda, Y.; Hatakeyama, S.; Murakami, N.; Yamahara, J.; Taniyama, T.; Hayashi, T.; Kitagawa, I. *Yakugaku Zasshi* **1993**, *113*, 460-467.
114. Sticher, O.; Soldati, F. *Planta Med.* **1979**, *36*, 30-42.
115. Soldati, F.; Sticher, O. *Planta Med.* **1980**, *39*, 348-357.

240

116. Meier, B.; Meier-Bratschi, A.; Dallenbach-Tölke, K.; Sticher, O. In *Advances in Chinese Medicinal Materials Research*; Chang, H.M.; Yeung, H.W.; Tso, W.-W.; Koo, A., Eds.; World Scientific Publ. Co.: Singapore, 1985; pp. 471-484.

117. Yamaguchi, H.; Kasai, R.; Matsuura, H.; Tanaka, O.; Fuwa, T. *Chem. Pharm. Bull.* **1988**, *36*, 3468-3473.

118. Guédon, D.; Abbe, P.; Cappelaere, N; Rames, N. *Ann. pharm. franç.* **1989**, *47*, 169-177.

119. Petersen, T.G.; Palmquist, B. *J. Chromatogr.* **1990**, *504*, 139-149.

120. Makhan'kov, V.V.; Samoshina, N.F.; Malinovskaya, G.V.; Atopkina, L.N.; Denisenko, V.A.; Isakov, V.V.; Kalinovskii, A.I.; Uvarona, N.I. *Chemistry of Natural Products* **1990**, *26*, 46-48.

121. Makhan'kov, V.V.; Samoshina, N.F.; Uvarova, N.I.; Elkyakov, G.B. *Chemistry of Natural Products* **1993**, *29*, 196-199.

122. Lang, W.-S.; Lou, Z.-C.; But, P.P.-H. *J. Chin. Pharm. Sci.* **1993**, *2*, 133-142.

123. Chuang, W.-C.; Sheu, S.-J. *J. Chromatogr. A* **1994**, *685*, 243-251.

124. Chuang, W.-C.; Wu, H.-K.; Sheu, S.-J.; Chiou, S.-H.; Chang, H.-C.; Chen, Y.-P. *Planta Med.* **1995**, *61*, 459-465.

125. Yamasaki, K.; Hashimoto, A.; Kokusenya, Y.; Miyamoto, T.; Nakai, H.; Sato, T. *Yakugaku Zasshi* **1995**, *115*, 62-71.

126. Kanazawa, H.; Nagata, Y.; Matsushima, Y.; Tomoda, M.; Takai, N. *Shoyakugaku Zasshi* **1989**, *43*, 121-128.

127. Kanazawa, H.; Nagata, Y.; Matsushima, Y.; Tomoda, M.; Takai, N. *J. Chromatogr.* **1993**, *630*, 408-414.

128. Bartolomé, R.; Masoliver, D.; Soler, M.; Vilageliu, J. *Alimentaria (Madrid)* **1987**, *24*, 73-76.

Chapter 17

Hawthorn (*Crataegus*): Biological Activity and New Strategies for Quality Control

O. Sticher[1] and B. Meier[2]

[1]Department of Pharmacy, Swiss Federal Institute of Technology (ETH) Zurich,
CH–8057 Zürich, Switzerland
[2]Zeller AG, Herbal Remedies, CH–8590 Romanshorn, Switzerland

Hawthorn consists of the dried tips of the flower-bearing branches of *Crataegus monogyna* or *Crataegus laevigata*. Phytomedicines based on the extract of the crude drug are indicated for the treatment of NYHA stage II heart failure patients. A survey of the botany, chemistry, pharmacology, and clinical uses, as well as of the quality control, of hawthorn and hawthorn-containing phytomedicines is given. Flavonoids and procyanidins are considered to be the main active compounds. Their structures, as well as their qualitative and quantitative determination in hawthorn and hawthorn-containing phytomedicines are presented. Research work on quality control and standardization conducted in the laboratories at ETH is discussed in detail.

Hawthorn is one of the most popular herbal drugs in central Europe. In Germany as well as in Switzerland there is a large number of phytomedicines based on the extract of hawthorn on the market. Older products are combinations, often with sedative herbal drugs. Modern preparations use hawthorn as a monodrug for the treatment of declining cardiac performance equivalent to stage II of the NYHA (New York Heart Association) classification, first proposed by the German Commission E. This monograph will be supplemented in the near future by a monograph of more international relevance edited by ESCOP (European Scientific Cooperation of Phytotherapie). The indications for hawthorn are similar in both monographs. The pharmacological profile of hawthorn is very promising. Several biochemical pathways are influenced by hawthorn extracts and no toxicological effects have been observed so far.

Compared to the well-documented pharmacological and clinical properties, quality control and standardization of hawthorn have been on a low level. The insufficient photometric method of DAB 10 has been used mainly for standardization. Meanwhile, more specific methods, especially for flavonoids have been established and will be discussed in this paper with the objective to adapt the level of the pharmaceutical quality of hawthorn preparations to the level of other popular herbal drugs like ginseng, ginkgo, St. John's wort and garlic.

Botany

The genus Crataegus (hawthorn; syn.: hawthorne, hedge thorn, may thorn, white thorn or whiterthorn) is classified in the tribe Crataegeae, subfamily Maloideae, of the Rosaceae family. It comprises about 150 to 1200 species, depending on the species concept and represents not only the largest but also a taxonomically difficult genus within the Maloideae. About 50 to 100 species occur in temperate areas of the Old World. The various species of Crataegus are shrubs of medium size or little trees with hard wood and thorny branches, bearing abundantly white flowers in May and revealing large numbers of red fruits in late summer. Due to numerous subspecies and hybridization it is difficult to classify them systematically. Some criteria for classification are the formation of the flowers, the shape of the leaves and how they are covered with hairs (*1,2*).

In phytotherapy, two species are of importance: *Crataegus monogyna* Jacquin and *Crataegus laevigata* (Poiret) De Candolle (syn.: *Crataegus oxyacantha* L.p.p. duct). In 1753, Linné classified both species as *C. oxyacantha* L., but subsequently they were subdivided by Jacquin by the end of the 18th century into *C. oxyacantha* L. (nowadays *C. laevigata*) and *C. monogyna* Jacq. (*3*).

C. monogyna, widely distributed in Europe, Asia and North Africa, grows in sites with high light intensity such as hedges and fences, as well as gardens and parks. Although it prospers on all kinds of soil, it prefers chalky, loamy soils. The leaves are laciniate, often 5- to 7-lobed and have a relatively dull surface (Figure 1). The flowers bear one style, and the fruits have one pit. The leaves of *C. laevigata* are shallowly 3-lobed or almost entire and have a very shiny surface. The flowers yield two styles, and the fruit has 2-3 pits. *C. laevigata* is spread throughout Europe and America, where it is also partially cultivated. It grows predominantly on dry soils, in hedges and, due to its tolerance of shade, also in continuous forests. *C. laevigata* begins to flower a week or two earlier than *C. monogyna*, and the evil-smelling scent of its flowers is much more distinct than the scent of *C. monogyna* (*2, 4-6*).

Hawthorn Monographs in European Pharmacopeias

The herbal drug "Crataegi folium cum flore" (Hawthorn leaves with flowers) is specified for example, in the German, Swiss and French pharmacopeias (DAB 10, Ph. Helv. VII, Ph. Franç. X). A monograph "Hawthorn Leaf with Flower" is in preparation for the European pharmacopeia (Ph. Eur. 3). Besides *C. monogyna* and *C. laevigata*, three other species are admitted in DAB 10: *C. pentagyna* Waldstein et Kitaibel ex Willdenow, *C. nigra* Waldstein et Kitaibel, and *C. azarolus* L. Due to their different TLC fingerprints, they are not listed in Ph. Helv. VII, Ph. Franç. 10, or in the draft of Ph. Eur. 3. The fruits and flowers of hawthorn are used therapeutically as well. The fruits are included in the German Pharmacon Codex (DAC 86) and drafted for Ph. Eur., the flowers are specified in Ph. Franç. X.

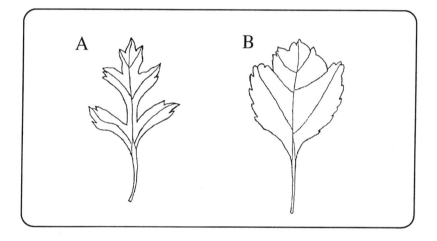

Figure 1. Leave shapes of *C. monogyna* (A) and *C. laevigata* (B) according to Byatt (*1*) and El-Gazzar (*7*).

Constituents of Hawthorn

Since the beginning of this century, the constitutents of various Crataegus species and of different parts of the plant have been investigated in detail. A series of classes of compounds have been detected, such as flavonoids, procyanidins, catechins, triterpenoid acids, aromatic carbonic acids, nitrogen containing compounds, as well as various ubiquitous compounds like fatty oil, fatty acids, essential oil, b-sitosterol, various sugars, sorbitol, vitamin B1, vitamin C, pectin, wax, carotenoids, chlorophyll, an anthocyan pigment, and oxalic acid. The pattern and content of constituents in plant material or extracts depend on various factors, such as the investigated species, the part of the plant, the date of collection, the soil, and the climate, as well as the solvent used for extraction. It is not surprising that varying or even contradictory results are reported in the literature (*8* and references cited therein). The most comprehensive studies were carried out with polyphenolic compounds, such as flavonoids and procyanidins, which are considered to contribute decisively to the pharmacological effects of hawthorn. In this paper only these two classes of compounds will be discussed in detail.

Flavonoids. Table I gives an overview of the main flavonoids isolated from various Crataegus species, representing a series of flavonol-O-glycosides, such as hyperoside, and flavone-C-glycosides, such as vitexin, vitexin-2"-O-rhamnoside, and acetylvitexin-2"-O-rhamnoside. Further constituents have been isolated or detected, such as kaempferol, quercetin-3-O-rhamnogalactoside, spiraeoside and isoquercitroside, crataeside, 8-methoxy-kaempferol-3-O-glucoside, luteolin-7-O-glucoside, orientin, isoorientin, isovitexin-2"-O-rhamnoside, orientin-2"-O-rhamnoside, isoorientin-2"-O-rhamnoside, schaftoside, isoschaftoside, neoschaftoside, neoisoschaftoside, vicenin-1, vicenin-2, vicenin-3, 8-methoxykaempferol-3-neohesperidoside, kaempferol-3-neohesperidoside, and an eriodictyol-glycoside. The main flavonoids are hyperoside, vitexin-2"-O-rhamnoside, and acetylvitexin-2"-O-rhamnoside. All other constitutents are only found in minor concentrations or in trace amounts (*8* and literature cited therein).

Most of the chemical investigations of hawthorn were made with *C. laevigata* and *C. monogyna*. Hyperoside and spiraeoside occur predominantly in the flowers, whereas the leaves contain mainly vitexin-2"-O-rhamnoside and/or acetylvitexin-2"-O-rhamnoside. Vitexin-2"-O-rhamnoside was observed as the main flavonoid in leaves of *C. laevigata*, while acetylvitexin-2"-O-rhamnoside prevails in leaves of *C. monogyna*. *C. laevigata* reveals only trace amounts of acetylvitexin-2"-O-rhamnoside. The flavonoid pattern of the other mentioned Crataegus species is similar but varies from species to species (*8* and literature cited therein).

Procyanidins. Procyanidins are the most common class of condensed proanthocyanidins. They consist of chains of catechin and/or epicatechin units linked by 4→8 or 4→6 bonds. The acid-lability of this interflavan C-C bond gives rise to the release of catechin or epicatechin and cyanidin, hence the name procyanidins. The flavan-3-ol skeleton has two centers of asymmetry, C-2 and C-3. From these arise pairs of monomeric diastereoisomers and a large number of possible configurations of the dimeric and oligomeric procyanidins. The absolute configuration of the monomers are usually 2R in all cases and 3S for (+)-catechin and 3R for (-)epicatechin.

Table I. Main Flavonoids of Hawthorn Leaves with Flowers

Compound	$R_1{}^a$	$R_2{}^a$	$R_3{}^a$	R_4
Quercetin	OH	H	H	OH
Hyperoside	O-gal	H	H	OH
Rutin	O-rut	H	H	OH
Vitexin	H	H	glc	H
Isovitexin	H	glc	H	H
Vitexin-2"-O-rhamnoside	H	H	glc-rha	H
Acetylvitexin-2"-O-rhamnoside	H		glc-rha-acetyl	H

^{a} glc = β-D-glucosyl
    rut = rutinose (α-L-rhamnosyl-1→6-β-D-glucosyl)
    gal = β-D-galactosyl
    rha = α-L-rhamnosyl

From the investigations of more than fifty plant species, it is known that one of the two diastereoisomeric flavan-3-ols, catechin or epicatechin, often clearly predominates in the plant, and, occasionally, the other isomer does not appear. As a consequence, the majority of the dimeric, oligomeric and polymeric procyanidins consist only of the corresponding monomer. When epicatechin predominates, the co-occurrence of the procyanidin dimers B-2 and B-5 and the trimer C-1 is the most frequently recognized pattern of procyanidins. This pattern was also found in hawthorn leaves and fruits. Beside catechin and epicatechin, the dimeric procyanidins B-1, B-2, B-4, B-5, the trimer C-1, and the tetramer D-1 were isolated from the leaves and fruits of Crataegus species. The content of procyanidins in the leaves varies from species to species and also during the vegetation period, revealing the highest concentration in late summer (*8* and literature cited therein).

4,8-all-2,3-cis-Procyanidin (B-2)

4,6-all-2,3-cis-Procyanidin (B-5)

Pharmacological Effects and Therapeutical Use

Overviews of the pharmacological effects of Crataegus extracts, as well as of the flavonoid and procyanidin fractions, have been published by several authors (e.g. *8-11*). For specific literature see Rehwald (*8*).

Dioscorides first mentioned the curative effect of hawthorn. Since the 14th century hawthorn was described in several herbal textbooks, but not as a cardioactive medicinal plant. The contemporary use of hawthorn for cardiovascular diseases started only in the second half of the this century. Due to its reliable action, its good success in therapy, and lack of side effects, Crataegus has obtained a proper place in phytotherapy and is firmly established in traditional and conventional medicine, in self-medication as well as in medicinal practises. In the German "Rote Liste 1994", 33 preparations containing only hawthorn extracts are listed, and the Swiss "Codex Galenica 1992" includes 11 mono- and about 25 multipreparations of Crataegus.

There is no final agreement concerning the question which constituents contribute to the well-known therapeutic effect of hawthorn preparations. During the past few decades, all classes of constituents were once regarded as principally responsible, and a series of pharmacological activities were demonstrated in numerous studies, either for extracts or for isolated fractions and pure compounds. The first attempt to find the active principle of hawthorn was made in 1927. Crataegic acid showed a cardiotonic effect and was, therefore, regarded as responsible for the pharmacological activity. Later it was shown that crataegic acid, a mixture of triterpenoid acids, produced dilatation of the coronary vessels, accompanied by a fall of the blood pressure. These findings could not be confirmed by other investigators.

It was shown in 1953 that aqueous and alcoholic preparations of Crataegus could lower blood pressure and induce muscle relaxation, suggesting the presence of choline-like compounds. In fact, choline and acetylcholine were detected, but these compounds are not characteristic for Crataegus and their content in the plant is too small to support them as the active constituents. The amines isolated from Crataegus caused hypertension and showed a positive inotropic activity in vitro on the guinea pig papillary muscle. But taken orally, which is the most common way of administration, the amine fraction of Crataegus cannot be responsible for the pharmacological effects due to rapid degradation by monoamine oxidases. Various authors claimed that the pharmacological effect of Crataegus is mainly based on a heptahydroxyflavan glycoside or a phlobaphene, respectively.

The classes of constituents discussed so far could not verify adequately their contribution to the therapeutic effect of hawthorn. Numerous studies and experiments have shown that flavonoids and procyanidins contribute significantly to the cardiac effect of hawthorn preparations, and that other constituents are probably only playing a supporting role. One of the main effects of Crataegus is the enhancement of the coronary blood flow and myocardial perfusion. Administration of hawthorn extracts to dogs showed the effect to be dose-dependent. Additional studies assigned the activity to the oligomeric procyanidins fraction. Other authors also found an increase in the coronary blood flow in rats and rabbits and concluded that the activity of Crataegus extracts is not due to a single group of constituents (flavonoids), but to the interaction between different groups of compounds. The mechanism of action is discussed as to be based on an inhibition of 3',5'-cyclic adenosine monophosphate phosphodiesterase. Computer-assisted model analysis showed that the Crataegus constituents catechin, vitexin and kaempferol exhibit similarities with the molecular structures of known phosphodiesterase inhibitors.

An improvement of the cardiac muscle contractility (positive inotropic effect) was shown in various animal models. This effect was obtained by the flavonoid and procyanidin fraction, but with different dose-response-curves. On the basis of computer-assisted model analysis, it was demonstrated that ursolic acid can interact with the digitaloid binding site for Na^+/K^+-ATPase. In addition, Crataegus exhibits a mild hypotensive effect. A reduction of the blood pressure in various animals was observed. A hypotensive effect in humans has also been reported. The mechanism of action is supposed to be an inhibition of angiotensin converting enzyme, as demonstrated for procyanidins. Furthermore, Crataegus showed an antiarrhytmic activity in rabbits and revealed a protective action in experimental arrhythmias in rats and rabbits. Also, an anti-ischaemic effect of Crataegus in animal models has been reported.

Apart from the cardiovascular effects, Crataegus shows activity on the central nervous system. A mild but significant sedative effect due to depressive action on the central nervous system was demonstrated in mice. A clear sedative effect indicating direct influence on the central nervous system was also shown in cats by the fraction of oligomeric procyanidins. A decrease of the body temperature and of the spontaneous activity, as well as a prolongation of the pentobarbital-induced sleep in mice, were found. Recently, antioxidant activities of different extracts from *C. monogyna* and of a series of constituents of hawthorn were reported. A good correlation between antioxidant activity and content of total phenolics, which indicated additive effects between the different classes of compounds, were found.

The activity profile, as well as the main indications for hawthorn preparations, can be summarized as shown in Table II.

Table II. Profile of Activity and Main Indications for Hawthorn Preparations

Profile of activity
- Weak positive inotropic, positive chronotropic and negative bathmotropic effects
- Increase of the coronary and myocardial circulation
- Vasodilatation in the peripheral arterial area
Indications
- Chronic heart failure defined as NYHA functional class II
- Deficiency of the coronary blood supply
- Mild forms of arrhythmias

Hawthorn shows important advantages compared with conventional inotropic substances, such as beta-agonists and cardiac glycosides. The effects achieved with hawthorn extracts (e.g. positive inotropic effect on the contraction amplitude) were more economic for the cell's energy pool due to only a moderate increase in oxygen consumption for both the contractile and the ionic processes. In contrast to cardiac glycosides, no toxic effects have been observed (*12*).

Recent Clinical Studies

For the evaluation of effectiveness and safety, controlled clinical studies are the method of choice. Although clinical studies with drugs of plant origin can impose some problems, several well-documented papers have been published within the last 15 years (1981-1994). Reviews on clinical trials have been published by Tauchert *et al.* (*13*) and by von Eiff (*14*).

The multifactorial activity of hawthorn (combined with the herb of passion flower, a drug with similar flavonoid pattern as hawthorn) has been shown by von Eiff *et al.* (*15*). Various parameters making the function of the heart easier have been influenced in a positive sense. In a randomized, double-blind, placebo-controlled, parallel study, in which patients were stratified according to physical performance, they received either hawthorn/passion flower extract or placebo. Exercise capacity and improvement in symptoms were evaluated in a total of 40 patients aged 53 to 86 years, with dyspnoea commensurate with an NYHA functional class II. Over a 6 week trial period, the patients received oral drops three times daily (2 ml of the preparation, equivalent to a daily dose of 1.56 g Crataegi folium cum flore, containing 15 mg of flavone-C-glycosides calculated as vitexin-2"-O-rhamnoside and 28 mg total procyandines, calculated as cyanidol). A 6-minute walking test and bicycle ergometer test were conducted to assess changes in exercise capacity. Blood lactate levels, heart rate and blood pressure were also measured during exercise. Also recorded were improvements in symptoms, particularly dyspnoea, and changes in urine and blood biochemical parameters.

The results showed that exercise capacity, measured in terms of a walking test, increased significantly in those patients receiving the hawthorn/passion flower extract ($p < 0.05$, two-tailed test). Maximum exercise capacity measured during a bicycle ergometer test increased by about 10% over the baseline in both groups, but changes between groups did not differ significantly. The subjective symptom, breathlessness, improved in both the extract and placebo groups by approximately 40% and 30%, respectively. Physical performance capacity at the aerobic transition threshold of 2 mmol lactate/l blood increased by 6.5% in the group receiving the extract, while it fell by 4.4% in the placebo group. This difference was, however, not statistically significant ($p > 0.05$). In the group receiving the extract there was a slight but significant fall in heart rate at rest and in mean diastolic blood pressure during exercise ($p < 0.05$, two-tailed test). There was a significant decrease in total plasma cholesterol levels in the group receiving the extract in comparison with the placebo group ($p < 0.01$, two-tailed test).

Positive results in the ergometer test system are dose dependent. A dose of 600 mg of extract (about 2 times more compared with the dosage of 1.56 g of the drug used by von Eiff *et al.* (*15*)) showed a significant effect in a study with 78 NYHA stage II heart failure patients (*16*). A dose of 900 mg of hawthorn extract has been compared with an adequate low dosage of captopril. None of the target parameters (ergometry, pressure-rate-product, score for five typical symptoms) showed any significant differences between the hawthorn preparation and the reference drug. 132 NYHA stage II heart failure patients have been included in this multicentered double-blind study (*17*).

Previous Analytical Work with Crataegus

Flavonoids. For the qualitative analysis (identity and purity control) of Crataegus plant material or Crataegus preparations, two methods are used: Thin-layer chromatography (TLC) on silica gel or reversed-phase sorbents and reversed-phase high-performance liquid chromatography (HPLC) fingerprint analysis. Both identify the flavonoid pattern and allow a fast and reliable identity and purity control. For identity control, TLC fingerprint analysis is probably the most wide-spread and most accepted method besides macroscopic and microscopic examination. It is found in pharmacopeias and other official monographs (Ph. Helv. VII, DAB 10, Ph. Franç. 10, DAC 86). TLC usually reveals sufficient information about the sample, but, nevertheless, reversed-phase HPLC fingerprint methods were developed as efficient completion to the TLC identity control (*18-24*). Usually, they are carried out on C-18 columns with mobile phases based on acetonitrile, water, and acetic acid, using isocratic or gradient elution.

Apart from checking identity and purity, quality control requires a quantitative determination of the active compounds or lead compounds. The usual method for the determination of flavonoids in Crataegus comprises the spectrophotometric analysis of a hydrolyzed extract after addition of aluminium chloride (e.g. DAB 10, DAC 86)

(Figure 2). This method goes back to Christ and Müller (25) and was applied until recently to nearly all flavonoid-containing herbal drugs. Glasl and Becker (26) demonstrated that the absorption of the analyzed solution is caused by colored pigments such as chlorophyll, carotenoids, and xanthophylls. These interferences, which occur especially in herbal drugs consisting of leaves, can be removed by extraction with hexane, though with a considerable loss of flavonoids (27). For plants with a high content of C-glycosides, such as Crataegus, this spectrophotometric method is still not suitable, because C-glycosides can not be hydrolyzed by acid and, therefore, are not extracted with ethyl acetate and not determined. By means of an HPLC fingerprint chromatogram (see later), it was shown in our laboratory that all the vitexin derivatives remain to a large extent in the aqueous solution which is discarded. This confirmed the previous results (20, 28). The wavelength of 425 nm is not the absorption maximum of the vitexin derivatives that have partially passed into the ethyl acetate. Since this method comprises only the flavonol-O-glycosides and not the flavone-C-glycosides, it is not accurate and yields unsatisfactory results.

Today, the method of choice for quantitative analysis of flavonoids is reversed-phase HPLC. Several HPLC methods for the quantitative determination of flavonoids in Crataegus are described (18, 20, 21-23, 30-32). They mostly consist of a quantification of individual or total flavonoids in a fingerprint chromatogram by means of external or internal standards. Until recently, Ph. Franç. 10 was the only pharmacopeia that used a reversed-phase HPLC method. Based on our work (see later), the Swiss (Ph. Helv. VII, supplement 1996) and the European (Ph. Eur. 3, draft) pharmacopeias also now contain a quantitative HPLC determination of flavonoids.

Procyanidins. A wide variety of analytical methods for procyanidins is currently being used. They have been the subject of numerous reviews (33-39). Qualitative analyses are often carried out using TLC on silica gel (37, 40-41). HPLC on reversed-phase material offers better separation and is a commonly applied method for qualitative analysis of procyanidins (for examples, see 42-45), but, nevertheless, the reported chromatograms lack a selective baseline separation and reveal remarkable matrix effects.

Quantitative analysis of procyanidins has several problems. Quantitative determination by HPLC is difficult because often the separation of the complex mixture is not sufficient for quantification, and/or reference standards for external standardization or calculation factors are not available. If a quantitative determination by HPLC is not feasible, a spectrophotometric analysis after a color reaction is possible (Figure 3). The reagents used are the non-specific Folin-Denis reagent, vanillin or p-dimethylaminocinnamaldehyde (37, 46-47). Owing to a certain specificity, the acid butanol method is generally the best assay. It is based on the release of cyanidin by heating with mineral acid in butanol. This method, originally published by Swain and Hillis (48), was made popular by Bate-Smith (49) and improved by Porter et al. (50). It is now proposed in Ph. Eur. 3 for hawthorn berries (draft).

The main problem with the spectrophotometric methods is that they suffer from lack of reproducibility. The color-generating reactions seem to not be quantitative. The reaction mechanisms and variables are not really known. It has to be kept in mind that the methods mentioned above are all conventional methods representing a

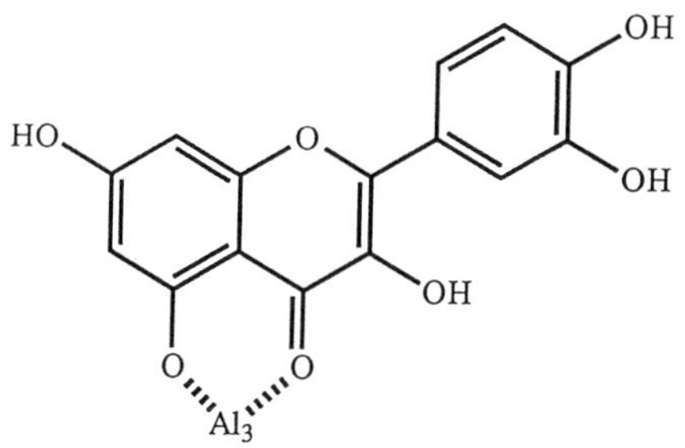

Figure 2. Spectrophotometric flavonoid assay according to DAB 10. The aluminium chelate complex of the flavonoid aglycones is determined after hydrolysis of the glycosides.

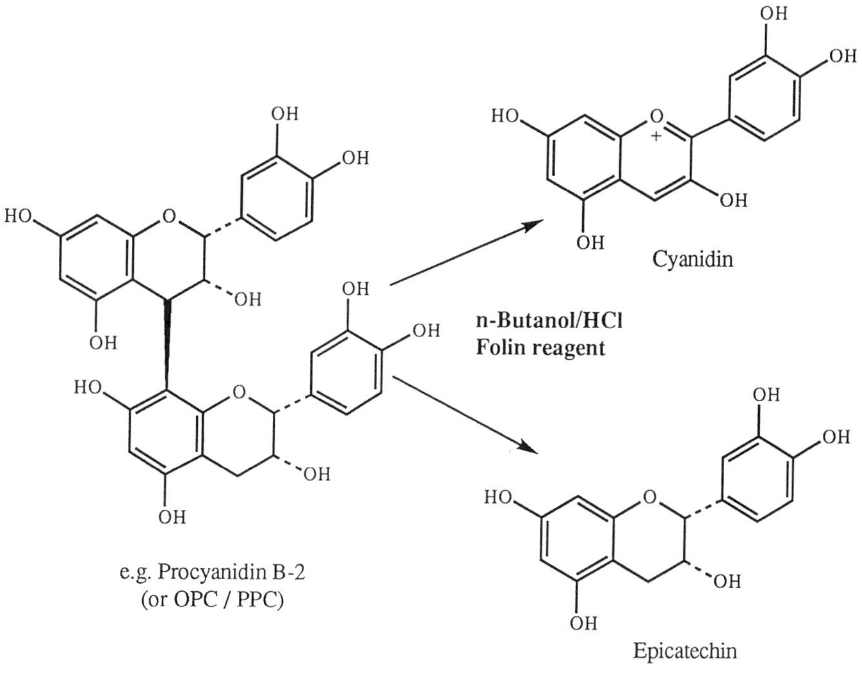

Figure 3. Principle of the spectrophotometric determination of procyanidins.

compromise between feasibility and accuracy and that the results for procyanidin content are meaningful and useful only together with a detailed instruction of the method that has been applied. Methods using micellar electrokinetic capillary chromatography (MECC) (*51*), HPLC with electrochemical detection, as well as LC-MS and LC/MS/MS (*52*) are under investigation.

Own Analytical Work on Crataegus

Hawthorn can be problematic for quality control since two groups of constituents, flavonoids and procyanidins, are considered to be the active principles. The standardization aims at either the flavonoids or the oligomeric procyanidins or both. Our primary goal was the development of suitable methods for quality control and standardization of the flavonoid glycosides occurring in hawthorn. They are pre-dominantly polar, hydrophilic substances. The columns most used for their separation are of the reversed-phase type. Since the majority of the reported chromatograms showed insufficient separation, the development of a selectively separating fingerprint method was elaborated. A basic modification, compared to previous HPLC methods, is the quantitative determination after acid hydrolysis. This step was considered necessary due to the great variety of flavonoid glycosides which can be reduced by hydrolysis to the major aglycones. The main flavonol-O-glycosides, hyperoside and rutin, can be hydrolyzed to quercetin, while the principal flavone-C-glycosides, vitexin-2"-O-rhamnoside and acetylvitexin-2"-O-rhamnoside, react to form vitexin. The availability of all glycosidic compounds for external standardization or calculation factors is not necessary. The commercially available standards, quercetin and vitexin, can be used for calibration curves for the standardization of Crataegus plant material and preparations. As a result, a simple, rapid, and reproducible method for the standardization of hawthorn extracts has been realized (*8, 53*).

Extraction and Sample Preparation. The work-up procedure consists of two basic steps: extraction and hydrolysis of the glycosides, and sample clean-up (Figure 4). The flavonol-O- and flavone-C-glycosides of hawthorn are hydrolized and analyzed separately. Mild hydrolytic conditions are sufficient for the hydrolysis of hyperoside and other flavonol-O-glycosides. For a complete hydrolysis of the vitexin derivatives, vitexin-2"-O-rhamnoside and its acetate, more hydrochloric acid and a longer reflux time are required. These conditions are too strong for the flavonol-O-glycosides and lead to a loss of 15-20% of quercetin. Sample clean-up was performed by filtration through a Bond Elut C-18 cartridge.

Qualitative Analysis. Five commercial samples of hawthorn leaves with flowers were examined for their flavonoid composition. They all showed rather uniform fingerprint chromatograms, with vitexin-2"-O-rhamnoside, hyperoside, and acetylvitexin-2"-O-rhamnoside as the main peaks in similar relative concentrations (Figure 5). Small amounts of vitexin and rutin were also found. Chlorogenic acid was detected in all of the samples.

Quantitative Analysis. Quantitative HPLC determinations of quercetin and vitexin/isovitexin after acid hydrolysis are shown in Figures 6 and 7, respectively.

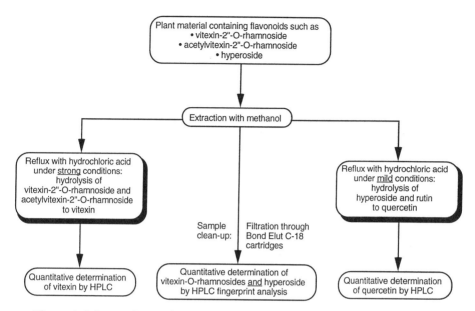

Figure 4. Scheme of extraction and sample preparation of hawthorn leaves with flowers for quantitative determination.

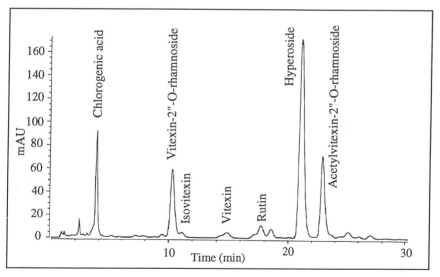

Figure 5. Typical HPLC fingerprint chromatogram of hawthorn leaves with flowers. Column: Hypersil ODS (5 mm) (100 x 4 mm i.d.); mobile phase: A = THF-acetonitrile-MeOH 92.4:3.4:4.2, B = 0.5% o-phosphoric acid; lin. gradient 0-12 min = 12% A in B, 12-25 min = 12% to 18% A in B, 25 -30 min = 18% A in B; flow: 1.0 ml/min; temperature: 25°C; detection: UV at 370, 336, 260nm.

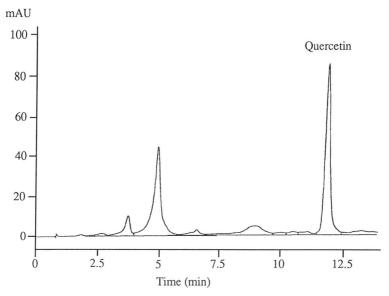

Figure 6. HPLC chromatogram of a hydrolyzed extract for the determination of quercetin. Column: Hypersil ODS (5 mm) (100 x 4 mm i.d.); mobile phase: A = MeOH, B = 0.5% o-phosphoric acid; gradient elution 0-15 min = 30% A/70% B to 55% A/45% B; flow: 1.0 ml/min; temperature: 25°C; detection: UV at 370, 260nm.

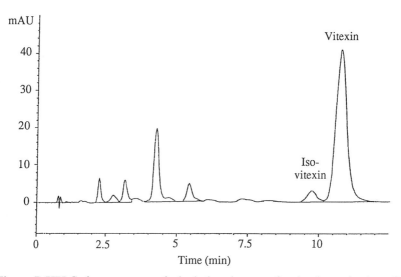

Figure 7. HPLC chromatogram of a hydrolyzed extract for the determination of vitexin and isovitexin. Column: Hypersil ODS (5 mm) (100 x 4 mm i.d.); mobile phase: A = THF-isopropanol-acetonitrile 10:8:3, B = 0.5% o-phosphoric acid; isocratic elution 0-13 min = 12% A/88% B; flow: 1.0 ml/min; temperature: 25°C; detection: UV at 336, 260nm.

Twelve batches of dried leaves and fruits of *Crataegus monogyna* at different plant development stages were examined by the hydrolysis method for their flavonoid content (Figure 8). The highest amount of total flavonoids was found in the leaves during the pre-flowering time (April) and the flowering time (beginning of May). Afterwards the total flavonoid content in the leaves decreased and remained at a rather constant level, comparable with the flowers. The fruits, containing mainly oligomeric and polymeric procyanidins, exhibited very low flavonoid concentrations, and their flavonoid determination and standardization was left out. The flowers and leaves revealed remarkable differences concerning the ratio of vitexin and querecetin: the ratio was 0.4 for the flowers, but 7 for the leaves, at the beginning of the collecting period, decreasing to 0.8 in October (*53*). These results are in accordance with results of other authors reporting the maximal flavonoid content for leaves during the pre-flowering and the flowering time in May (*31,54*), with increasing amounts of flavonol-O-glycosides towards the end of the vegetation period. Taking into consideration these results, it is recommended to harvest both the leaves and the flowers during the flowering time on one hand to get the highest content of total flavonoids and, on the other hand, a well-balanced proportion of flavonol-O- and flavone-C-glycosides.

Validation of the Quantitative Hydrolysis Method. The newly elaborated hydrolysis method was validated thoroughly. Hydrolysis conditions, modes of extraction, calibration curves, precision, and accuracy were examined. Additionally, the hydrolysis method was compared with three other methods.

In a first step the hydrolysis of the reference standards vitexin-2"-O-rhamnoside and hyperoside was examined. Subsequently, the conditions were transferred to the plant material and adapted once again. This attempt showed, as seen earlier, that flavonol-O-glycosides and flavone-C-glycosides require different hydrolysis conditions (Figure 9). Chromatogram 2 shows that hyperoside and rutin are compeletely hydrolyzed while vitexin-2"-O-rhamnoside can still be detected, since the conditions are too mild for a complete hydrolysis of vitexin-2"-O-rhamnoside to vitexin. Acetylvitexin-2"-O-rhamnoside is completely hydrolyzed to vitexin-2"-O-rhamnoside. On the other hand, chromatogram 3 shows that vitexin-2"-O-rhamnoside and acetylvitexin-2"-O-rhamnoside are completely hydrolyzed, since the corresponding peaks can not be detected anymore. Not only vitexin but also isovitexin resulted from the acid hydrolysis due to a Wessely-Moser rearrangement.

The preparation of the plant material was performed by continuous, exhausting extraction in a Soxhlet apparatus (Figure 10). Since this kind of extraction is exhaustive (in the second and third extraction less than 1% of the main first extraction was obtained) it leads to a complete extraction of the flavonoids and is suitable for a quantitative determination. Other extraction methods like reflux or turbo extraction were tested but they revealed only incomplete extraction of the flavonoids, even if repeated several times. The resulting flavonoid content was at least 15% lower.

Five different samples of hawthorn leaves with flowers were quantified by the hydrolysis method described above in order to obtain some information on its precision and accuracy as well as about the uniformity of content and ratio of vitexin and quercetin. All samples showed similar flavonoid contents with comparable ratios of vitexin to quercetin (Figure 11). The relative standard deviations ($n = 3$ or 6) were between 1% and 3% in all samples.

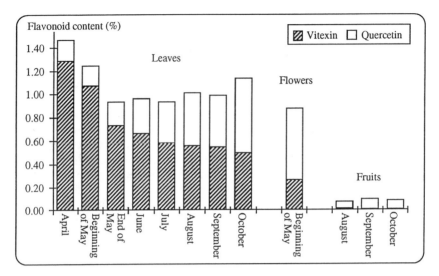

Figure 8. Flavonoid determination of *Crataegus monogyna* at different stages of development. For chromatographic conditions, see Figures 6 and 7.

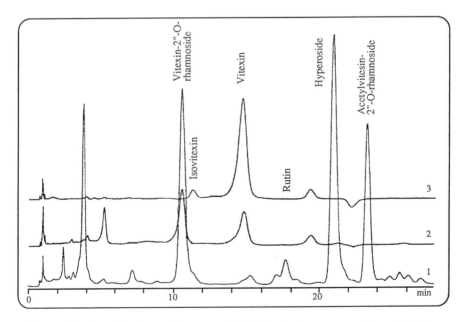

Figure 9. Control of completeness of hydrolysis by means of an HPLC fingerprint chromatogram: 1: unhydrolyzed extract, 2: hydrolyzed extract [hydrolysis conditions for quercetin derivatives (quercetin showed a retention time > 40 min under the selected chromatographic conditions)], 3: hydrolyzed extract [hydrolysis conditions for vitexin derivatives]. For chromatographic conditions, see Figure 5.

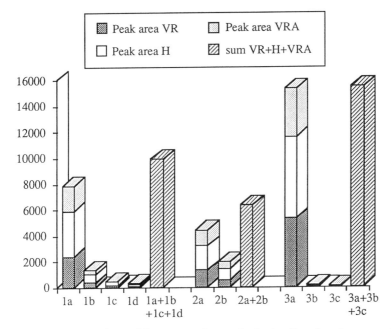

Figure 10. Comparison of three extraction methods: 1 reflux, 2 turbo extraction, 3 Soxhlet. a = first, b = second, c = third, d = forth extraction. VR = vitexin-2"-O-rhamnoside, VRA = acetylvitexin-2"-O-rhamnoside, H = hyperoside.

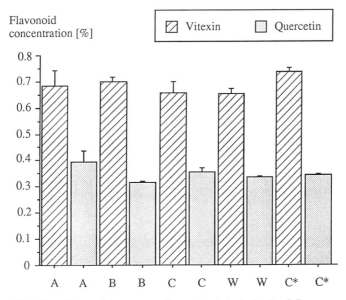

Figure 11. Mean flavonoid contents (± standard deviation) of five commercial samples of hawthorn leaves with flowers (samples A, B, C, W, and C*), determined by HPLC after hydrolysis.

The recoveries were determined by twice adding two different amounts of each vitexin and vitexin-2"-O-rhamnoside and two different amounts of querecetin and hyperoside at the beginning of the sample extraction. For vitexin-2"-O-rhamnoside and hyperoside, the theoretically expected amounts of vitexin and quercetin, assuming a complete hydrolysis, were calculated. The recoveries were, on an average, 96.4% for vitexin and 91.5% for quercetin. This indicates almost complete hydrolysis and only little loss of compounds during the sample preparation.

To investigate its validity, the hydrolysis method (method 1) was finally compared to three other methods, namely the HPLC fingerprint method (method 2) described above, an HPLC method used by the French Pharmacopeia (method 3), and the spectrophotometric determination described in DAB 10 (method 4) (Figure 12). Methods 1 and 2 consider the whole spectrum of flavonoids—the flavonol-O-glycosides and the flavone-C-glycosides—and yield similar amounts of total flavonoids. Nevertheless, method 1 should be preferred, since the flavonoid pattern is reduced with a simple hydrolysis procedure to two compounds; therefore, it is possible to perform the quantitative determination with only two reference standards, whereas for the fingerprint method more reference compounds and calculating factors are needed. Although method 3 is very similar to method 2, the obtained flavonoid contents are clearly lower. There is a significant loss of about 30% of hyperoside compared to methods 1 and 2, mostly due to incomplete extraction of the plant material. Compared to methods 1 and 2 there is a significant decrease of 50-60% of vitexin-2"-O-rhamnoside resulting from the non-consideration of the acetylated vitexin-2"-O-rhamnoside. Method 4, which determines the total flavonoid content, calculated as hyperoside, revealed results in the range of those of method 3. But it is—as often criticized—not specific and neglects the flavone-C-glycosides. It should not be applied to herbal drugs that have a considerable content of flavone-C-glycosides (8, 53).

Finally, four different commercially available products containing Crataegus extracts were quantified by the hydrolysis method and compared to each other (Figure 13). The preparations were: a) Faros 300 (Lichtwer Pharma, Berlin, Germany) PZN-4830052, Batch 93060100, b) Crataegutt novo (Dr. Willmar Schwabe, Karlsruhe, Germany) PZN-2505313, Batch 2331093, c) Crataegutt forte (Dr. Willmar Schwabe, Karlsruhe, Germany) PZN-3443086, Batch 0391093, d) Valverde Herzdragees (Ciba-Geigy, Basel, Switzerland), IKS No. 478360018, Batch 900333. In these four preparations, remarkable differences concerning the flavonoid content were found. Faros 300 yielded clearly the highest and Crataegutt novo an extremely low content, whereas Crataegutt forte and Valverde Herzdragees exhibited similar, medium flavonoid concentrations. Even after including the daily dosage, which is nine capsules in the case of Crataegutt novo compared to three for the other products, the results are still considerably lower for this preparation.

Conclusions

Besides general aspects on the botany, chemistry, and pharmacology of hawthorn and clinical uses of phytomedicines based on this plant, the objective of this paper was to show that accurate methods of assaying and standardization are a prerequisite to insuring the reproducibility and efficacy of a given phytomedicine. By using the

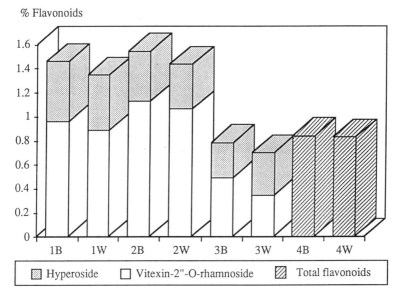

Figure 12. Flavonoid determination of two samples (B and W) by four different methods: 1 = HPLC after acid hydrolysis, 2 = quantitative HPLC fingerprint, 3 = HPLC according to Ph. Franç. X, 4 = according to DAB 10.

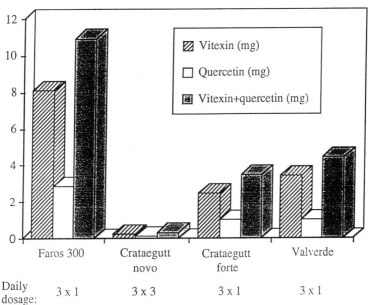

Figure 13. Comparison of the flavonoid content of four commercial preparations, determinded by HPLC after hydrolysis.

HPLC methods presented, every step in the production of hawthorn phytomedicines can be controlled, beginning with quality control of the raw material, continuing with inprocess controls during the production of standardized extracts, and ending with the quality control of the commercial products.

Hawthorn has been established within the last ten years as a valuable alternative to synthetic compounds in the therapy of the NYHA stage II heart failure patients. Unfortunately, herbal remedies have not been accepted until now in the textbooks of pharmacology. Hawthorn should be introduced in the therapeutic treatment of heart insufficiency as the starting drug. The clinical results show that hawthorn has to be ranked above the ACE-inhibitors as starting drug in the treatment of the declining cardiac performance. Therefore, the scientific base is available.

The clinical studies of the last 15 years show a trend to higher dosages of hawthorn up to extract amounts corresponding to 5 g of the drug Crataegi folium cum flore. Unfortunately, the analytical description of the extracts tested in the studies is not on the level of the HPLC-method described in this paper. Only a few results for comparison are available. Therefore, it is difficult to correlate daily flavonoid dosage with therapeutic activity. Relating proposals (Commission E, ESCOP) have to be validated in future. Furthermore, the analysis of procyanidins has to be developed in the same manner as for flavonoids. Until now, a convincing method is still missing. The workshop on "Quality control of hawthorn" (proceedings in preparation) showed that research is promising some progress in the near future.

Acknowledgments

The authors wish to thank the former Ph.D. student Anne Rehwald for performing the laboratory work on Crataegus and also for reviewing various general aspects in her Ph.D. thesis. The authors are also extremely grateful to Annemarie Suter for editorial assistance.

Literature Cited

1. Byatt, J. *Watsonia* **1975,** *10,* 253-264.
2. Christensen, K. *Revision of Crataegus Sect. Crataegus and Nothosect. Crataeguineae (Rosaceae-Maloideae) in the Old World.*; Systematic Botany Monographs, The American Society of Plant Taxonomists: Ann Arbor, MI, 1992; Vol. 35, pp 1-18.
3. Byatt, J. *Bot. J. Linn. Soc.* **1974,** *69,* 15-21.
4. Hegi, G. *Illustrierte Flora von Mitteleuropa*; J.F. Lehmanns Verlag: München, 1908-1931; Vol. IV, Part 2, pp 725-739.
5. Allen, D. *Watsonia* **1980,** *13,* 119-120.
6. Bauer, I.; Hölscher, U. In *Hagers Handbuch der Pharmazeutischen Praxis*; Hänsel, R.; Keller, K.; Rimpler, H; Schneider, G., Eds.; Springer Verlag: Berlin Heidelberg, 1992; 5th ed., Vol. 4, Drogen A-D; pp 1040-1062.
7. El-Gazzar, A. *Bot. Jahrb. Syst.* **1980,** *101,* 457-469.
8. Rehwald, A. *Ph. D. Thesis*, ETH Zurich, No. 10959, 1995.
9. Ammon, H.P.T.; Händel, M. *Planta Med.* **1981,** *43,* 105-120, 209-239 and 313-322.
10. Leukel-Lenz, A. *Ph. D. Thesis*, Philips-Universität, Marburg, 1988.

11. Ammon, H.P.T.; Kaul, R. *Dtsch. Apoth. Ztg.* **1994**, *134*, 2433-2436, 2521-2535 and 2631-2636.
12. Pöpping S.; Fischer Y.; Kammermeier H. *Münch. med. Wschr.* **1994**, *136*, Suppl. 1, S39-S46.
13. Tauchert, M.; Siegel, G.; Schulz, V., Eds. *Münch. med. Wschr.* **1994**, *136*, Suppl. 1, S3-S73.
14. von Eiff M. *Untersuchungen zur Pharmazeutischen Qualität, Unbedenklichkeit und therapeutischen Wirksamkeit eines Crataegus/Passiflora-Extraktes*; Thesis, Universität Basel, 1994 .
15. von Eiff, M.; Brunner, H.; Haegeli, A.; Kreuter, U.; Martina, B.; Meier, B.; Schaffner, W. *Acta Therapeutica* **1994**, *20*, 47-66.
16. Schmidt U.; Kuhn U.; Ploch M.; Hübner W.-D. *Phytomedicine* **1994**, *1*, 17-24.
17. Tauchert M.; Ploch M.; Hübner W.-D. *Münch. med. Wschr.* **1994**, *136*, Suppl. 1, S27-S33.
18. Forni, G. *Fitoter.* **1980**, *51*, 13-33.
19. Tittel, G.; Wagner, H. In *Flavonoids and Bioflavonoids, 1981*; Farkas, L.; Gabor, M.; Kállay, F.; Wagner, H., Eds.; Elsevier Scientific Publishing Company: Amsterdam, 1982; pp 299-310.
20. Wagner, H.; Tittel, G.; Bladt, S. *Dtsch. Apoth. Ztg.* **1983**, *123*, 515-521.
21. Ficarra, P.; Ficarra, R.; Tommasini, A.; De Pasquale, A.; Guarniera Fenech, C.; Iauk, L. *Farmaco Ed. Prat.* **1984**, *39*, 148-157 and 342-354.
22. Ficarra, P.; Ficarra, R.; De Pasquale, A.; Monforte, M.; Calabro, M. *Farmaco* **1990**, *45*, 247-255.
23. Pietta, P.; Manera, E.; Ceva, P. *J. Chromatogr.* **1986**, *357*, 233-238.
24. Pietta, P.; Mauri, P.; Manera, E.; Ceva, P.; Rava, A. *Chromatographia* **1989**, *27*, 509-512.
25. Christ, B.; Müller, K. *Arch. Pharm.* **1960**, *293*, 1033-1042.
26. Glasl, H.; Becker, U. *Dtsch. Apoth. Ztg.* **1984**, *124*, 2147-2152.
27. Dallenbach-Tölke, K.; Nyiredy, Sz.; Meier, B.; Sticher, O. *Planta Med.* **1987**, *53*, 189-192.
28. Glasl, H. *Fresenius Ztschr. Anal. Chem.* **1985**, *321*, 325-330.
29. Hiermann, A.; Kartnig, T. *Sci. Pharm.* **1984**, *52*, 30-35.
30. Lamaison, J.; Carnat, A. *Pharm. Acta Helv.* **1990**, *65*, 315-320.
31. Lamaison, J., Carnat, A. *Plant. Méd. Phytothér.* **1991**, *25*, 12-16.
32. Kartnig, T.; Kögl, G.; Heydel, B. *Planta Med.* **1993**, *59*, 537-538.
33. Tempel, A. *J. Chem. Ecol.* **1982**, *8*, 1289-1298.
34. Deshpande, S.; Cheryan, M.; Salunkhe, D. *CRC Critical Reviews in Food Science and Nutrition*; CRC Press, Inc.: Cleveland, OH, 1986; Vol. 24, Issue 4, pp 401-449.
35. Hagerman, A.; Butler, L. *J. Chem. Ecol.* **1989**, *15*, 1795-1810.
36. Karchesy, J. In *Chemistry and Significance of Condensed Tannins*; Hemingway, R.; Krachesy, J. Eds.; Plenum Press: New York, 1989; pp 197-202.
37. Porter, L. In *Plant Phenolics*; Harborne, J., Ed.; Methods in Plant Biochemistry; Academic Press: London, 1989, Vol. 1; pp 389-419.
38. Macheix, J.; Fleuriet, A.; Billot, J. *Fruit Phenolics*; CRC Press Inc.: Boca Raton, FL, 1990; pp 81-90.
39. Scalbert, A. In *Plant Polyphenols: Synthesis, Properties, Significance*; Hemingway, R.; Laks, P., Eds.; Plenum Press: New York, 1992; pp 259-280.
40. Hölzl, J.; Strauch, A. *Planta Med.* **1977**, *32*, 141-153.
41. Schrall, R.; Becker, H. *Planta Med.* **1977**, *32*, 297-307.
42. Lea, A. *J. Chromatogr.* **1980**, *194*, 62-68.
43. Lea, A. *J. Chromatogr.* **1982**, *238*, 253-257.
44. Vande Casteele, K.; Geiger, H.; De Loose, R.; Van Sumere, C. *J. Chromatogr.* **1983**, *259*, 291-300.
45 Perez-Ilzarbe, F.; Martinez, V.; Hernandez, T.; Estrella, I. *J. Liquid Chromatogr.* **1992**, *15*, 637-646.

46. Goldstein, J; Swain, T. *Nature* **1963,** *198,* 587-588.
47. Butler, L.; Price, M.; Brotherton, J. *J. Agric. Food. Chem.* **1982,** *30,* 1087-1089.
48. Swain, T.; Hillis, W. *J. Sci. Food Agric.* **1959,** *10,* 63-68.
49. Bate-Smith, E. *Phytochem.* **1973,** *12,* 907-912.
50. Porter, L.; Hrstich, L.; Chan, B. *Phytochem.* **1986,** *25,* 223-230.
51 Kreimeyer, J.; Petereit, F.; Nahrstedt, A. Workshop: *Crataegus - Pharmazeutische Qualität und Wirksamkeit,* ETH Zurich, October 11, 1996.
52. Rohr, G. Workshop: *Crataegus - Pharmazeutische Qualität und Wirksamkeit,* ETH Zurich, October 11, 1996.
53. Rehwald, A.; Meier, B.; Sticher, O. *J. Chromatogr. A.* **1994,** *677,* 25-33.
54. Kartnig, T.; Hiermann, A.; Azzam, S. *Sci. Pharm.* **1987,** *55,* 95-100.

Chapter 18

Influence of the Flavonolignan Silibinin of Milk Thistle on Hepatocytes and Kidney Cells

J. Sonnenbichler, I. Sonnenbichler, and F. Scalera

Max Planck Institut für Biochemie, D–82152 Martinsried, Germany

The pharmacologically active components of the plant *Silybum marianum* are the flavonolignans, silibinin and silichristin. It has been shown that these compounds can be used successfully in therapy to promote faster regeneration of diseased liver. The biochemical mechanism for this cell-regenerating power has been elucidated. It has been demonstrated that silibinin stimulates the activity of the DNA-dependent RNA-polymerase I, thus causing an increase in rRNA synthesis and an accelerated formation of intact ribosomes. The consequence of this stimulation is a general increase in the rate of synthesis of all cellular proteins. Molecular modelling revealed that silibinin may imitate a steroid hormone by binding specifically to polymerase I, thus stimulating the enzyme activity. The molecular mechanism described has been demonstrated in experiments with rat and mice liver *in vivo*, with hepatocyte cultures, isolated liver nuclei, and purified enzyme and receptor proteins *in vitro*. The increase in protein synthesis offers a good explanation for the liver-regenerating power of the plant extract. Similar results have been recently found with human and monkey kidney cells.

Extracts of the flowers and leaves of *Silybum marianum* (milk thistle or St. Mary's thistle), Figure 1, have been used for centuries to treat liver diseases. In the 1960s, scientists began to isolate the most important ingredients of the extracts of milk thistle, and their chemical structures were elucidated in the study-groups of Pelter and Hänsel (*1*), as well as of Wagner et al. (*2, 3*). The efforts at isolation first lead to a mixture, which was named silymarin. It was with this mixture that most of the clinical studies were carried out. We know today that the constituents are the compounds shown in Figure 2, and that silibinin makes up approximately 60% of the main components. These compounds are flavonolignans and are produced in the plant under conditions of strong solar radiation.

Every physician knows the abundance of clinical pictures associated with the term "liver diseases". These clinical pictures are just as varied as are the causes of the diseases—starting with viral infections and continuing on to toxic damage, for example,

264

Figure 1. *Silybum marianum.*

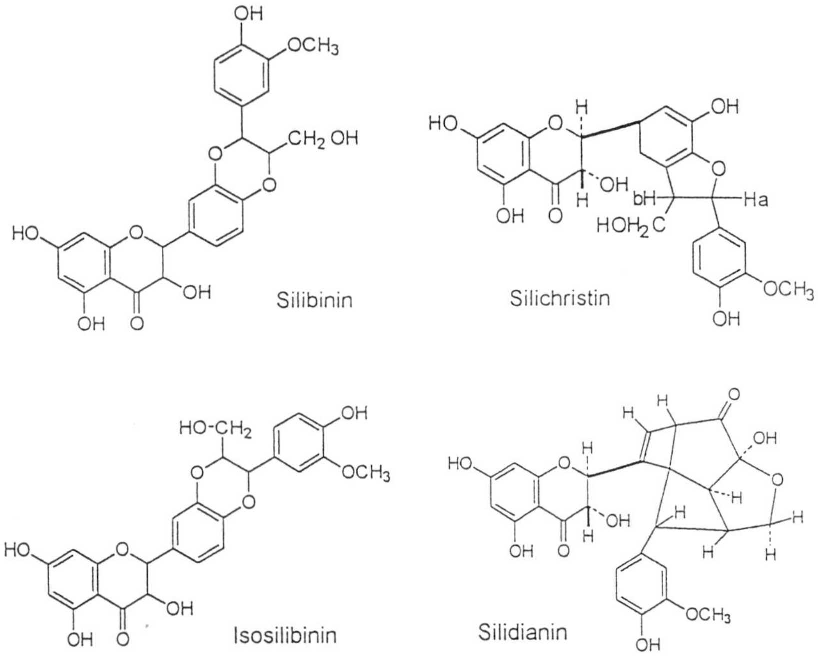

Silibinin

Silichristin

Isosilibinin

Silidianin

Figure 2. Constituents of Silymarin.

from alcohol intoxication. It is, therefore, not surprising that one faces considerable difficulties in selecting a suitable group of patients for double-blind clinical trials. Nevertheless, the therapeutic efficacy of silymarin has been convincingly verified. Fintelmann and Albert (4) observed already in 1980 that after toxic liver damage, administration of silymarin results in a significantly accelerated normalization of GOT (glutamate-oxaloacetate transaminase) and GPT (glutamate-pyruvate transaminase), Figure 3. A 1989, a double-blind study by Feher (5) likewise demonstrated that after administration of silymarin to patients with alcohol-induced liver diseases, a distinctly more rapid normalization of the transaminases and bilirubin takes place.

A multicenter study by Ferenci and colleagues (6) in Vienna found that the average survival rate of cirrhotic patients in the Child A stage could be markedly lengthened if they were given silymarin, Figure 4. Berenquer and Carrasco (7) observed an accelerated normalization in the plasma albumin values in patients with chronic inflammatory liver diseases after treatment with silymarin. One could summarize all these findings under the term "increased regenerative ability".

Studies on the Mechanism of Action of Silibinin

Irrespective of the noxious agents causing the damage (i.e. virus or toxin), in biochemical terms, "regenerative ability" means the restoration of damaged cellular components. Above all, defective macromolecules with biochemically meaningful functions must be replaced in order to restore normal cell function. Besides nucleic acids, these include, above all, proteins as building blocks of cell walls, organelles and as enzymes which in the end determine the entire primary metabolism. The synthesis of proteins is managed according to the scheme in Figure 5.

In the past 20 years, we have been able to clarify in considerable detail how the flavonolignans of milk thistle influence this mechanism. Among other things, it was necessary to measure the rate of synthesis of the various macromolecules. This is done generally in the following manner: a specific building block of the macromolecule is radioactive-labelled and then employed in the experiment. For example, in the case of quantitative determination of protein biosynthesis, this would involve labelling an amino acid, or in the case of nucleic acid measurements, a nucleotide unit would be thus labelled. After isolation of the high-molecular weight products, the time-dependent incorporation of the precursor can then be measured.

We carried out experiments on the livers of rats and mice *in vivo*, as well as with isolated cells - particularly with hepatocyte cultures, with isolated organelles (e.g., with cell nuclei *in vitro*), and finally with isolated enzymes and receptor proteins. The following was found under the influence of silibinin:

1. Protein biosynthesis, measured in units of time, proceeds approximately 25-30% faster under the influence of the flavonolignan silibinin compared to controls (8), Figures 6 and 7.

2. After determining the specific radioactivity of hundreds of different newly synthesized cellular proteins, it could be seen that the rate of synthesis of all the cellular proteins is increased equally without any preference or *de novo* synthesis of certain proteins.

3. We found that preceding this stimulation of protein synthesis is a great increase in the rate of RNA synthesis (9), Figure 8. We differentiate here three species of RNAs, which are indicated in Figure 9. Although the rates of synthesis of amino acid-activating tRNAs and of the heterogenic m-RNAs are not influenced by silibinin, the synthesis of the rRNAs

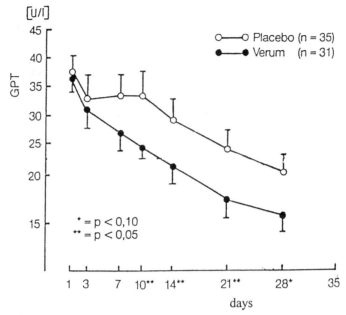

Figure 3. Time dependent normalization of a liver-specific enzyme after toxic liver disease in men, in the presence and absense of silymarin.
(Reproduced with permission from reference 4. Copyright 1980 Therapiewoche.)

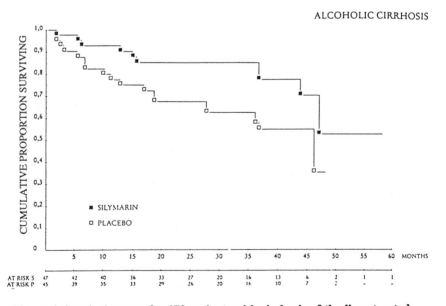

Figure 4. Survival curves for 170 patients with cirrhosis of the liver treated with silymarin or placebo, data analyzed according to the etiology of liver disease. Placebo vs. silymarin in alcoholic cirrhosis: Wilcox-Breslov test, p=0.011; Mantel-Cox test, p=0.012.

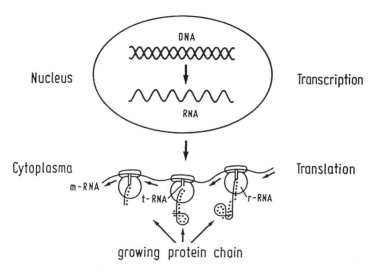

Figure 5. Scheme for the protein biosynthesis in eucaryotes including the transcriptional and translational process.

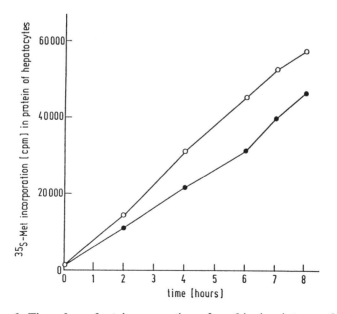

Figure 6. Time dependent incorporation of methionine into synthesized proteins in isolated rat hepatocytes in the presence of 10 ug/ml silibinin (○) and without silibinin (●).

Figure 7. Incorporation of ^{14}C-leucine into rat liver proteins 1.5 h after application of silibinin *in vivo,* and 3 h before sacrifice. S=10 mg/kg silibinin, C=control.

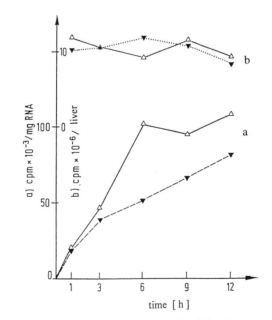

Figure 8. Time dependent incorporation of uridine into newly synthesized RNA in rat livers *in vivo* with (Δ)and without (▲) silibinin application (a). Precursor pools (b).

(Reproduced with permission from reference 10. Copyright 1984 Walter de Gruyter.)

is strongly stimulated (*10*), Table I. These rRNAs consist of a 5.8S, 18S and 28S-RNA species and are essential building blocks of the ribosomes—those "sewing machines" on which the proteins are synthesized.

Table I. Influence of silibinin on the rate of synthesis of different RNA species in rat liver *in vivo* (%).

total RNA	+99
rRNA (5.8S)	+33
rRNA (18S)	+74
rRNA (28S)	+29
tRNA	+1
mRNA (from nuclei)	-7

4. It was also shown that the synthesis of the mature—those with the associated ribosomal proteins—more complex ribosomes is distinctly increased, Figure 10.

Upon reflection of these findings, it becomes clear that with the increase in "sewing machines"—that is to say, the ribosomes—also the protein synthesis of the entire cell is increased, which then accelerates the cell regeneration.

5. Finally, the cause for the stimulation was found to be a specific enzyme, namely DNA-dependent RNA-polymerase I, whose activity is greatly increased by silibinin (*11*), Figure 11. This enzyme catalyzes the transcription of rRNA, and silibinin acts as a positive effector on this biocatalyst, but specifically with eucaryotes. With procaryotic RNA-polymerases (Gram-positive and Gram-negative bacteria), we could not find such a stimulation (*12*). In the case of partially hepatectomized rats even the DNA-replication is stimulated (*13*).

In enzymology one can recognize negative and positive effectors, for example in the case of pacemaker enzymes. It is astonishing though why a plant metabolite can exert such a specific effect on the molecular biology of an animal cell. In order to clarify this question, we investigated around 30 structurally related compounds with respect to these properties. Only a few of the tested substances exhibited a similar effect (*14*), Table II. Structural comparisons and molecular modelling of the biologically active compounds revealed a similarity to steroids, which is indicated in Figure 12. In fact, it is known that besides their specific ability for genetic induction (silibinin does not do this!), steroids can stimulate also the activity of RNA-polymerase I, i.e. they can stimulate the general rate of rRNA-transcription. In cooperation with Prof. Jungblut of the Max-Planck-Institute for Endocrinology, Braunschweig, we were able to demonstrate that silibinin binds competitively to a purified, isolated steroid receptor, albeit at a reduced affinity (*15*), Figure 13. Silibinin thus imitates a regulator of the cell itself, resulting in a stimulation of the entire cellular protein synthesis. We can now summarize the mechanism of action in Figure 14.

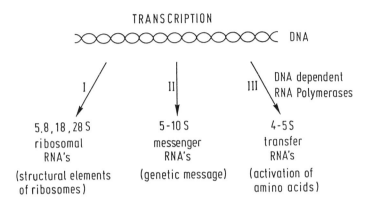

Figure 9. Different RNA types that are transcribed from the DNA template.

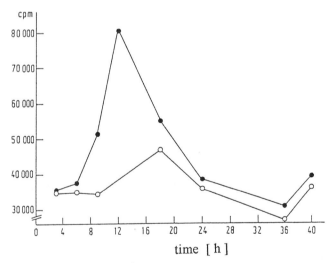

time [h]

Figure 10. Incorporation of ^{14}C-uridine into RNA of complex ribosomes. ^{14}C-Orotic acid (50 µCi) was injected into Wistar-rats (80-120 g) i.p. 3 h before sacrifice. Silibinin-hemisuccinate (1.5 mg) (●) was previously injected at different times. Controls (○) were treated only with solvent.

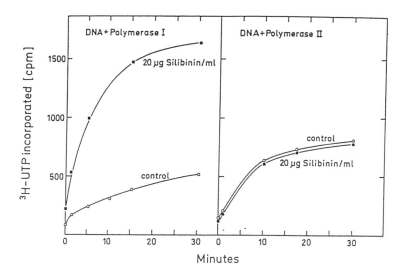

Figure 11. Transcription activities of the isolated and purified DNA-dependent RNA-polymerases *in vitro* with and without addition of silibinin, with calf thymus DNA as the template.
(Reproduced with permission from reference 11. Copyright 1977 Walter de Gruyter.)

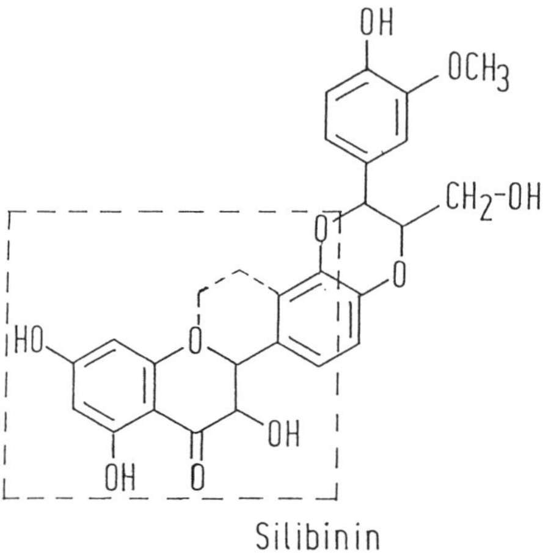

Silibinin

Figure 12. Comparison of silibinin to the steroid skeleton. The steroid skeleton is enclosed in the dashed box.

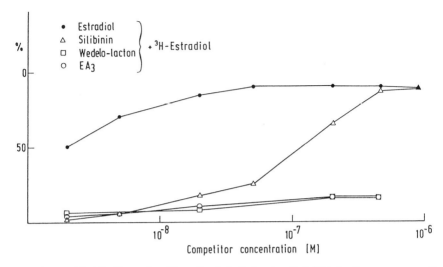

Figure 13. Binding of radioactive labelled estradiol (%) in the presence of unlabelled estradiol and unlabelled silibinin as a competitor to the isolated and purified steroid-receptor from pig uteri. Wedelo-lactone and EA3 are inactive controls.

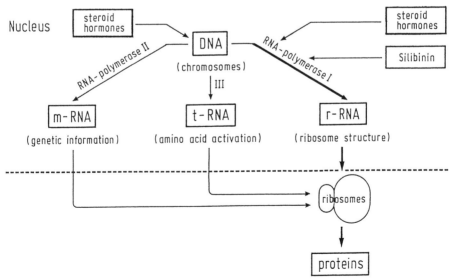

Figure 14. Summary scheme for the interaction of silibinin with the molecular biology of the cell.

Table II. Effects of various flavonoids on protein and RNA synthesis in liver cells (+ stimulation, - inhibition, -0 no effect, ( ) marginal effects).

Flavanones:		**Flavonoles:**	
Pinocembrin	0	Gossypetin	(+)
Isokuranetin	0	Galangin	0
Homoeryodictyol	0	Fisetin	(+)
Naringenin	0	Kämpferol	(+)
Hesperitin	++	Luteolin	0
Flavanone	0	Morin	(+)
		Rhamnetin	0
		Quercetin	0
Flavanonoles			
Taxifolin	++	**Flavonolignans**	
Flavones		Silibinin	++
Acacetin	0	Silidianin	0
Apigenin	0	Silichristin	+
		Isosilibinin	(-)
Isoflavones			
Prunetin	0	**Others:**	
Irigenin	(+)	Benzalacetophenone	0
Genistein	-	Chalcone	0
		Cynarin	0
Catechin	++	Curcumin	0

This mechanism applies for the period of time in which silibinin is available to the cell. The absorption of current preparations after oral administration is approximately 50-60%. We were able to confirm this in rats with radioactive-labelled silibinin, Figure 15. Two maxima of silibinin concentrations occur in the liver via enterohepatic circulation. Then, after approximately 40 hours, the flavonolignan is excreted in the form of glucoside and sulfate metabolites (16).

Considering the biochemical mechanism described, it would be expected that the stimulation of protein biosynthesis also occurs in non-hepatocytes, even though the flavonoid concentrations are by far highest in the hepatocytes. In fact, we have been able to ascertain in the past months that also in cell cultures of human and monkey kidney cells, silibinin causes increased protein synthesis, nucleic acid synthesis and also increased replication of dividing cells by approximately 25-30%, Table III. Already, positive clinical findings have also been obtained in this regard. Our experiments with liver and kidney cells demonstrated that only silibinin and silichristin exhibit the stimulatory effect, but not isosilibinin or silidianin, Table IV.

We have also worked with tumor cells, particularly human hepatoma cells (Alexander cells), rat hepatoma cells (Raji cells), Burkitt lymphoma cells, and HeLa cells, Figure 16. No stimulation was found in any of these cases. An obvious explanation would be that the transcription and translation in malignant cell lines already proceed at maximal rates and cannot be accelerated any further.

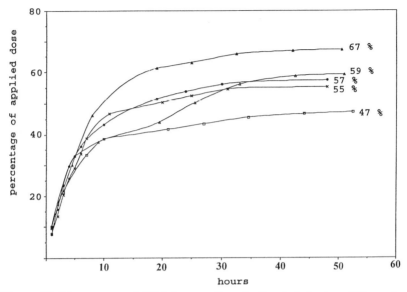

Figure 15. Time course of silibinin excretion via the choleductus of five rats in percentage of applied drug.

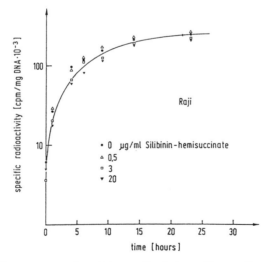

Figure 16. Time course of protein synthesis in cultures of malignant cell lines (for example, Raji-cells from rat hepatoma) in the presence of different concentrations of silibinin-hemisuccinate.

Table III. Effect of various silibinin-hemisuccinate concentrations on the cell line Vero from green monkey kidneys. Four days after transfer of 7.5×10^4 cells/ml in clusters with 400 µl medium 199, the cells were counted. Silibinin-hemisuccinate was added three days beforehand.

silibinin-hemisuccinate (µg/mL)	final cell counts	cell vitality (%)	growth (%)
0	2.86×10^5	92	
	2.86×10^5	93	100
10	3.30×10^5	93	
	3.26×10^5	94	116
20	3.50×10^5	94	
	3.44×10^5	94	123
50	2.60×10^5	91	
	2.58×10^5	92	92

Table IV. Effect of 10 ug/ml silibinin-hemisuccinate, silidianin, silichristin and iso-silibinin on the kidney cell line Vero. Details described in Table III.

probe	final cell counts	cell vitality (%)	growth (%)
control	4.26×10^5	95	
	4.34×10^5	96	100
control + 1% ethanol	4.26×10^5	96	
	4.28×10^5	96	99
silibinin hemi-succinate	4.96×10^5	96	
	4.98×10^5	96	116
silidianin	4.10×10^5	96	
	4.92×10^5	96	94
silichristin	4.92×10^5	96	
	4.88×10^5	96	114
isosilibinin	3.90×10^5	96	
	3.72×10^5	95	89

Additional mechanisms of action have been proposed for the flavonolignans of milk thistle, and some have been verified in experiments. For example, these polyphenols have the ability to intercept radicals, such as radicals formed from oxidative alcohol degradation or from other noxious agents in the liver which can cause cell damage. Data for the scavenger function of the silymarin components have been obtained in *in vitro* experiments primarily by the study-groups of Feher (*17*), Mira (*18*), Valenzuela and Guerra (*19*), and György (*20*). However, the flavonoid concentrations required in this case are so high, that, in my opinion, they cannot be achieved in the organism through oral application of silymarin preparations.

Furthermore, we know that in high concentrations silibinin diminishes the membrane permeability of the hepatocytes, so that many toxins can hardly then penetrate the cell (*21*). For example, the absorption of amanitine, the poison of the green death-head, is slowed down so much that the excretion of the toxin takes place before the actual toxic effect can set in. One can also employ this possibility for the successful therapy against mushroom poisoning (*22*). The silibinin concentrations required in such a case are around 10-20 times higher than those needed for stimulation of protein synthesis and can be achieved only via intravenous administration.

Conclusion

If one considers all of the biochemical effects described here, then the clinical results obtained with silibinin and silymarin can be explained from the view of the molecular biologist and confirm that the flavonolignans of milk thistle—above all, silibinin and silichristin—are indeed interesting drugs.

Acknowledgement. Silibinin (Legalon) and the flavonolignans were a gift of Dr. Madaus AG, Köln, FRG, which is gratefully mentioned.

Literature cited

1. Pelter, A.; Hänsel, R. *Tetrahedron* **1968,** 25, 2911
2. Wagner, H.; Hörhammer, R.; Münster, R. *Naturwissenschaften* **1965,** 52, 305
3. Wagner, H.; Seligmann, O.; Hörhammer, L.; Seitz, M.; Sonnenbichler, J. *Tetrahedron Letters* **1971,** 22, 1895
4. Fintelmann, V.; Albert, A. *Therapiewoche* **1980,** 30, 5589
5. Feher, J. *Orvosi Hetilap* **1989,** 130, 2723
6. Ferenci, P.; Fragosics, B.; Dittrich, H.; Frank, H.; Benda, L.; Lochs, H.; Meryn, S.; Base, W.; Schneider, B. *J. of Hepatology* **1989,** 2, 105
7. Berenquer, J.; Carrasco, D. *Münchner Med. Wochenschr.* **1977,** 119
8. Sonnenbichler, J.; Zetl I. *Plant Flavonoids in Biology and Medicine II* **1986,** 369
9. Sonnenbichler, J.; Mattersberger, J.; Rosen , H. *Hoppe-Seyler's Z. Physiol. Chem.* **1976,** 357, 1171
10. Sonnenbichler, J.; Zetl, I. *Hoppe-Seyler's Z. Physiol. Chem.* **1984,** 365, 555
11. Machicao, F.; Sonnenbichler, J. *Hoppe-Seyler's Z. Physiol. Chem.* **1977,** 358, 141
12. Schnabel, R.; Sonnenbichler, J.; Zillig, W. *FEBS Letters* **1982,** 50, 400
13. Sonnenbichler, J.; Goldberg, M.; Hane, L.; Madubunyi, I.; Vogl, S.; Zetl, I. *Biochemical Pharmacology* **1986,** 35, 538

14. Sonnenbichler, J.; Pohl, A. *Hoppe-Seyler's Z. Physiol. Chem.* **1980,** *361*, 1757
15 Sonnenbichler, J.; Zetl, I. In *Progress in Clinical and Biological Research*, Cody, V.; Middleton, E.; Harborne, J.B., Eds., Alan R. Liss, Inc. 1986, Vol. 213; p. 319
16. Sonnenbichler, J.; Mattersberger, J.; Hanser, G. *Hoppe-Seyler's Z. Physiol. Chem.* **1980,** 361, 175
17. Feher, J. *Free Rad. Res. Commun.* **1987,** *3*, 373
18. Mira, M.L. *Free Rad. Res. Commun.* **1987,** *4*, 125
19. Valenzuela, A.; Guerra, R. *Experientia* **1986,** *42*, 139
20. György, I. *Radiat. Phys. Chem.* **1992,** *39*, 81
21. Sonnenbichler, J.; Sonnenbichler, I.; Scalera, F. in press.
22. Hruby, K.; Fuhrmann, M.; Csomos, G.; Thaler, H. *Wiener Klin. Wochenschr.* **1983,** *95*, 225

Chapter 19

Phytotherapeutic Immunomodulation as a Treatment Modality in Oncology: Lessons from Research with Mistletoe

H.-J. Gabius[1] and S. Gabius[2]

[1]Institute für Physiologische Chemie, Tieräztliche Fakultät, Ludwig-Maximilians-Univeristät, Veternärstr. 13, D–80539 München, Germany
[2]Hämatologisch-onkologische Schwerpunktpraxis, Sternstr. 12, D–83022 Rosenheim, Germany

The reason to introduce mistletoe (*Viscum album* L.) preparations to so-called alternative cancer therapy was perceived in its entirety by the spiritual intuition of Rudolf Steiner, without any clinical support. Presently, their application is counted among the heavily advertised array of unproven methods in oncology. During the past decade we have shown that a purified constituent of mistletoe, namely the galactoside-specific lectin, can be referred to as biological response modifier, when applied in vitro or in vivo within a narrow low-dose range (1-2 ng/kg body weight). Although the characterization of the lectin's biochemical and cell biological properties expectedly proved to be straightforward, the scientific interpretation of its potency as elicitor of immune responses, e.g. enhanced cytokine secretion (IL-1, IL-6, TNF-α), has clearly not yet reached the same stage of maturity. It is now essential to responsibly and carefully evaluate the potential clinical uncertainty of the lectin-dependent immunological reactions in adequate model systems prior to advocating the design of fully controlled clinical trials, since so-called hemopoietic cytokines can be stimulators of growth for diverse tumor cell types in vitro and in vivo. The superficial use of scientific terminology without detailed and convincing proof of efficacy and safety should not delude physicians, pharmacists, and patients into ignoring the current discrepancy between facts and fiction.

An enormous variety of formulas and mixtures inundates the market for treatment of cancer. Judged on the basis of the often showy advertising, which billboards the methods as cures emphatically recommended by alleged experts, it seems to be surprising that various cancer types still pose a therapeutic problem. Official statements by cancer societies and critical reviews of the literature place emphasis on the fact that these preparations have so far not at all lived up to their promises (*1-3*). The scientific chain of reasoning should not be compromised (*4-6*).

Taking mistletoe as a graphic example, the background on its introduction, the ensuing therapeutic recommendations, the clinical experience with extract preparations, and the scientific perspective are outlined to understand why dealing diligently with this matter should not be considered as unimportant.

Mistletoe as a Spiritual Entity

The anecdotal nature of annotations in documents from the Middle Ages casts doubt on the assumption that mistletoe preparations have had a tradition as anticancer agents (7). Indeed, no reference to any clinical experience ("das bloße Probieren und Experimentieren") was even required for R. Steiner to boldly denote mistletoe as the future remedy for cancer (8-10). He intuitively defined physically inexplicable forces from this so-called highly spiritual entity as a fundamental cure which allegedly works on levels beyond the physical body (8-11). Experimentation is explicitly excluded as a tool to enhance our level of understanding. In contrast, the perception of the spiritual world, guided by the masterly intuition of the prepared mind, categorically replaces any empirical reasoning. To familiarize the reader with Steiner's concept (12, 13), examples of advice dogmatically deducted from the anthroposophical system are the use of lead, sugar, and honey in the treatment of atherosclerosis and the strong warning to consume potatoes, in order to avoid egoism and materialism. Having deliberately left the area of scientific, self-critical search for truth and entered a dogmatic, pseudo-religious sphere, it is a startling phenomenon that scientific nomenclature is often used instead of the inherent wording of the anthroposophical system of belief (14). However, this attempt to square the circle exposes the extract preparations and their application to the commonly accepted rules of quality assessment and control.

Mistletoe Extracts as Commercial Products

Various proprietary extracts are on the market in Germany, taking advantage of special, legally benevolent regulations for anthroposophic, homeopathic, and phytotherapeutic preparations. The manufacturing processes of the individual mistletoe-derived products differ notably and can e.g. include vigorous shaking for "homeopathic dynamization", bacterial fermentation, or rhythmic warming. In addition to the differences in processing of the starting material, the variability of its biochemical composition makes it nearly impossible to guarantee a product of reproducible content. The nature of the host tree, the climatic conditions, and the time of harvest are among the factors with proven impact on the composition. Nonetheless, the manufacturers unanimously claim profound efficiency in cancer treatment. In contrast to R. Steiner's purely spiritual concept ("werdende Natur"), pharmacological effects by substances ("gewordene Natur"), especially on the level of the immune system, are made responsible for the purported success. Due to the biochemical complexity of the extracts with compounds such as viscotoxins, lectins, polysaccharides, alkaloids, and flavonoids (for reviews see 7, 15, 16), it is understandable, albeit scientifically questionable, that the manufacturers attribute efficacy to a concerted and harmoniously orchestrated action of the mixture. The nature of the postulated synergism remains elusive.

The apparent positive attitude concerning its clinical value is not at all supported by critical reviews of the literature on application of commercial extract preparations. The average quality of seemingly positive reports is judged to be "disappointingly poor" and does not justify the widespread and uncontrolled use in the treatment of cancer patients (17, 18). The lack of tumor-inhibiting efficacy was also noted with a commercial product in four animal tumor models. Furthermore, in vitro studies revealed a higher degree of susceptibility of normal cells than carcinoma cells to toxic mistletoe lectins (19, 20). The latter report clearly shows that a cell-type-specific chemotherapy with these two proteins, which belong to the class of potent AB-toxins (21, 22), is not feasible. Having first been detected by Krüpe, by virtue of the capacity of mistletoe extract to agglutinate erythrocytes (23), and then preparatively enriched by galactose-dependent desorption from extract-exposed B-type erythrocytes (24), the abundant galactoside-specific agglutinin (VAA-I) was then purified independently by three groups (25-27). Due to systematic evaluation of various types of affinity ligands and spacers in chromatography, its isolation has been optimized, yielding approximately 1 mg of lectin from 12 g dried mistletoe leaves (28). In view of the emerging role of lectins in immune regulation—its present status outlined elsewhere (29, 30)—the availability of pure material made it possible to determine whether the lectin may be involved in any regulatory processes. As already emphasized, the inherent toxicity of the lectin precluded any assays at concentrations which will inevitably cause cell death. Therefore, our interest focused on the ng/mL (ng/kg body weight) range of concentration (dose). It is noteworthy that the N-acetylgalactosamine-specific lectin (VAA-II), which is also present in extracts with pronounced host tree dependence (16), was simultaneously tested.

Mistletoe Lectin as Biological Response Modifier

To assess any lectin-dependent effects, endotoxin-free preparations were tested in vitro and in vivo after subcutaneous injections. Immunomodulatory responses, such as enhancement of phagocytic activity of granulocytes or of NK cell cytotoxicity tested in vitro, were seen with blood samples from rabbits and human volunteers, if treated within a narrow low-dose range in the order of 1 ng lectin/kg body weight (31). The second lectin was considerably less active in this respect (31). When lectin-containing extracts were injected to reach the described active lectin quantity, similar reactions were seen (31, 32). Deliberate lectin depletion of the extract removed the active principle from the mixture, as the otherwise unchanged solutions then displayed suppressive potency for immune parameters (31, 32). It is thus fair to assume that the galactoside-specific lectin, given in a narrow dose range, can alter certain parameters of the immune system. Further studies have extended the panel of such parameters to, e.g., the number of helper T cells and the extent of CD25 expression for breast cancer patients (33). However, it is essential to keep in mind that these parameter alterations have not been proven as reliable indicators of prognosis (34). Although they reveal a measurable response to lectin application, it is either premature or irresponsible to repeatedly refer at this stage to extract application as "method of choice for a scientifically acceptable immunotherapy" (35), as commented upon in detail elsewhere (36, 37).

To at least be able to partially explain the effects on the level of the organism, the binding capacity of the lectin to glycoconjugates and to cells capable of deciphering lectin-triggered biosignaling was analyzed. Especially blood group H/B-like structures, as well as the P_1 determinant, are strongly reactive with the lectin, which also accommodates N-acetyllactosamine- and Le^d-terminating antennae of complex-type glycoproteins into its binding site (38-42). Similar to animal lectins which bind galactosides, the application of sophisticated NMR-techniques in combination with computer-assisted molecular modeling has raised evidence that in solution the ligand fits snugly into the plant agglutinin's binding site without notable deviation from its global minimum conformation in the free state (43). Judged on the basis of the rather broad panel of galactosides, which also includes the a-2,6-sialylated derivative of N-acetyllactosamine as common glycoprotein-linked epitope, it is obvious that the lectin can in principle find ligands on diverse cell types. This low-stringency ligand selectivity can account for its proposed ecological role as a defense molecule for the plant (44).

To convey a message into the cell's interior, binding to surface epitopes is only one, albeit essential, factor. The spatial rearrangement of the surface ligands, i.e. cluster formation, should then follow to entail an intracellular signaling cascade (45). Lectins form cross-linked complexes of long-range order—as often at least as bivalent molecules are predestined to—whose formation is implicated in elicitation of cellular activities such as mitogenesis (46). Complexes of constant stoichiometry with a model ligand can also be established by this mistletoe lectin (48). Their formation on cell surfaces is linked to the occurrence of positive cooperativity at a lectin concentration in the range of ng/mL x 10^5 cells (47). Intracellularly, increases in Ca^{2+}-mobilization, in phosphorylation of a 28 kDa protein, and in production of phosphatidylinositol-4,5-biphosphate were detected for a human monocytic leukemia cell line as a standard model (48). These changes can precede further responses, e.g. an enhanced secretion of cytokines, namely interleukin-1, interleukin-6 and tumor necrosis factor-α (49). This reaction is strictly dependent on recognitive protein-sugar interaction, as shown in respective controls (49). By altering mediator levels in the complex cytokine network, the lectin can rightfully be denoted as "biological response modifier" (50, 51). Current strides in unveiling the uncertainty of such endogenous mediators are welcome sparks to ignite efforts to clarify the role of altered cytokine availability to exclude any potential harm for the patients. It is not at all disgraceful to admit that we are presently ignorant of a complete comprehension of this immune factors network.

Lectin-elicited Immunomodulation: Friend or Foe?

Long-term cultured murine model systems with metastatic capacity provide the opportunity to test anti-tumor efficacy under experimentally favorable conditions. Although distant in their properties to the clinically apparent tumors, assays in such test systems are a reasonable first step to initiate the evaluation of a novel approach. Lectin application at an immunomodulatory dose reduced the extent of metastasis formation in lung or liver and reduced tumor growth for L1-sarcoma cells and RAW117-H10 lymphosarcoma cells (52, 53). However, a curative effect was not observed. At any rate, such effects in a limited number of model systems adapted to in vitro growth over a long time should not be overestimated.

Presently, the notion is gaining acceptance that cytokines should not merely be considered as anti-tumor agents, and that denominations such as tumor necrosis factor-α should be viewed cautiously (54, 55). Starting with a euphoric period of high hopes, we have entered an academic, rapidly burgeoning stage in which an understanding of the complexity of the cytokine network and the inherently contextual meaning of the interplay of the individual factors is being sought. As likewise known from normal cell types, multifunctional pleiotropic growth factors can trigger opposite activities from diverse tumor cell types, cautioning against the vision that, e.g., the lectin-dependent enhancement of cytokine secretion is principally and solely beneficial (56-59).

Since tumors have an enormous potential to adjust to their environment and maintain their growth potential, it is not surprising that this adaptive process also includes taking advantage of host immune reactions, a process that is generally referred to as the "flip side of immunity" (60, 61). The still ill-defined complexity of the network harbors the potential of unanticipated local and systemic side effects. Therefore, immunosuppression, not augmentation, may trigger clinically favorable effects which can establish an advantageous factor associated with chemotherapy or transplantation-linked immunosuppressive therapy (61, 62). To emphasize that immunomodulation can indeed have a bad side that *must not be irresponsibly neglected or carelessly played down*, we point out that common human cancer types have been convincingly shown to be stimulated by auto- or paracrine mechanisms. The set of lectin-triggered cytokines increases estrogen synthesis in breast cancer cells (63) and proliferation of ovarian cancer cells (64, 65); interleukin-6 and tumor necrosis factor-α can additively stimulate the proliferation of non-Hodgkin's lymphoma cells (66); two interleukins promote growth of colorectal carcinoma cells (67); and interleukin-6 can positively affect growth of renal cell carcinoma, cervical carcinoma, hairy cell leukemia, and myeloma cells in vitro (68-70).

Although these in vitro results certainly have a limited range of relevance for the in vivo situation, it is unquestionable that a further increase of cytokine production in a situation, where this factor is adversely correlated to prognosis, is not advised. For example, an increased interleukin-6 level is supposed to be a negative prognostic indicator in patients with ovarian carcinoma, metastatic renal cell carcinoma, non-small cell lung carcinoma, and multiple myeloma (71-74). Inherent heterogeneity of cytokine receptor expression and responsiveness to the lectin are quite troublesome factors to be carefully reckoned with to attain the long-range goal of a custom-made therapeutic program for the individual patient (75). Especially in advanced stages, where unconventional methods are often applied as supposedly harmless "placebos," a connection to cachexia should not be excluded. It is actually known from experimental systems that interleukin-1, interleukin-6, and tumor necrosis factor-α are involved in progression of tumor cachexia, and that interleukin-6 can be implicated in the nutritional status of patients with malignant mesothelioma and esophageal squamous cell carcinoma (76-79). Furthermore, promotion of angiogenesis by tumor necrosis factor-α at low doses can aggravate the clinical perspective (80).

The essential role of interleukin-6 for B cell neoplasms in vivo, which can be counteracted by cytokine-specific or cytokine receptor-specific antibodies of clinical potential, and the in vivo demonstration of its growth-stimulatory effect for advanced-stage human melanoma cells, attest to the importance of considering all aspects of the functions of pleiotropic cytokines (*81-83*). Although there is a publication bias towards reporting individual cases displaying an unfavorable response—due to the lack to unequivocal proof of relevance from large study groups—it should make us think that the clinical course of two patients with advanced solid tumors appeared to worsen during treatment with recombinant interleukin-6 (*84*). Similar casuistic evidence led to the recommendation to avoid mistletoe application for leukemia and lymphoma patients (*85*). Notably, in response to a critical commentary, representatives of one commercial distributor publicly shared this view (*86*).

Conclusions

As emphasized elsewhere (*37*), "there is no reason why the burden of proof for clinical efficiency and lack of harmful effects should be less in this area than for any drug in mainstream medicine." The indisputable *primum non nocere* (first of all, do not harm) must not be compromised by any treatment modalities. In order to provide an optimal standard of quality, the evaluation of lectin-dependent immunomodulation requires extensive work in in vitro systems with murine and human tumor material, in order to dissect responses in this microenvironment for individual tumor types and, if favorable, in strictly controlled clinical trials with close-meshed tumor marker monitoring. Furthermore, each tumor type should necessarily be analyzed as a separate entity with no unwarranted generalizations. It is our firm conviction that a great deal of experimental and, if promising, clinical work lies ahead to reliably separate the good, neutral, or even bad sides of the complex, pleiotropic, and presently ill-defined, interaction of the lectin-dependent immune parameters with the individual tumor. As commentaries on other phytopreparations, homeopathic practice, and Kampo (Chinese herbal products used in Japan) medicine similarly reveal (*87-90*), the notion of a risk-free application of the suggestively called "gentle medicines" is untenable. In the interest of the patients, this notion should be replaced by credible, generally accepted facts. It appears logical to us that evidence for efficacy and safety should precede the general, uncontrolled application and not *vice versa*. The immanent ethical judgement has unequivocally been given by Werning in a noteworthy commentary (*91*).

Literature Cited

1. Hauser, S. P. *Curr. Opinion Oncol.* **1993**, *5*, 646-654.
2. Burkhard, B. *Onkologe* **1995**, *1*, 583-589.
3. Drings, P.; Brittinger, G.; Gaedicke, G.; Heimpel, H.; Hossfeld, D. K.; Huber, C.; Meuer, S.; Wannenmacher, M.; Winkler, K. *Onkologie* **1995**, *18*, 158-162.
4. Windeler, J. In *Unkonventionelle medizinische Verfahren*; Oepen, I., Ed.; G. Fischer Verlag, Stuttgart, 1993; pp. 104-128.
5. Bock, K. D. *Onkologe* **1995**, *1*, 541-547.
6. Sewing, K. F. *Neue Jurist. Wschr.* **1995**, *48*, 2400-2402.

284

7. Becker, H.; Schmoll gen. Eisenwerth, H. *Mistel - Arzneipflanze, Brauchtum und Kunstmotiv im Jugendstil*; Wissenschaftliche Verlagsgesellschaft, Stuttgart, 1986.
8. Ullrich, H. *Dt. Ärztebl.* **1988**, *85*, B1301-1307.
9. Wolff, O. *Anthroposophisch orientierte Medizin und ihre Heilmittel*; Freies Geistleben, Stuttgart, 1990.
10. Stratmann, F. In *Unkonventionelle medizinische Verfahren*; Oepen, I., Ed.; G. Fischer Verlag, Stuttgart 1993; pp. 68-95.
11. Gabius, H.-J.; André, S; Kaltner, H.; Siebert, H.-C.; von der Lieth, C.-W.; Gabius, S. Z. *ärztl. Fortbild.* **1996**, *90*, 103-110.
12. Marx, H. H. *Z. Allg. Med.* **1994**, *70*, 31-34.
13. Kühne, P. *Novalis* **1996**, *50*, 10-11.
14. Gabius, H.-J.; Gabius, S. *Dt. Ärztebl.* **1996**, *93*, A810.
15. Franz, H. *Pharmazie* **1985**, *40*, 97-104.
16. Luther, P.; Becker, H. *Die Mistel. Botanik, Lektine, medizinische Anwendung*; Springer Verlag, Berlin, 1987.
17. Hauser, S. P. *Therapiewoche* **1993**, *53*, 76-81.
18. Kleijnen, J.; Knipschild, P. *Phytomedicine* **1994**, *1*, 255-260.
19. Berger, M.; Schmähl, D. *J. Cancer Res. Clin. Oncol.* **1983**, *105*, 262-265.
20. Schumacher, K. *Med. Klinik* **1986**, *81*, 423-428.
21. Barbieri, L.; Batelli, M. G.; Stirpe, F. *Biochim. Biophys. Acta* **1993**, *1154*, 237-282.
22. Read, R. J.; Stein, P. E. *Curr. Opinion Immunol.* **1993**, *3*, 853-860.
23. Krüpe, M. *Blutgruppenspezifische pflanzliche Antikörper (Phytoagglutinine)*; Enke Verlag, Stuttgart, 1956.
24. Luther, P.; Prokop, O.; Köhler, W. *Z. Immun. Forsch.* **1973**, *146*, 29-35.
25. Luther, P.; Theise, H.; Chatterjee, B.; Karduck, D.; Uhlenbruck, G. *Int. J. Biochem.* **1990**, *11*, 428-435.
26. Franz, H.; Ziska, P.; Kindt, A. *Biochem. J.* **1981**, *195*, 481-484.
27. Olsnes, S.; Stirpe, F.; Sandvig, K.; Pihl, A. *J. Biol. Chem.* **1982**, *257*, 13263-13270.
28. Gabius, H.-J. *Anal. Biochem.* **1990**, *189*, 91-94.
29. Gabius, H.-J.; Gabius, S. Eds. *Glycosciences: Status and Perspectives*; Chapman & Hall, London, Weinheim, 1997.
30. Gabius, H.-J. *Eur. J. Biochem.* **1997**, in press.
31. Hajto, T.; Hostanska, K.; Gabius, H.-J. *Cancer Res.* **1989**, *49*, 4803-4808.
32. Gabius, H.-J.; Hostanska, K.; Trittin, A.; Gabius, S.; Hajto, T. *Therapeutikon* **1990**, *4*, 38-45.
33. Beuth, J.; Ko, H. L.; Gabius, H.-J.; Burrichter, M.; Oette, K.; Pulverer, G. *Clin. Investig.* **1992**, *70*, 658-661.
34. Sauer, H. *Münch. Med. Wschr.* **1996**, *138*, 159.
35. Beuth, J.; Ko, H. L.; Pulverer, G. *Therapiewoche* **1995**, *45*, 1594-1599.
36. Gabius, H.-J.; Kayser, K.; Gabius, S. *Naturwissenschaften* **1995**, *82*, 533-543.
37. Gabius, H.-J.; Gabius, S. *Onkologie* **1996**, *19*, 74.
38. Lee, R. T.; Gabius, H.-J.; Lee, Y. C. *J. Biol. Chem.* **1992**, *267*, 23722-23727.
39. Wu, A. M.; Chin, L. K.; Franz, H.; Pfüller, U.; Herp, A. *Biochim. Biophys. Acta* **1992**, *1117*, 232-234.
40. Debray, H.; Montreuil, J.; Franz, H. *Glycoconjugate J.* **1994**, *11*, 550-557.
41. Lee, R. T.; Gabius, H.-J.; Lee, Y. C. *Carbohydr. Res.* **1994**, *254*, 269-276.
42. Galanina, O. E.; Kaltner, H.; Khraltsova, L. S.; Bovin, N. V.; Gabius, H.-J. submitted for publication

43. Siebert, H. C.; von der Lieth, C.-W.; Gilleron, M.; Reuter, G.; Wittmann, J.; Vliegenthart, J. F. G.; Gabius, H.-J. In: *Glycosciences: Status and Perspectives*; Gabius, H.-J.; Gabius, S., Eds.; Chapman & Hall, London, Weinheim, 1997; pp. 291-310.
44. Rüdiger, H. In: *Glycosciences: Status and Perspectives*; Gabius, H.-J.; Gabius, S., Eds.; Chapman & Hall, London, Weinheim, 1997; pp. 415-438.
45. Villalobo, A.; Horcajadas, J. A.; André, S.; Gabius, H.-J. In: *Glycosciences: Status and Perspectives*; Gabius, H.-J.; Gabius, S., Eds.; Chapman & Hall, London, Weinheim, 1997; pp. 485-496.
46. Brewer, F. *CHEMTRACTS* 1996, *6*, 165-179.
47. Gupta, D.; Kaltner, H.; Dong, X.; Gabius, H.-J.; Brewer, F. *Glycobiology* 1996, *6*, 843-849.
48. Gabius, H.-J.; Walzel, H.; Joshi, S. S.; Kruip, J.; Kojima, S.; Gerke, V.; Kratzin, H.; Gabius, S. *Anticancer Res.* 1992, *12*, 669-676.
49. Hajto, T.; Hostanska, K.; Frei, K.; Rordorf, C.; Gabius, H.-J. *Cancer Res.* 1990, *50*, 3322-3326.
50. Gabius, S.; Joshi, S. S.; Kayser, K.; Gabius, H.-J. *Int. J. Oncol.* 1992, *1*, 705-708.
51. Gabius, H.-J.; Gabius, S.; Joshi, S. S.; Koch, B.; Schröder, M.; Manzke, W. M.; Westerhausen, M. *Planta Med.* 1994, *60*, 2-7.
52. Joshi, S. S.; Komanduri, K. C.; Gabius, S.; Gabius, H.-J. In: *Lectins and Cancer*; Gabius, H.-J.; Gabius, S., Eds.; Springer Verlag, Heidelberg, 1991; pp. 207-216.
53. Beuth, J.; Ko, H. L; Gabius, H.-J.; Pulverer, G. *In vivo* 1991, *5*, 29-32.
54. Malik, S. T. A. Sem. *Cancer Biol.* 1992, *3*, 27-33.
55. Vassalli, P. *Curr. Biol.* 1993, *3*, 607-610.
56. Sporn, M. B.; Robert, A. B. *Nature* 1988, *332*, 217-219.
57. Balkwill, F. Sem. *Cancer Biol.* 1992, *3*, 1-2.
58. Michiel, D. F.; Oppenheim, J. J. Sem. *Cancer Biol.* 1992, *3*, 3-15.
59. Quesenberry, P. *J. Exp. Hematol.* 1993, *21*, 835-836.
60. Prehn, R. T.; Prehn, L. M. *Arch. Surg.* 1989, *124*, 102-105.
61. Prehn, R. T. *Cancer Res.* 1994, *54*, 908-914.
62. Stewart, F.; Tsai, S.-C. J.; Grayson, H.; Henderson, R.; Opelz, G. *Lancet* 1995, *346*, 796-798.
63. Reed, M. J.; Ghilchik, M. W. *Lancet* 1995, *346*, 1422-1423.
64. Wu, S.; Rodabaugh, K.; Martinez-Maza, O.; Watson, J. M.; Silberstein, D. S.; Boyer, C. M.; Peters, W. P.; Weinberg, J. B.; Berek, J. S.; Bast, R. C. *Am. J. Obstet. Gynecol.* 1992, *166*, 997-1007.
65. Wu, S.; Boyer, C. M.; Whitaker, R. S.; Berchuck, A.; Wiener, J. R.; Weinberg, J. B.; Bast, R. C. *Cancer Res.* 1993, *53*, 1939-1944.
66. Voorzanger, N.; Touitou, R.; Garcia, E.; Delecluse, H.-J.; Rousset, F.; Joab, I.; Favrot, M. C.; Blay, J.-Y. *Cancer Res.* 1996, *56*, 5499-5505.
67. Lahm, H.; Petral-Malec, D.; Yilmaz-Ceyhan, A.; Fischer, J. R.; Lorenzoni, M.; Givel. J. C.; Odartchenko, N. *Eur. J. Cancer* 1991, *28A*, 1894-1899.
68. Akira, S.; Kishimoto, T. Sem. *Cancer Biol.* 1992, *3*, 17-26.
69. Barut, B.; Chauhan, D.; Uchiyama, H.; Anderson, K. C. *J. Clin. Invest.* 1993, *92*, 2346-2352.
70. Eustace, D.; Han, X.; Gooding, R.; Rowbottom, A.; Riches, P.; Heyderman, E. *Gynecol. Oncol.* 1993, *50*, 15-19.
71. Zhang, X. G.; Klein, B.; Bataille, R. *Blood* 1989, *74*, 11-13.
72. Berek, J. S.; Chung, C.; Kaldi, K.; Watson, J. M.; Knox, R. M.; Martinez-Maza, O. *Am. J. Obstet. Gynecol.* 1991, *164*, 1038-1043.

73. Blay, J.Y.; Negrier, S.; Combaret, V.; Attali, S.; Goillot, E.; Merrouche, Y.; Mercatello, A.; Ravault, A.; Tourani, J.-M.; Moskovtchenko, J.-F.; Philip, T.; Favrot, M. *Cancer Res.* **1992**, *52*, 3317-3322.

74. Katsumata, N.; Eguchi, K.; Fukuda, M.; Yamamoto, N.; Ohe, Y.; Oshita, F.; Tamura, T.; Shinkai, T.; Saijo, N. *Clin. Cancer Res.* **1996**, *2*, 553-559.

75. Gabius, S.; Kayser, K.; Gabius, H.-J. *Münch. Med. Wschr.* **1995**, *137*, 602-606.

76. Gelin, J.; Moldawer, L. L.; Lönnroth, C.; Sherry, B.; Chizzonite, R.; Lundholm, K. *Cancer Res.* **1991**, *51*, 415-421.

77. Strassmann, G.; Jacob, C. O.; Evans, R.; Beall, D.; Fong, M. *J. Immunol.* **1992**, *148*, 3674-3678.

78. Fitzpatrick, D. R.; Manning, L. S.; Musk, A. W.; Robinson, B. W. S.; Bielefeldt-Ohmann, H. *Cancer Treatment Rep.* **1995**, *21*, 273-288.

79. Oka, M.; Yamamoto, K.; Takahashi, M.; Hakozaki, M.; Abe, T.; Iizuku, N.; Hazama, S.; Hazama, S.; Hirazawa, K.; Hayashi, H.; Tanyoku, A.; Hirose, K.; Ishihara, T.; Suzuki, T. *Cancer Res.* **1996**, *56*, 2776-2780.

80. Fajardo, L. F.; Kwan, H. H.; Kowalski, J.; Prionas, S. D.; Allison, A. C. *Am. J. Pathol.* **1992**, *140*, 539-544.

81. Hilbert, D. M.; Kopf, M.; Mock, B. A.; Köhler, G.; Rudikoff, S. *J. Exp. Med.* **1995**, *182*, 243-248.

82. Tsunenari, T.; Akamatsu, K.-I.; Kaiho, S.-I.; Sato, K.; Tsuchiya, M.; Koishihara, Y.; Kishimoto, T.; Ohsugi, Y. *Anticancer Res.* **1996**, *16*, 2537-2544.

83. Lu, C.; Sheehan, C.; Rak, J. W.; Chambers, C. A.; Hozumi, N.; Kerbel, R. S. *Clin. Cancer Res.* **1996**, *2*, 1417-1425.

84. Ravoet, C.; De Greve, J.; Vandewoude, K.; Kerger, J.; Sculier, J.-P.; Lacor, P.; Stryckmans, P.; Piccart, M. *Lancet* **1994**, *344*, 1576-1577.

85. Heimpel, H. *Tumordiagn. Ther.* **1995**, *16*, 205.

86. Mayer, H. *Pharmazeut. Ztg.* **1996**, *141*. 2882-2883.

87. De Smet, P. A. G. M.; Keller, K.; Hänsel, R.; Chandler, R. F., Eds. *Adverse Effects of Herbal Drugs*; Springer Verlag, Berlin, 1992.

88. Arzneimittelkommission der deutschen Ärzteschaft. Dt. Ärztebl. **1996**, *93*, B2135.

89. Ernst, E. *Schweiz. Med. Wschr.* **1996**, *126*, 1677-1679.

90. Okada, F. *Lancet* **1996**, 348, 5-6.

91. Werning, C. *Med. Mon. Pharm.* **1995**, *18*, 113.

Chapter 20

Chemistry and Biology of *Hypericum perforatum* (St. John's Wort)

Hans D. Reuter

Siebengebirgsallee 24, D–50939 Köln, Germany

Hypericum perforatum, besides the hypericins (protohypericin, pseudohypericin, cyclopseudohypericin, hypericin), contains additional supposed biologically active compounds such as hyperoside, rutin, quercitin, and chlorogenic acid. As mechanisms for the antidepressant action of Hypericium, an increase in the number of neurotransmitters, an inhibition of type A monoamine oxidase (MAO), an inhibition of catechol-O-methyl-transferase (COMT), a modulation of cytokine expressivity, hormonal effects, and photodynamic effects have been discussed. It is still not possible to assign the antidepressant effects of *Hypericum perforatum* to specific constituents. Following the inhibition of MAO and COMT by the hypericins, it would seem unlikely that inhibition of these enzymes is the main active antidepressant principle of Hypericum. Some experiments suggest that the flavonoids are also involved in the action of the total extract. From recent investigations it can be concluded that the dopaminergic system is involved in the activity of Hypericum extract.

St. John's wort has been used for the treatment of psychiatric disorders for centuries since the times of Paracelsus (1493-1541). St. John's wort was then described as "arnica for the nerves." Today St. John's wort is one of the most important psychotropic drugs. It is widely used in Germany and central Europe for the treatment of psychovegetative disturbances, light depressive states, anxiety and/or nervous restlessness, according to the drug monograph of the German drug regulatory authorities *(1)*.

St. John's wort originates from the wild in Europe and western Asia. The drug consists of the dried flowering tops. Particularly noteworthy are the yellow to yellow-brown flowers which under certain circumstances are still present in the form of cymes and whose petals are covered with numerous dark dots or streaks. Fifty to sixty stamens of each flower are usually fused into three groups. The often shriveled and folded green to brownish-green smooth, ovate-elliptic leaves are up to 3.5 cm long, with clearly visible translucent dots. These dots represent oil glands, filled with voatile oil. Because of the refractive index of the oil glands differing from that of the other cells, the leaves seem to be perforated if they are looked at against the light. From this phenomenon the designation *Hypericum perforatum* is derived.

Constituents of Hypericum Herb

The constituents of Hypericum herb (Figure 1) are 0.05-0.3% hypericin and hypericin-like substances, notably pseudohypericin, isohypericin and the protohypericins. The nothern European broad-leaved variety *perforatum* tends to have less of the hypericins than the southern European variety *angustifolium* DC., the concentration correlating well with the leaves oil gland counts. The special structure of the hypericins consisting of a system of conjugated double bonds is responsible for their red color and of the rather severe phototoxic skin reactions which may occur upon exposure to ultraviolet light after applying Hypericum oil topically *(2)*.

Among the antibiotic substances there are about 3% hyperforin, a phloroglucinol derivative structurally related to the bitter substances of hops *(2)*.

Furthermore, there are flavonoids, especially hyperoside and rutin, and 2,3,6,7-tetrahydroxy-xanthone respectively. Hypericum also contains biflavones, particularly I3,II8-biapigenin (0.26%), amentoflavone and small amounts of procyanidins. The essential oil (0.05 to 0.3%) consists of n-alkanes, especially $C_{29}H_{60}$, α-pinene and other monoterpenes. Additionally, there are up to 10% condensed tannins *(2)*.

When making extracts it must be remembered that hypericin and the pseudohypericin are released in very different amounts. Thus, in a hot infusion, 22% of pseudohypericin is released within the first 10 minutes, whereas only 6% of the hypericin is released. In the subsequent 60 minutes a further 23% of the pseudohypericin, compared to a further 7% of the hypericin, is released. Experiments with extraction at various temperatures show that maximum release of pseudohypericin is achieved at 60-80°C, above which the yield decreases. Basically, this also applies to hypericin, but both the amount released and the temperature sensitivity are lower. With increased methanol concentration, the amount of hypericin extracted increases. Pseudohypericin increases with 80% methanol and decreases with 100% methanol to 30% of the 80% methanol value *(3)*.

Controlled Clinical Trials

Despite the long use of Hypericum preparations, systematic investigation for its effectiveness against depression has been neglected until approximately 10 years ago. In contrast to the development of synthetic drugs, studies into the clinical efficacy and tolerability of Hypericum have been performed predominantely while pharmacological investigations were conducted merely as supplementary measures. To date, there have been about 30 controlled trials with Hypericum extracts, involving over 1500 depressive patients with daily dosages of 900 mg for between 28 and 42 days. They have confirmed the antidepressant action of Hypericum extracts, especially that of LI 160 or Jarsin. The extract LI 160 is characterized phytochemically as follows: 100 mg extract containing 300 µg hypericins. Trials comparing the effects of this preparation and maprotilin showed

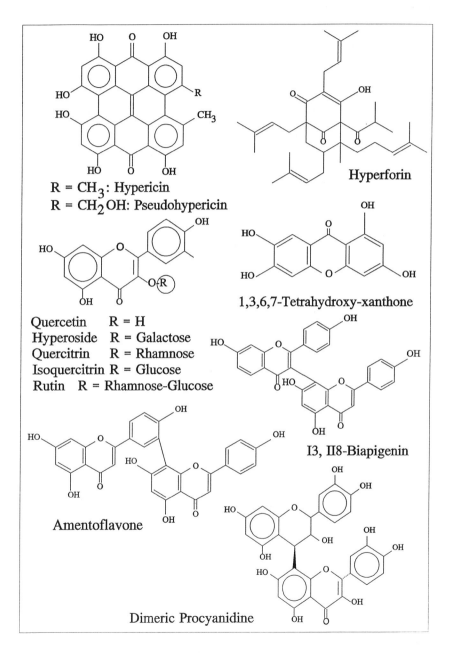

R = CH₃: Hypericin
R = CH₂OH: Pseudohypericin

Hyperforin

1,3,6,7-Tetrahydroxy-xanthone

Quercetin   R = H
Hyperoside   R = Galactose
Quercitrin   R = Rhamnose
Isoquercitrin   R = Glucose
Rutin   R = Rhamnose-Glucose

I3, II8-Biapigenin

Amentoflavone

Dimeric Procyanidine

Figure 1. Constituents of Hyperici herba.

similar depression-relieving and mood-lifting effects in both products, whereas maprotilin showed a more pronounced sedative action *(4, 5)*. Jarsin, with its action spectrum of slightly stimulative and very slightly sedative components, is very similar to imipramin *(6)*.

Despite proof of the therapeutic efficacy of St. John's wort in patients with mild depressions the mechanism of the antidepressant effect is still unclear.The latest issue of the Comission E drug monograph of the German drug regulatory authorities dates from 1984 *(1)*. This monograph recommends as a "medium daily dose for internal use: 2-4 g dried extract or 0.2-1.0 mg total hypericin in other forms of administration". It thus regards hypericin as the definitive active constituent. However, due to the subsequent availability of improved analytical and pharmacological experiments, we must now assume that the hypericins (proto-hypericins, pseudohypericins, cyclopseudohypericins, hypericin) are not the only important active constituents in *Hypericum perforatum*. As active substances also hyperoside, rutin, quercitin, hyperforin and chlorogenic acid must be considered.

Proposed Mechanisms for the Antidepressant Action of Hypericum and Animal Models

The following mechanisms are being discussed for the antidepressant action of St. John's wort:
- Increase in the number of neurotransmitters *(7)*
- Inhibition of type A monoamine oxidase *(8)*
- Inhibition of catechol-O-methyltransferase *(9)*
- Modulation of cytokine expressivity *(10)*
- Hormonal effects *(12)*
- Photodynamic effects, as in light therapy *(13)*

The following animal experiments have been performed for the determination of the antidepressant action of St John's wort:
- Exploration behaviour in unfamiliar environments (mice) *(14)*
- Ethanol sleep period (mice) *(14)*
- Action of reserpine (mice) *(14)*
- Behaviour on the waterwheel (aggression of isolated male mice) *(14)*
- Clonidine depression (gerbils) *(14)*
- Forced swimming test (Porsolt) *(12)*

The Amine Theory: Inhibition of MAO and COMT by Hypericum Extract Fractions

One of the theories concerning the mechanism of action of Hypericum extracts is the so-called amine theory *(15)*. It is based on the fact that in endogeneous depression an insufficient monoaminergic postsynaptic excitation--preferably by epinephrine and serotonin--is seen. The therapeutic effects of Hypericum extracts could be explained by either blocking the reuptake of serotonin and/or norepinephrine into the presynaptic nerve endings and/or by inhibiting the decomposition of epinephrine. By inhibiting the enzymes responsible for this process the levels of epinephrine could be increased. By the successive action of MAO and COMT, norepinephrine is transformed either to 3,4-

dihydroxymandelic acid or to vanillin-mandelic acid (Figure 2). The function of norepinephrine in the synapses, representing the contact place between the nerve cells and e.g. the sensory cells is demonstrated in Figure 3. Norepinephrine is located in the presynaptic vesicles and from there diffuses into the synaptic gap. At the postsynaptic membrane it induces an alteration of its permeability for ions which on their part modify the bioelectric polarisation of the membrane. By the enzymatic decomposition of the neurotransmitter, which also may consist of dopamine, serotonin, and GABA (gamma-aminobutyric acid) the neurotransmitter effect is rapidly abolished and a repolarisation of the membrane takes place.

Thiede *(9)* has investigated the inhibiting potential of several fractions of a Hypericum extract. The fractions 1-6 were obtained by extraction with different solvents resulting in the isolation of the components of the essential oil (Table I).

Table I. Solvents used for the preparation of extracts from 100 mg *Hypericum perforatum* and composition of the fractions obtained. Fractions obtained are characterized by the total amounts in mg and in % of the total amounts present in the original extract *(9)*.

Fraction 1: Solvent: petrol ether:ether 1:1 yields 37 mg (2%) lipid-soluble components such as n-alkenes, pinene and other compounds of the essential oil
Fraction 2: Solvent: ether yields 66 mg (3%) lipid soluble components
Fraction 3: Solvent: ether:acetone = 8:2 yields 34 mg (2%) lipid-soluble components as fraction 1
Fraction 4: Solvent: acetone yields 47 mg (2%), little hypericin and flavonols
Fraction 5: Solvent: methanol:acetone (1:1) = yields 189 mg (8%) hypericin, little flavonols
Fraction 6: solvent: methanol yields 1824 mg (83%) flavonoids, flavone glycosides, xanthones

By comparing the composition of the fractions with their ability to inhibit MAO and COMT, it becomes evident that St John's wort extract only inhibits MAO and COMT activity at relatively high concentrations (Table II). The COMT inhibition is mainly due to the flavonoids or flavone glycosides. The MAO inhibition, due to as yet unidentified constituents, can be observed in all high concentrations of St John's wort extract fractions. The MAO inhibition caused by isolated hypericin is considerably less than that of the total extract. Only at therapeutically irrelevant high doses of total extract (10^{-3}-10^{-4} M) can an essential inhibition of MAO, but not of COMT, be seen. From these findings it can be concluded that the effect of the extract on norepinephrine metabolism is not only due to hypericin, and that inhibition of MAO and COMT is not the main mechanism by which Hypericum extracts exert their therapeutic effects *(9)*.

292

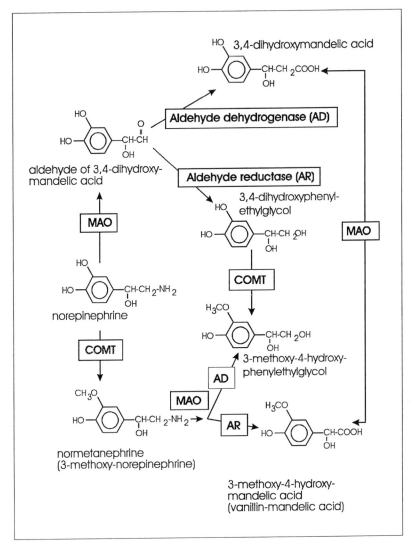

Figure 2. Decomposition of norepinephrine (MAO = monoamine oxidase, COMT = catecholamine-O-methyltransferase).

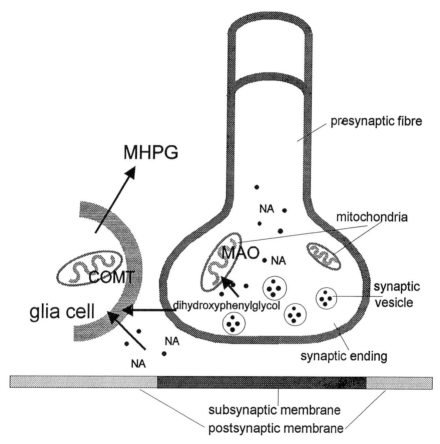

Figure 3. Metabolism of epinephrine in the synapses.

**Table II. Inhibition (%) of MAO (M) and COMT (C) in vitro by extracts from
Hypericum perforatum. For extracts and fractions, the mean mol mass of substance
mixtures was taken as that of hypericin, i.e. 502 *(9)*.**

Concentration (M)	10^{-3}		10^{-4}		10^{-5}		10^{-6}	
	M	*C*	*M*	*C*	*M*	*C*	*M*	*C*
Solvent	12	0	0	0	0	0	0	0
Hypericin	25	0	2	0	0	0	0	0
Hypericum extract	68	-	27	35	0	0	0	0
Extract fraction 1	39	11	16	7	5	6	0	0
Extract fraction 2	39	0	0	0	0	0	0	0
Extract fraction 3	71	10	29	0	13	0	0	0
Extract fraction 4	51	0	4	0	0	0	0	0
Extract fraction 5	59	83	5	16	0	0	0	0
Extract fraction 6	42	93	5	21	0	4	0	0

Results from experiments by Bladt and Wagner *(8)* on the MAO inhibition of
hypericins and St John's wort fractions are essentially the same as those found by Thiede
(9). Of the various St. John's wort constituents quercetin, luteolin and kaempherol, found
in concentrations of about 10^{-5} M, demonstrated the highest inhibitory action (80-95%).
Similar inhibition was shown by xanthones (70-90%) and hydroxyxanthones (60-80%).
1,3 Dihydroxyxanthone showed inhibition in concentrations of 10^{-3} M (70%) to 10^{-6} M
(40%). The phloroglucin derivative, hyperforine, which occurs in St. John's wort in
concentrations of about 3% showed only weak inhibition of less than 20% at a
concentration of 10^{-5} M. At this same molar concentration, quercitrin was the only polar
compound to demonstrate inhibition (65%) whilst the phenol carboxylic acids showed no
effect. The total hypericin fraction only showed an inhibitory effect (70%) at a high
concentration of 10^{-3} M. At 10^{-4} M both the total hypericin extract and pure hypericin
showed less than 10% MAO inhibition, contrary to the 27% inhibition by the extract
reported by others (Table II). These findings again establish that hypericins cannot be
regarded as inhibitors of MAO.

While the amine-theory can explain the mode of action of the classical
antidepressants, it is insufficient to explain the therapeutic effect of Hypericum extracts.
Especially the latency period of several weeks between drug application and
antidepressant effect can not be explained on the basis of this theory.

Effect of Hypericum Extract on the Expression of Serotonin Receptors

In addition to an increased availability of the neurotransmitter, i.e. by inhibition of MAO,
the antidepressant effects of Hypericum extracts could be explained by a decrease in the
number of serotonin receptors. Müller and Rossol *(7)* investigated the effect of
Hypericum extract on the expression of serotonin receptors using a neuroblastoma cell
line to establish a model for the regulation of neurotransmitters by immunologically
active compounds such as cytokines. The cells were incubated with the Hypericum

extract LI 160 for 2, 4, 6, 8 and 10 hours, then washed. Examination of the cells by means of fluorescence microscopy showed maximum reduction of the expression of the serotonin receptors after 6 hours of incubation. Also by diluted extract solutions a distinct reduction of receptor expression occured. From the LI160-induced reduction of the serotonin receptor availability, resulting in a decreased reuptake of the neurotransmitter into the cell, the authors conclude a possible mechanism of the antidepressant effect of the Hypericum extract. Contrary to the classical antidepressants, which bind to the receptor, it is suggested that neighboring epitopes of the serotonin receptors are occupied by the constituents of Hypericum, with subsequent masking of the receptors.

The Macrophage Theory of Depression

Besides direct effects on neurotransmitter metabolism, indirect effects mediated by immune modulators have been proposed for the antidepressant effects of Hypericum extract. The immune system and the nervous system are the only biological systems which are equipped with the ability of a memory. These systems are characterized by numerous regulatory interactions and biochemical correspondence. Important molecules are produced together by both systems, excreted and recognized as active substances such as endorphins, ACTH (adrenocorticotropic hormone), ADH (antidiuretic hormone) and others. Provided that the constituents of the Hypericum extract can not penetrate the blood-cerebrospinal fluid-barrier, their primary point of attack would be located outside of the central nervous system. The effect must then be mediated by CSF-penetrating substances. In this case effects on the cells of the immune system, such as lymphocytes and monocytes, could be expected. Taking into consideration these interactions between the immune system and the nervous system, the following mechanism could be supposed: stimulation of monocytes induces release of IL-6 (interleukin-6) which in turn influences, via the corticotropin releasing hormone and the lymphocytes, the release of ACTH. ACTH then induces the release of cortisol which converts the normal emotional state into a depressive one (10).

In his macrophage theory of depression, Smith (16) has suggested that in predisposed subjects, monocyte-interleukins are able to induce all symptoms of depressive states. Therefore, various specific or unspecific stimuli of monocuclear cells, such as infections, allergies, and estrogens, would induce an increase of interleukine-levels resulting in mental changes (Figure 4).

Based on these considerations Thiele and coworkers (10) investigated the effect of Hypericum extract on the stimulated cytokine expression in vitro in a whole blood culture system. The assessment of the PHA(phytohemagglutinin)- and lipopoly-saccharide-stimulated and unstimulated release of the cytokines interleukin 1ß, interleukin 6 and tumor-ncrose-factor-α (TNF-α) was performed with anticoagulated blood from 4 healthy and 4 depressive patients. Hypericum extract LI 160 significantly depressed the release of IL-6 stimulated by phytohemagglutinin and lipopolysaccharide by 92, 89, 90 and 97% in healthy persons and by 39, 95, 87 and 94% in depressive patients. In contrast to the stimulated release of IL-6, the release of IL-1ß and TNF-α was influenced to a lesser extent. Furthermore, investigations of Müller and Rossol (7) have shown an increased expression of IL-1 on neurodermal cells, pointing out that also IL-1 directly takes part in the neuronal control.

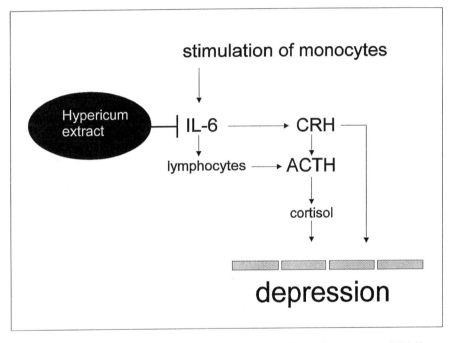

Figure 4. Modulation of cytokine expression by the Hypericum extract LI 160.

Central Effects of Hypericum Extracts, Being Mediated by the Dopaminergic System

In vivo, Okpanyi and Weischer *(14)* were able to demonstrate a prolongation of an anaesthesia by ethanol, an increase of the spontaneous motility, a partial reserpin antagonism, and a suppression of the aggression behaviour of isolated male mice. However, these experiments have been critisized in so far as the Hypericum extract used had not been characterized phytochemically, the extract was applied in high doses, 50% propanediol was used as solvent, and hypericin as reference substance had not been applied by the oral route such as the Hypericum extract, but was given intraperitoneally.

More recently a series of pharmacological trials were performed to confirm the assumption that Hypericum extracts exert a central effect and that these effects are mediated by the dopaminergic system.

Investigations of Winterhoff and coworkers *(11)* have shown that Hypericum extract in mice induces an increase in body temperature. The increase is significant one hour after the application of 250 and 500 mg Hypericum extract per kg body weight. The temperature remains elevated above the initial temperature for 4 hours. These changes of body temperature indicate that oral Hypericum extract exerts central effects.

These indications of a central effect of Hypericum were further investigated in other experimental approaches by Winterhoff and coworkers *(12)*. A standard model for the assessment of the central effects of drugs is to measure the influence on the ketamine induced sleeping time. Pretreatment of female mice with 50, 200 and 500 mg/kg Hypericum extract results in a dose-related shortening of the sleeping time. The effect was significant with 200 mg extract and highly significant with 500 mg. A somewhat smaller effect was seen after pretreating the animals with concentrations of pure hypericin of 0.15, 0.6 and 1.5 mg/kg. The mean duration of anaesthesia by the extract was shortened by 4.3%, 18.9% and 36.5%, by hypericin at concentrations corresponding to that of the extract by 2%, 9% and 22.4%. The same effect as by 500 mg Hypericum extract was seen with 20 mg/kg bupropione. Pretreatment of the animals with the unspecific dopamine receptor blocker haloperidol and with the specific D_2-receptor sulpiride completely abolished the effects of Hypericum extract and bupropione.

A more specific method for testing the antidepressant effect of drugs is the forced swimming test according to Porsolt *(17)*. The method is based on the observation that a rat, when forced to swim in a situation from which there is no escape, will become immobile making only those movements neccessary to keep its head above the water. Hypericum extract caused a reduction in immobility time which was significant at 125 mg/kg. Further increasing the concentration did not result in an stronger inhibition. Again 20 mg bupoprione proved to be active *(12)*.

From the fact that the anti-immobility effect of Hypericum extract observed during the forced swimming test and the Hypericum-induced decrease of the ketamine induced sleeping time are antagonized by the unspecific dopamine receptor blocker haloperidol as well as by the specific D_2-receptor sulpiride, it can be concluded that the dopaminergic system is involved in the activity of the Hypericum extract LI 160. These results are in good accordance with the findings of Winterhoff et al. *(12)* that prolactin levels are reduced after repeated application of Hypericum extract. This also reflects a dopaminergic effect.

Conclusions

Summarizing the results of the various pharmacological investigations done so far the following conclusions can be made:
1. The inhibition of monoamine oxidase does not seem to be the main principle of the antidepressant effect of Hypericum extract. Based on the composition of the extract, the hypericins do not influence MAO activity at therapeutically used doses in a clinically relevant manner. Other constituents, such as the flavonoids, have a distinctly stronger effect on MAO.
2. Hypericum extract decreases the expresison of serotonin reseptors in the CNS. The constituents which are responsible for this effect have not been identified so far.
3. Besides direct effects on neurotransmitter metabolism, indirect effects mediated by immune modulators seem to play a role in the antidepressant effect of Hypericum extract. Thus, Hypericum extract significantly depressed the stimulated release of IL-6, which in turn influences, via the corticotropin releasing hormone and the lymphocytes, the release of ACTH. ACTH then induces the release of cortisol, which converts the normal emotional state into a depressive one.
4. From the results of investigating the effects of Hypericum extract on the ketamine induced sleeping time and the forced swimming test it can be concluded that the dopaminergic system is involved in the activity of Hypericum extract.
5. Further investigations are necessary to identify the constituents of Hypericum which are responsible for its antidepressant effects.

Literature cited

1. Kommission E; Monographie "Johanniskraut - Hyperici herba"; *Bundesanzeiger* Nr. 228, 5.12.1984.
2. Wichtl, M. Medpharm Scientific Publ. Stuttgart **1994**, pp. 273-275.
3. Niesel, St.*Inauguraldissertation* 1992, Freie Universität Berlin.
4. Harrer, G., Hübner, W. D., Podzuweit H. *Nervenheilk.*, **1993**, *12*, 297-30.
5. Johnson, D., Ksciuk, H., Woelk, H., Sauerwein-Giese, E., Frauendorf, A. *Nervenheilk.*, **1993**,*12*, 328-330.
6. Vorbach, E. U., Hübner, W. D., Arnoldt, K., H. *Nervenheilk.*, **1993**, *12*, 290-296.
7. Müller, W. E. G., Rossol, R. *Nervenheil.*, **1993**, *12*,357-358.
8. Bladt, S., Wagner, H. *Nervenheilk.*, **1993**, *12*,349-52.
9. Thiede, H. M., Walper, A. *Nervenheil.*, **1993**, *12*, 346-348.
10. Thiele, B., Brink, I., Ploch, M. *Nervenheilk.*, **1993**, *12*, 353-356.
11. Winterhoff, H., Hambrügge, M., Vahlensieck, U. *Nervenheilk.*, **1993**, *12*, 341-345.
12. Winterhoff, H., Butterweck, V., Nahrstedt, A., Gumbinger, H. G., Schulz, V., Erping, S., Boßhammer, F., Wieligman, A. In: Loew, D., Rietbrock, N. (Eds.); *Phytopharmaka in Forschung und klinischer Anwendung.* Steinkopff Verlag Darmstadt **1995**; pp. 38-56.
13. Martinez, B., Kasper, S., Ruhrmann, S., Möller, H. J. *Nervenheilk.*, **1993**, *12*, 302-307.
14. Okpanyi, S. N., Weischer, M. L. *Arzneim-Forsch/Drug Res* **1987**, *37*, 10-13.
15. Kretschmar, R., Stille, G., Psychopharmaka. In: Estler, C.-J. (Ed.) *Pharmakologie und Toxikologie.* Schattauer Verlag Stuttgart, New York , **1995**, p. 203.
16. Smith, R. S. *Med. Hypotheses* **1991**, *35*, 298-306.
17. Porsolt, R. D. *Eur. J. Pharmacol.* **1978**, *47*, 379-391.

Chapter 21

Vitex agnus-castus (Chaste Tree): Pharmacological and Clinical Data

H. Winterhoff

Institute of Pharmocology and Toxicology, University of Münster, Domagkstrasse 12, D–48149 Münster, Germany

The medicinal use of extracts from *Vitex agnus-castus* has a long tradition. The use in „disturbances of the female genital system" was first reported more than two thousand years ago. Nowadays ethanolic extracts from the fruit are used for the treatment of mastopathy, premenstrual syndrome and luteal insufficiency—complaints caused by a mild or a latent hyperprolactinemia. The mode of action of *Vitex agnus-castus* extracts was studied in vitro using primary cultures from rat pituitaries as a test system. A dose dependent reduction of prolactin secretion was observed. The effect could be blocked by the dopamine-receptor antagonist, haloperidol, confirming a dopamine agonistic activity of the drug. The prolactin lowering effect of *Vitex agnus-castus* was also found in in vivo experiments in rats. Clinical studies in patients with premenstrual syndrome, luteal insufficiency, and mastopathy show a simultaneous decrease in the clinical syndrome score and prolactin levels.

Vitex agnus-castus is a shrub belonging to the family of Verbenaceae which is widespread in the Mediterranean region and in Asia. As is the case for a considerable number of medicinal plants described, e.g. in the German Pharmacopoeia, the plant was used anciently as an herbal remedy.

The medicinal use of *Vitex agnus-castus* was mentioned by Dioskurides, Hippokrates, Plinius, and Galen. Even at that time the plant was namely used for the treatment of "dysregulations of the female genital system." Moreover, the shrub was a symbol of chastity. The Greek goddess Hera is said to have been born under an Agnus-castus shrub. The vestal virgins carried branches of *Vitex agnus-castus*. Accordingly, in the Middle Ages at the robing ceremony of novices, the way to the monastery was sprinkled with flowers from *Vitex agnus-castus*. The fruit of the plant, which smell and taste like pepper, was used by the monks as a spice and as an anaphrodisiac. In addition, the fruit was said to be useful not only in diverse disturbances of the female genital system but were also used as a lactogen, as an anorectic, as a hypnotic, and for dyspepsia (*1*).

The fruit is also the part of the plant which is used nowadays. In the monograph of the Commission E, a daily dosage of 30-40 mg of an ethanolic extract of the fruit of *Vitex agnus-castus* is recommended for the treatment of the premenstrual syndrome, luteal phase defects and other disturbances of the menstrual cycle, as well for the treatment of mastodynia (*2*).

Though such symptoms have been treated with extracts of *Vitex agnus-castus* for many years, the mode of action was not determined until much later. One prerequisite for this was the discovery of the occurrence and the physiological meaning of prolactin in men about 30 years ago. Shortly after this finding, the pathophysiogical meaning of hyperprolactinemia was discovered. In a considerable number of women with anovulatory (no ovum discharged) cycles and sterility, increased basal prolactin levels were found. Subsequently, hyperprolactinemia turned out to be rather widespread in women and to be one of the most common reasons for infertility (*3*). In women with manifest hyperprolactinemia, often a correlation between the severity of the clinical findings and the prolactin levels is found (*4*).

Later it became apparent that even minor changes of prolactin synthesis and/or secretion cause pathophysiological changes in the women concerned. In patients with such a "latent hyperprolactinemia," basal prolactin levels in the normal range are found; whereas, the prolactin increase following thyrotropin releasing hormone (TRH) stimulation is augmented. Furthermore, the nocturnal prolactin rise increases and/or the progesterone levels are found to decrease. The clinical signs are mastopathy, disturbances of the menstrual cycle, or even sterility. In 50% of the patients suffering from luteal phase defects, a manifest or only a latent hyperprolactinemia is diagnosed. In a study from 1992, Aisaka et al. diagnosed in 62% of 753 infertile women a latent hyperprolactinemia identified by a diminished progesterone phase of the cycle, an augmented prolactin increase following TRH application, or an increased nocturnal prolactin rise (*5*).

Pituitary prolactin secretion is regulated mainly negatively by dopamine—the prolactin inhibiting factor—which is mediated mainly by D_2 receptors. Various peptides, such as TRH and VIP (vasoactive intestinal peptide), stimulate prolactin release in pharmacological dosages, but their physiological role is still controversial.

Shortly after finding that hyperprolactinemia was the cause for a substantial part of unexplained sterility, a treatment for this pathological state was discovered— bromocriptine, a dopamine-agonistic drug derived from ergot. Several thousand children, in the meantime, owe their life to the treatment of their mothers with bromocriptine and other dopamine agonistic drugs. While dopamine agonists are indicated in medical treatment of manifest hyperprolactinemia, *Vitex agnus-castus* may be useful in treatment of latent hyperprolactinemia.

In mild forms of hyperprolactinemia, all the rather nonspecific symptoms are found which are indications for *Vitex agnus-castus*. Correspondingly, the efficacy of *Vitex agnus-castus* preparations, e.g. in patients with luteal insufficiency, was reported in the older literature, suggesting a prolactin lowering effect by a dopaminergic activity.

A number of experimental and clinical studies support the assumption that *Vitex agnus-castus* reduces the circulating prolactin levels by a dopaminergic mode of action.

In Vitro Studies

As a test system primary rat pituitary cell cultures have been used. For the in vitro investigations a hydroethanolic extract from *Vitex agnus castus* and the commercial preparation, Mastodynon[R] (which also contains, besides the main constituent *Vitex agnus-castus* extract, diluted extracts from *Iris, Lilium, Cyclamen, Caulophyllum,* and *Ignatia*) were freed from ethanol under reduced pressure, and the resulting precipitate was removed. As shown in Figure 1, the water soluble constituents from *Vitex agnus-castus* as well as those from Mastodynon[R] reduced both the basal and the TRH-stimulated prolactin secretion. The effect was comparable to that of the reference substance dopamine (*6, 7, 9*).

Since the vitality of the cells proved to be unchanged, toxic effects as the reason for the reduced prolactin secretion could be excluded (*7*). Subsequently, diverse concentrations of both preparations were tested in the same test system. A dose-dependent reduction of prolactin release was observed (Figure 2, open bars). The effect was markedly reduced by the dopamine receptor antagonist, haloperidol (Figure 2, black bars). This confirms the assumption that *Vitex agnus-castus* exerts a dopaminergic activity. The effect of the preparation Mastodynon [R] was comparable, and it was shown that *Vitex agnus-castus* is responsible for its activity.

By repeating these experiments, the authors confirmed a dose-dependent effect of the extract. Moreover, the inhibition of prolactin secretion seemed to be a specific activity as neither FSH- nor LH- secretion were changed by the *Vitex agnus-castus* extract (*7*).

The effects of *Vitex agnus-castus* were reinvestigated by Sliutz et al. in 1993 using the same in vitro test system. They also found a dose dependent reduction of prolactin release under basal conditions, as well as following TRH stimulation—maximal inhibition by *Vitex agnus-castus* extract was found at a concentration of 460 µg/mL. As a reference substance, lisuride was used. The effects of the plant extract, as well as that of lisuride, could be blocked by haloperidol, confirming the hypothesis that *Vitex agnus-castus* acts as a dopaminergic drug (*8*).

The D_2 receptor, which mediates the effects of dopamine on prolactin release, is amply expressed in the corpus striatum of mammals. Hence, plasma membrane preparations of striatal tissue can be used as a source for this receptor. In vitro studies with such membrane preparations, using radioactive sulpiride (a D_2 receptor antagonist) as a ligand, showed a very high potency of constituents from *Vitex agnus-castus* to displace the ligand from this receptor. In addition, a marked affinity to the D_1 receptor was found (*9*).

To check the relevance of these in vitro findings, the effect of *Vitex agnus-castus* on prolactin release was studied in vivo.

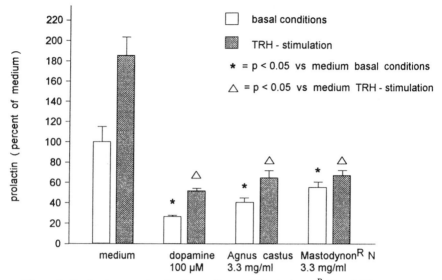

Figure 1. Prolactin secretion in vitro: effects of Mastodynon[R] N and *Vitex agnus-castus* on primary cultured rat pituitary cells. (Reproduced from Ref. 9 with permission, Copyright 1995, Steinkopff Verlag, Darmstadt.)

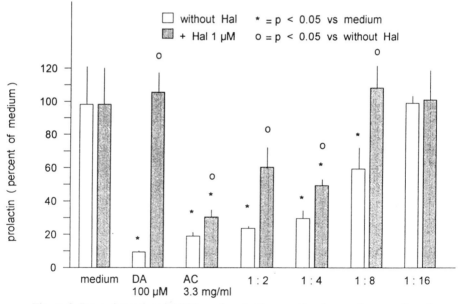

Figure 2. Dose-dependent inhibition of prolactin secretion from primary cultured rat pituitary cells by the water soluble constituents of *Vitex agnus-castus* (AC) and suppression of the effects by the nonspecific dopamine (DA) antagonist, haloperidol (Hal). (Reproduced from Ref. 9 with permission, Copyright 1995, Steinkopff Verlag , Darmstadt.)

Animal Experiments

Animals with inserted jugular vein catheters have been studied. In such animals blood samples can be drawn repeatedly without causing pronounced stress. This is of specific importance when studying effects on prolactin levels, since prolactin is a "stress hormone," being released in response to any stressful situation.

Pretreatment values show the basal prolactin values (Figure 3). *Vitex agnus-castus* extract or physiological saline were injected intravenously, and 15 minutes later the rats were exposed to an ether stress. As shown in the figure, ether exposure causes a quick and pronounced increase in prolactin levels. This increase was clearly less pronounced in the *Agnus-castus* group, indicating an inhibition of prolactin secretion in vivo as well. In a similar experiment, the prolactin increase following handling-stress was completely abolished by pretreatment with the plant extract. Both experiments clearly indicate an inhibition of prolactin secretion in vivo (*6, 9*).

As indirect proof for the prolactin lowering effect of *Vitex agnus-castus,* the inhibition of lactation was studied. This experimental design was used by Flückiger et al. (*10*) to study the effects of bromocriptine. Postpartum lactating rats were injected subcutaneously once daily from day 5 to day 8 with physiological saline, 5 mg/kg bromocriptine, or with 2x50 mg of the water soluble constituents of *Vitex agnus-castus*. In the pubs, milkspots can be seen when the stomach is filled with milk. The extract from *Vitex agnus-castus,* as well as bromocriptine, caused an increase in the number of pubs without milk spots and an increase in mortality of the pubs, presumably as a consequence of a reduced milk production in the mothers (Figure 4). This finding is not only in good accordance with the prolactin lowering effect of *Vitex agnus-castus* measured radioimmunulogically in vivo, it also confirms that indeed the biological activity was simultaneously reduced (*11*). The effects presented thus far can be explained by a direct dopaminergic action of the drug, as well as by an indirect effect mediated by an increased hypothalamic dopamine secretion.

Studies in hypothalamus-lesioned rats should resolve this question. Animals in which the hypothalamus is lesioned electrochemically cannot release hypothalamic dopamine. Consequently, they have high basal prolactin levels. In such animals, the intravenous injection of a purified fraction of a *Vitex agnus-castus* extract caused a pronounced and long lasting decrease in prolactin levels. This experiment shows clearly that *Vitex agnus-castus* exerts dopaminergic activity at the pituitary level; however, an indirect effect, by increasing the endogenous dopamine secretion, could be ruled out (*9*).

After demonstrating the activity of *Vitex agnus-castus* by both in vitro and in vivo experiments, it was most important to prove the clinical efficacy.

Clinical Studies

A considerable number of older clinical studies were performed with *Vitex agnus-castus* preparations in patients with diverse disturbances of the menstrual cycle.

In one study, 49 patients with *secondary amenorrhoe* were treated over a period from six weeks up to 46 months. In 40 of the patients menstruation reappeared, while only in nine patients did the treatment fail. In the same study with 59 patients with

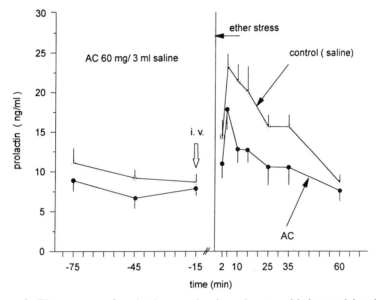

Figure 3. Time course of prolactin secretion in male rats with inserted jugular catheters. After three pretreatment values (-75 to -15 min), the animals received 60 mg *Vitex agnus-castus* (AC) extract or physiological saline i.v. After an additional 15 min, the animals were exposed to an ether stress for 30 s. The clearly less pronounced prolactin increase following *Agnus castus* indicates an inhibitory effect under in vivo conditions. (Reproduced from Ref. 9 with permission, Copyright 1995, Steinkopff Verlag, Darmstadt.)

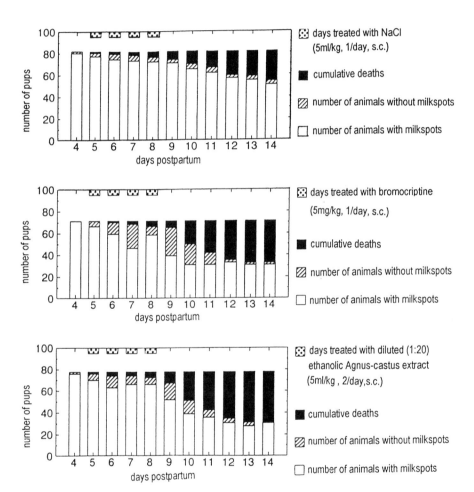

Figure 4. Treatment of lactating rats with bromocriptine (5 mg/kg) or *Vitex agnus castus* (50 mg/kg) on days 5-8 postpartum. Cumulative number of pups with and without milk spots, as well as dead animals. (Reproduced from Ref. 11 with permission, Copyright 1991, Hippokrates-Verlag, Stuttgart.)

oligomenorrhoe, 88% of the patients experienced clearly shortened intervals between bleedings. A comparable therapeutic outcome was reported for the treatment of *polymenorrhoe* with *Vitex agnus-castus*. In 62 of 70 women, the treatment proved to be effective (*12*).

Comparable positive results were also obtained in patients with *premenstrual syndrome* (PMS). In a total of 36 women, a number of physical and psychological symptoms, e.g. headache, tension of the breast, tiredness, anxiousness, and depression were judged before and following three months of treatment. In addition, the length of the luteal phase was determined as the period of increased temperature. The period of the luteal phase was lengthened significantly from 5.5 ± 3.8 d to 11.4 ±1.5 d, while the symptom score reduced significantly under the treatment (*12*). Further investigations were performed by several practitioners with a total of 1542 patients with PMS. The patients were treated for an average of 166 days, with 33 % of the patients reported to be completely cured, while in 57 % the symptoms were clearly reduced. Only 4% reported no improvement; for another 5% no data could be obtained; and 2.1% of the patients terminated the treatment because of unwanted side effects (*13*). None of the studies were double-blinded, nor were prolactin levels determined. Though such studies have to be completed by additional investigations, the pronounced therapeutic success gives hints of activity for the extract.

In recent studies, hormonal parameters have been determined in addition to general symptoms. In one study, 45 infertile normoprolactinemic women were treated over a period of at least three menstrual cycles with 40 drops of an ethanolic extract from *Vitex agnus-castus* daily. All patients had pathologically low progesterone levels at day 20. The outcome of the treatment was judged by means of the frequency of increased progesterone levels and on the extension of the hyperthermic phase. In 39 patients the treatment proved to be effective, and seven of the patients even became pregnant during the treatment (*14*).

In a double-blind study, a total of 93 patients with mastopathy were treated over a period of 4 cycles with a dose of 30 drops of Mastodynon[R] twice daily. The complaints score was significantly reduced in the verum group, and the prolactin levels were also reduced (*15*).

In 13 patients with disturbances of the menstrual cycle, a three-month treatment with Mastodynon[R] caused a clear decrease of the prolactin levels as well as a normalization of the cycle (*16*).

Patients with luteal phase defects due to a latent hyperprolactinemia were treated for three months with a daily dosage of 20 mg of an ethanolic extract of *Vitex agnus-castus* (n = 17) or placebo (n=20). This relatively low dose was chosen corresponding to the first monograph on *Vitex agnus-castus*. The latent prolactinemia was confirmed by an augmented prolactin increase following TRH injection, a shortened luteal phase, and low midcycle progesterone levels. After the treatment period a clear difference between the placebo and the verum group became obvious—in the verum group the TRH-induced prolactin increase was less pronounced, the length of the luteal phase was increased significantly, and the progesterone levels increased (*17*).

A good efficacy was reported repeatedly for premenstrual mastodynia, which is often caused by a latent hyperprolactinemia and associated with corpus luteum insufficiency. In those patients the latent hyperprolactinemia often causes complaints at the end of the luteal phase when the inhibitory effect of progesterone is removed and the prolactin secretion is increased. Such patients were treated for a three-month period with Mastodynon [R] or placebo in a multicenter study. Whereas placebo caused only minor, insignificant changes of the prolactin levels, treatment with the verum resulted in a significant decrease in prolactin levels, confirming the activity of the extract (9).

Twenty healthy male subjects were treated with placebo or with three different doses of an ethanolic extract from *Vitex agnus-castus* corresponding to 120, 240 or 480 mg drug/day successively over 14-day periods. The daily dose was given in three administrations, with a wash-out phase of at least one week. Basal and TRH-stimulated prolactin values were slightly reduced following the higher doses, whereas following the low dose an increased basal and stimulated prolactin level was determined (18, 19). At the moment it is uncertain whether the extract indeed contains agonistic and antagonistic compounds, as the authors suppose, or whether the effect in males cannot fully be compared to that in females. Anyway, in spite of the higher doses used, no toxic signs were observed.

Active Constituents

One problem that is not yet answered is the identification of the active principle, although numerous publications deal with the composition of the fruit from *Vitex agnus-castus*. Flavonoids and the iridoids, agnuside and aucubin, have been identified. The fruit contains 0.7-1.2% volatile oil, consisting mainly of the sesquiterpenes, β-caryophyllene and germacrene B, and of the monoterpenes, α-pinene, sabinene, β-phellandrene, and variable amounts of 1,8-cineole (20). However, these substances cannot be responsible for the biological activity, since water-soluble constituents proved to be responsible for the activity (9). Although a fractionation of the water soluble fraction was performed and three active fractions were obtained, yet in none of these fractions could single compounds be identified. The most active fraction proved to be thermolabile, with the activity being lost during storage. This is in clear contrast to findings with the original extract, which does not show any loss of activity under the same conditions. Maybe this will turn out as a further example of a phytopharmacon in which "inactive" constituents prove to be important by improving the stability or bioavailabilty of the extract.

Literature Cited

1. Böhnert,K.J.; Hahn,G. *Erfahrungsheilkunde* **1990**, *9*, 494-502.
2. German Federal Gazette **1992**, No.226.
3. Flückiger,E.; Del Pozo,E.; von Werder, K. In: *Prolactin physiology, pharmacology and clinical findings*; Flückiger,E.; Del Pozo,E.; von Werder, K. Eds., Springer Verlag: Berlin 1982.
4. Schneider, H.P.G.; Bohnet, H.G. *Gynäkologie* **1981**, *14*, 104-118.
5. Aisaka, K.; Yoshida, K.; Mori, H. *Hormone Res.* **1992**, *37* (Suppl. 1), 41-47.
6. Jarry, H.; Leonhardt, S.; Wuttke, W.; Behr, B.; Gorkow, C. *Z. Phytother.* **1991**, *12*, 77-82.

308

7. Jarry, H.; Leonhardt, S.; Gorkow, C.; Wuttke, W. *Exp. Clin. Endocrinol.* **1994**, *102*, 448-454.
8. Sliutz, G.; Speiser, P.; Schultz, M.; Spona, J.; Zeilinger, R. *Hormone Metab. Res.* **1993**, *25*, 253-255.
9. Wuttke, W.; Gorkow, C.; Jarry, H. In: *Phytopharmaka in Forschung und klinischer Anwendung I;* Loew, D.; Rietbrock, N. Eds. Steinkopff Verlag: Darmstadt,1995.
10. Flückiger, E.; Briner, U.; Bürki, H.R.; Marbach, P.; Wagner, H.R.; Doepfner, W.; *Experientia* **1979**, *35*, 1677-1678.
11. Winterhoff, H.; Gorkow, C.; Behr, B. *Z. Phytother.* **1991**, *12*, 175-179.
12. *Hager's Handbuch der Pharmazeutischen Praxis*, Hänsel, R.; Keller, K.; Rimpler, H.; Schneider, G. Eds. 5th Ed. Springer Verlag: Berlin, 1994; Vol 6., pp. 1184-1192.
13. Dittmar, F.W.; Böhnert, K.J.; Peeters, M.; Albrecht, M.; Lamertz, M.; Schmitz, U. *TW Gynäkologie* **1992**, *5*, 60-68.
14. Propping, D.; Katzorke, T.; Belkien, L. *Therapiewoche* **1988**, *38*, 2992-3001.
15. Kubista, E.; Müller, G.; Spona, J. *Gynäk. Rdsch.* **1986**, *26*, 65-79.
16. Roeder, D. *Z. Phytother.* **1994**, *15*, 157-163.
17. Milewicz, A.; Gejdel, E.; Sworen, H.; Sienkiewicz, K.; Jedrzejak, J.; Teucher,T.; Schmitz, H. *Arzneim.-Forsch. /Drug Res.* **1993**, *43*, 752-756.
18. Loew, D.; Gorkow, C.; Schrödter, A.; Rietbrock, S.; Merz, P.G.; Schnieders, M.; Sieder, C. *Z. Phytother.* **1996**, *17*, 237-243.
19. Merz, P.G.; Gorkow, C.; Schrödter, A.; Rietbrock, S.; Sieder, C.; Loew, D.; Dericks-Tan, J.S.E.; Taubert, H.D. *J. Exp. Clin. Endocrinol. Diabetes* 1997 (in press).
20. Zwaving, J.H.; Bos, R. *Planta Med.* **1996**, *62*, 83-84.

INDEXES

Author Index

Subject Index